Applied Probability and Statistics (Continued)

CHAKRAVARTI, LAHA and ROY · Handbook of Methods of Applied Statistics, Vol. I
CHAKRAVARTI, LAHA and ROY · Handbook of Methods of Applied Statistics, Vol. II
CHERNOFF and MOSES · Elementary Decision Theory
CHIANG · Introduction to Stochastic Processes in Biostatistics
CLELLAND, deCANI, BROWN, BURSK, and MURRAY · Basic Statistics with Business Applications
COCHRAN · Sampling Techniques, *Second Edition*
COCHRAN and COX · Experimental Designs, *Second Edition*
COX · Planning of Experiments
COX and MILLER · The Theory of Stochastic Processes
DAVID · Order Statistics
DEMING · Sample Design in Business Research
DODGE and ROMIG · Sampling Inspection Tables, *Second Edition*
DRAPER and SMITH · Applied Regression Analysis
GOLDBERGER · Econometric Theory
GUTTMAN and WILKS · Introductory Engineering Statistics
HALD · Statistical Tables and Formulas
HALD · Statistical Theory with Engineering Applications
HOEL · Elementary Statistics, *Second Edition*
HUANG · Regression and Econometric Methods
JOHNSON and LEONE · Statistics and Experimental Design: In Engineering and the Physical Sciences, Volumes I and II
LANCASTER · The Chi Squared Distribution
MILTON · Rank Order Probabilities: Two-Sample Normal Shift Alternatives
PRABHU · Queues and Inventories: A Study of Their Basic Stochastic Processes
SARHAN and GREENBERG · Contributions to Order Statistics
SEAL · Stochastic Theory of a Risk Business
WILLIAMS · Regression Analysis
WOLD and JURÉEN · Demand Analysis
WONNACOTT and WONNACOTT · Econometrics
YOUDEN · Statistical Methods for Chemists

Tracts on Probability and Statistics

BILLINGSLEY · Ergodic Theory and Information
BILLINGSLEY · Convergence of Probability Measures
CRAMÉR and LEADBETTER · Stationary and Related Stochastic Processes
RIORDAN · Combinatorial Identities
TAKÁCS · Combinatorial Methods in the Theory of Stochastic Processes

This is a text on regression analysis and its application in economics and related fields. It systematically organizes classical and recent developments in the uses of the least-squares method and its variants in estimation and testing of behavioral and physical relationships.

The material includes the important elements of simple and multiple regression, multivariate regression, and simultaneous equations, with the emphasis on:

- thorough treatment of the various tests in regression analysis, including tests of hypotheses about regression co-efficients, structural stability, and linear restrictions in general
- up-to-date discussion of basic econometric problems as extensions of the standard regression model, such as generalized least squares, errors in the variables, and multicollinearity
- a concise introduction to multivariate regression, nonlinear regression, setwise least squares, identification, and estimation and application of simultaneous-equation models.

Regression and Econometric Methods

About the author

DAVID S. HUANG is Professor of Economics at Southern Methodist University.

After receiving a Ph.D. degree in 1961 from The University of Washington, he taught at The University of Texas and carried on research for the Social Systems Research Institute at The University of Wisconsin. In 1967 he was a visiting professor of Business Economics at the Graduate School of Business, The University of Chicago.

Dr. Huang is the author of INTRODUCTION TO THE USE OF MATHEMATICS IN ECONOMIC ANALYSIS (Wiley, 1964). He has also produced two monographs and has published extensively in leading journals.

Regression and Econometric Methods

DAVID S. HUANG
Southern Methodist University

John Wiley & Sons, Inc.
New York . London . Sydney . Toronto

By the same author

Introduction to the Use of Mathematics in Economic Analysis, Wiley, New York, 1964

Library of Congress Catalogue Card Number: 75-115650

ISBN 0-471-41754-8

Printed in the United States of America

10 9 8 7 6 5 4 3 2 1

Preface

This book is for students and researchers in economics and in other behavioral disciplines where the use of the concepts and techniques of regression analysis is rapidly becoming commonplace. It is intended for students who have had, at least, one good statistics course and who have some mathematical maturity (for instance, having completed calculus). The emphasis is on application and intuitive appreciation, so that the proofs and derivations are supplied only to the extent essential for the orderly development of the topics. To use a statistical procedure properly and effectively in research, it is important to understand the underlying assumptions that lead to the procedure. Thus my emphasis has also been on a thorough discussion of the assumptions of the various models. On learning the matrix notations, the student may adopt a cookbook approach to the use of this text.

There is ample material for a one-semester course in regression analysis or in econometrics for first-year graduates and well-prepared seniors. The book also may be used as an econometrics text for generalists in doctoral programs in economics or business where certain training in quantitative methods is required of every candidate. The book adapts a three-step format.

1. The first three chapters are introductory and provide the fundamentals of simple regression analysis. The material is easy, but a full appreciation of it is necessary for an understanding of the subsequent chapters.

2. Chapters 4 to 7, at an intermediate level, form the backbone of the book. I recommend that these chapters be covered in some detail in a one-semester course in regression analysis or in econometrics.

3. Chapters 8 to 10 are concerned with multivariate regression analysis and its extensions. The last three chapters may be treated with varying degrees of intensity in a one-semester course. However, these chapters may be used along with other books if the present volume covers a course that lasts longer than one semester.

DAVID S. HUANG

Dallas, Texas, 1969

Acknowledgments

In writing this book, I cited the work or obtained the assistance of many people. I am grateful to Douglas G. Chapman, who inspired my interest in applied statistics, and to Guy H. Orcutt, who encouraged me in economic research. Much of what I know and practice has been learned from Arthur S. Goldberger and Arnold Zellner; they have been of great help in my professional work. It was an educational experience to have had Franklin M. Fisher read the drafts of this book. His comments helped clarify many concepts and strengthened the exposition. T. Dudley Wallace read the final draft, caught errors, and made suggestions for improvement. Michael D. McCarthy and Michael G. McCracken read and commented on the first four chapters in the first-draft stage. I thank all of these scholars, but the responsibility for any errors and defects is mine.

I am indebted to the Literary Executor of the late Sir Ronald A. Fisher, F.R.S., to Dr. Frank Yates, F.R.S., and to Oliver & Boyd Ltd., Edinburgh for permission to reprint a part of Table III from their book *Statistical Tables for Biological, Agricultural and Medical Research.* I am also indebted to many journals and individuals for permissions to reproduce their publications in various forms. Proper acknowledgments for these permissions are given in the text. I thank Mrs. Laura Vissing and Mrs. Martha Wells for assistance in the preparation of the manuscript. Finally, I thank my wife, Ruth, and my children, Milton and Lena, for being cheerful and inspiring while this book was being written.

D. S. H.

Contents

CHAPTER 1

INTRODUCTION

Much of this book is concerned with regression analysis in economics. Regression analysis is a body of statistical theory and techniques. As a branch of statistics, it is useful in scientific efforts that usually include the test and estimation of behavioral hypotheses.

An example in economics would be that of analyzing production possibilities in an industry. Here the analyst might hypothesize that the underlying production process is that of the Cobb-Douglas production function or that of the more general constant-elasticity-of-substitution variety and might use available data to estimate the parameters of the hypothesized relation. After the parameters are estimated it may be possible to test if the elasticity of output with respect to a certain productive factor is of certain size or to test if there exist constant returns to scale in the industry. Regression analysis can be extremely helpful in these endeavors.

Another example might be in the area of consumer behavior, for example, in connection with the purchase of automobiles. It could be that one has a set of cross-section data that represents information in any one year on purchase or nonpurchase, disposable income, and asset holdings pertaining to individual households of a certain region, and interest may center on how the income and the assets influence purchase behavior. Such a question, in part, may be answered by fitting a regression equation to the variables under analysis. In this connection, we recall the area of work in economics where regression analysis merely serves the function of describing data involving many variables. Some of the well-known works sponsored by the National Bureau of Economic Research emphasize the approach of extracting statistical regularities from a complex of data collected for a large number of variables and, thereby, suggest plausible laws of economic behavior. Wesley Mitchell's *Business Cycles* is a case in point. Whether as a tool for inference or as a descriptive device the body of knowledge on regression analysis and on its applications has grown by leaps and bounds in recent years.

Frequently, the term regression analysis is used interchangeably with the analysis of dependence relations and also with least-squares analysis. By a dependence relation we mean a relation among two or more variables such that the variation of one of the variables is determined, with or without error, by the remaining variables; statistically, the dependent variable has a probability distribution conditional on the other variables. Also, we observe that regression theory and techniques are applicable to a wide variety of forms of dependence relation. Recently it has become well known that relations nonlinear in the parameters can be adequately handled by the method of least squares. As will be seen later, the least-squares method is but one of the several procedures that can be used in analyzing dependence relations. As such, it will be a mistake to consider regression analysis as solely composed of the least-squares analysis. On the other hand, however, the method of least squares is simple to use and is applicable in a very large number of situations, so that the whole area of the literature on regression analysis is *almost* filled with theories and procedures related to or implied by the method of least squares. This last point will be reflected in a large part of the present book.

This is a book about the method of least squares and its extensions. One of the basic objectives here is to organize in a systematic fashion the classical and the recent developments in the uses of the least-squares method and its variants in the estimation and the testing of behavioral and physical relationships. Although the standard body of knowledge in regression analysis is predicated on a set of classical assumptions, economists have built on or extended these assumptions to suit economics problems. The result is a flourishing in a part of the econometric methodology where, for instance, the recent developments make it possible to consider the estimation and the related analysis of several dependence relations simultaneously with the aid of the least-squares procedure.

The plan of this book is as follows. In the next chapter we provide some basic discussions of a model and relate it to economic and regression models. Chapter 3 is concerned with the assumptions and procedures of simple regression analysis. In this chapter we put down the general framework of the univariate regression model—the framework that underlies the development throughout Chapters 4, 5, and 6. These last three chapters provide the fundamentals of the multiple regression model (still a univariate model) and there, as in the subsequent chapters, matrix notations are extensively used. Many special problems arise in the application of the standard regression models to economic and business problems. Some of these problems and related techniques are selected for discussion in Chapter 7. Another special problem, which has received increasing attention in the recent years, is one that falls in the general area of multivariate regression analysis. In Chapters 8, 9, and 10, we introduce the multivariate regression model and take up its natural extension in the forms of simultaneous equation models that are much discussed in the econometrics literature.

CHAPTER 2

SOME FUNDAMENTALS OF MODELS

2.1. The Model Explained

A model, as used in a physical sense, is a miniature of a real object, such as a building, a landscape, etc. Presumably, the miniature has all the features of the corresponding real object, but this requirement of a model is rarely fulfilled in practice. A model of an office building will most likely be lacking an elevator. Also, a model of an airplane set in a wind tunnel lacks the collection of precision instruments that are necessarily present in a real airplane. A model, then, for all practical purposes and from a physical point of view, may be taken to possess most of the essential features of the real object that it purports to represent. Of course, the essential features that are present in a model, will be determined by the purposes for which the model is built. Thus, a commercial builder who is trying to sell a particular design of office building will construct a model of the building portraying mainly the outward features of the building, whereas the aerospace engineer interested in testing the effects of turbulance and of stress of airstream on an airplane will build a model plane with features that most adequately embody the designs of the wing and body structure of the real airplane.

It seems clear, from this discussion, that a model usually embodies the essential features of its real counterpart and that the features included in the model are suited to specific purposes. Put differently, a model is generally not a true and complete representation of its real counterpart, and model builders can put in different essential features of the real object in the model as their interests dictate.

Not all models are sized-down versions of real physical objects. In its more general use, a model can range from a set of subjective or objective statements about some aspect of human behavior to a simplified version of a complex electrical system. Thus, Orcutt (1960, p. 897) writes: "A model of something is a representation of it designed to incorporate those features deemed to be significant for one or more specific purposes. In some cases such features are directly observable · · · . In other cases models incorporate more subtle features such as how the thing modeled responds to stimuli or otherwise behaves."

2.2. Economic Models

Much similarity to the preceding discussion can be claimed for economic models. In an objective sense, an economic model is a simplified relief of a part or whole of an aspect of economic life. In subjective usage, an economic model is a set of propositions or hypotheses purported to describe and/or predict economic behavior of men and nations. As is familiar, an economic model can be cast in simple English or can be expressed in some other language, such as mathematics. Whatever language used, an economic model generally possesses three ingredients: basic behavior units, variables, and relations. And, because of the nature of the ingredients and because most economic phenomena are quantifiable, economic models are frequently represented by sets of mathematical formulas or, sometimes, logical statements. For instance, as a simple descriptive device,

$$X = f(K, L) \tag{2.1}$$

where X is physical output of an industry, and K and L, respectively, are capital and labor inputs of the industry is an economic model saying what the production possibilities, in general, are in the industry. Here, the behavior units can be the firms of the industry, the variables in the model are X, K, and L, and f defines the relation between the output and the input. Also, as a predictive relation we might have

$$Q = g(P_s/P_0, Y, A) \tag{2.2}$$

as the result of making behavioral assumptions about the consumer's choices given his income and other circumstances. In this relation, Q is the quantity of a commodity demanded by a consumer, P_0 is the price of the commodity, P_s is the price of the closest substitute, Y is his income, and A is his asset holdings. In this relation, the behavior unit is the consumer; the variables are his income and assets, the prices of the commodities in question, and the quantity he chooses to purchase of the first commodity. Notice that the form of the relations is left open and that the relations are exact—exact in the sense that whatever functional forms f and g might have, K and L explain completely the magnitudes of X, while the variables in the right-hand side of the Q relation make perfect account of the variation in Q. These illustrative observations lead us to consider a few types of economic models.

Given that the relational form of an economic model can be specified, the model may be looked at in two ways: one, for exactness of specified form, and two, for amenability to statistical estimation and testing. With respect to the first, two types of economic models exist: exact models and inexact models. As mentioned above, the Expressions 2.1 and 2.2 are examples of exact models. It is often the case that when specifying these relations the economist might wish to allow for the possibility that the variation in a variable cannot be explained away perfectly by a handful of other variables, primarily, because of the complexity of the real world and the economist's lack of complete knowledge. In this case, one usually adds an explanatory variable called disturbance, say u, in the right-hand side of the relation, thus, making the model inexact. Therefore, if it is thought that capital and labor inputs alone would not explain the level of output because of the exclusion of entrepreneurship, or because technological advance is difficult to quantify, or because the model builder wishes not to be bothered by accounting for factors other than labor and capital, and so on, (2.1) may be rewritten as

$$X = f(K, L, u) \tag{2.3}$$

Similarly, (2.2) may be recast as

$$Q = g(P_s/P_0, Y, A, v) \tag{2.4}$$

In these last two expressions, u and v are referred to as catchall terms or disturbances. Now these expressions show that the disturbances are present and, in other cases, we may write (2.3) and (2.4) as

$$X = f(K, L) + u \tag{2.3a}$$

and

$$Q = g(P_s/P_0, Y, A) + v \tag{3.4a}$$

where the disturbances u and v are additive to their respective equations. Whether the disturbance is additive or otherwise to an equation is a matter that a model builder must decide, but a significant point here is the recognition that the theoretical or hypothesized relation, such as $X = f(K, L)$ here, may not agree with actual observation and that the relation is subject to error. Models such as (2.3), (2.4), (2.3a), and (2.4a) are referred as to error-in-the-equation models or models with equation errors.*

Models that admit errors or disturbances in their behavioral equations, in general, are amenable to statistical test. Now, error in an equation arises either because the knowledge concerning the behavior to be modeled is imperfect (as to the functional form and the variables to be used), or because practical considerations make it necessary to limit attention to a number of crucial variables. For instance, the hypothesized production function $f(K, L)$, in truth, may be

$$X = f(K, L) + f_1(X_1, X_2, \ldots, X_k) \tag{2.5}$$

* Equation error is distinct from what is called measurement error, which involves the question of whether each variable in a model can be measured accurately.

where $X_1, X_2, \ldots, X_k$ are additional variables that account for that part of the variation in X left unaccounted for by $f(K, L)$. If so, $X = f(K, L)$ is an approximation to (2.5) and is a wrong equation for the production function. It may be that $X_1, X_2, \ldots, X_k$ cannot be observed, so that the model builder is forced to use $f(K, L)$ as the approximation. Or it may be that the analyst believes the term $f_1(X_1, X_2, \ldots, X_k)$ to be like a random variable that does not give a systematic account of the variation in X. In any case, models with equation error, such as (2.3a), cannot be tested statistically until the disturbance terms are assigned stochastic properties. For instance in a research process one may prefer to replace $f_1(X_1, X_2, \ldots, X_k)$ by u and to let the letter be a random variable with a probability distribution.

Sometimes, theoretical models in economics are formulated without consideration as to whether the variables in the models are all measurable, in the sense that quantitative or qualitative observations on these variables can be actually or potentially made. These variables are usually defined as abstract concepts (such as relating to the process of introspection) and are used in logical operations that lead to certain theoretical propositions. This type of model is nonstatistical, since there is no way to obtain data for the models to have statistical implications.

Many economic models that we see today are pseudostatistical models, very much in the sense of what Hicks calls models with econometric implications, although they are not strictly econometric models. Models like those given by (2.1) and (2.2) might be called pseudostatistical because the variables in the models are observable and because only a minor additional specification is necessary for these models to qualify as statistical models. The additional specification usually consists of appending to the already specified functional form a disturbance term that has well-defined statistical properties. Thus the so-called exact models we discussed earlier in this section can become statistical.

Economic models that have a well-specified functional form and statistically defined disturbance terms are statistical models. (Outside of the economics world, one has statistical models in the fields of biology, chemistry, etc.) Economic models that qualify as statistical models are called econometric models. An econometric model can consist of one or more equations or relations.

2.3. Regression Models

An econometric model consisting of one equation may be called a regression model, since the term regression means a dependence relation.

But, just as econometric models are a subclass of economic models, regression models are a subclass of statistical models. Thus, a regression model is a statistical dependence relation. Obviously, one can have regression models in various physical and behavioral sciences. We said that the term regression meant a dependence relation. This definition is really too broad to help us understand the more technical meaning of the term regression as it is used in modern context.

The tracing of statistical literature shows that originally the term regression refers to a phenomenon in which the heights of the sons of tall fathers and those of the sons of short fathers tend to move toward the overall average heights. This happens as follows. Consider a collection of the heights of a number of sons and classify these heights according to the stature of the fathers. We find that tall fathers tend to have tall sons, and short fathers short sons; furthermore, the average height of the sons corresponding to a group of tall fathers is less than the average of the fathers' heights while, for a group of short fathers, the sons' average height is greater than their fathers' average height. Thus, there is a tendency for the tall sons and short sons alike to move toward the average height of all men. This last phenomenon was named by Galton as a law of regression, retrogression to an earlier stage of development, or a movement toward mediocrity. Of course, the phenomenon on heights is made possible by the biological facts that there are upper and lower limits to the human stature, so that the retrogression interpretation can be fallacious.

The modern interpretation of the term regression is that of dependence on the average. We say that regression phenomenon exists where fathers are divided into groups according to their heights and the sons' heights are compared with the fathers' on a group to group basis. In this kind of comparison, it can be expected that the heights of the sons whose fathers belong to height Class A will tend to depend on the average heights in Class A except for some random deviation; the heights of the sons whose fathers belong to height Class B will tend to depend on the average height of Class B, and so on. Also, it can be expected that a tall father will tend to have sons whose average heights is greater than the average of the sons of a short father. There are several interpretations to what we call dependence on the average, and these interpretations will become clear later. For now, we may say that a regression model is a statistical model that embodies assumptions and/or hypotheses about a dependent relationship between one variable (dependent variable) and one or more other variables. The model is usually cast in the form of a well-specified equation in which all the variables have well defined statistical properties.

Regression analysis represents a statistical assessment of a dependent relationship. Methods available for the analysis are numerous, although the least-squares method is the most frequently used. Much of our effort in this book is concerned with the application of the least-squares method to simple as well as to complicated regression models. However, it is well to realize that the least-squares method is not the only method for regression analysis. The usuage of regression equation and least-squares equation as interchangeable is due primarily to the dominance of the least-squares technique among the various available techniques of regression analysis.

CHAPTER 3

SIMPLE CORRELATION AND REGRESSION

The basic purpose of this chapter is to explain the principles and procedures of simple regression analysis with emphasis on the standard assumptions and their meaning, the computational procedures, the interpretation of computed results, and the research strategy related to the analysis. Before embarking on this task, however, we treat simple correlation briefly as a prelude to regression analysis, primarily, for the purpose of contrasting the differing assumptions of the two analyses.

As will be apparent later, the concepts and techniques discussed in this chapter are fundamental to the subsequent development of this book. A thorough understanding of simple regression analysis is crucial to a successful study of multiple regression analysis and related statistical problems so that the student is advised to study this chapter very carefully.

3.1. Simple Linear Correlation Analysis

We do not intend to make an exhaustive survey of simple (or two-variable) linear correlation analysis in this section. Here, we are interested in presenting some fundamental elements of the analysis as they will relate to our development of simple linear regression analysis. Furthermore, a basic knowledge of correlation analysis is assumed. It would be safe to say that the two analyses are concerned with the degree of dependence between two variables but that the assumptions about directions of causation and others are quite different. Also, regression analysis is a more flexible tool for research work than is correlation analysis.

3.1.1. Correlation Between Two Variables

If a researcher is interested in the degree of association or the degree of mutual dependence between any two variables, say, X and Y, and if he has a number of observations (or values) on X and Y, he most likely will calculate the so-called sample correlation coefficient r as follows

$$r = \frac{\sum_{i=1}^{n}(X_i - \bar{X})(Y_i - \bar{Y})}{\sqrt{\sum_{i=1}^{n}(X_i - \bar{X})^2 \sum_{i=1}^{n}(Y_i - \bar{Y})^2}} \tag{3.1}$$

where X_i and Y_i are the ith observations on X and Y, respectively, n is the sample size, and $\bar{X}$ and $\bar{Y}$ are sample means of the variables. Here, the researcher does not ask whether X depends on Y or vice versa; he is interested in the degree of covariation between the variables. In other words, the researcher, by virtue of computing the r, may be thought of as having reasons to believe that the question of the direction of causation between X and Y does not exist or is not important.

Although there are occasions on which one is merely interested in the empirical correlation coefficient, usually the point of calculating r as in (3.1) is to guess the value of the true correlation coefficient between two variables. The population correlation coefficient ρ between two variables, say X and Y, is defined as

$$\rho = \frac{\sigma_{12}}{\sigma_1 \sigma_1} \tag{3.2}$$

where σ_1 is the standard deviation of X, σ_2 is the standard deviation of Y, and σ_{12} is the covariance of X and Y which is defined as $E(X - \mu_1)(Y - \mu_2)$ with the understanding that μ_1 and μ_2, respectively, are the means of X and Y. It follows that ρ can be seen as one of the parameters characterizing the joint probability distribution of the two variables. The interpretation of the sample correlation coefficient, then, should be dictated by the nature of the joint distribution between the two variables in question.

An interesting case of the joint probability distribution between two variables is the bivariate normal distribution. Let X and Y be the two random variables with the following joint distribution:

$$f(X, Y) = \frac{1}{2\pi\sigma_1\sigma_2\sqrt{1 - \rho^2}} e^{-Q(X,Y)} \tag{3.3}$$

where

$$Q(X, Y) = \frac{1}{2(1 - \rho^2)}\left[\frac{(X - \mu_1)^2}{\sigma_1^2} - \frac{2\rho(X - \mu_1)(Y - \mu_2)}{\sigma_1\sigma_2} + \frac{(Y - \mu_2)^2}{\sigma_2^2}\right] \tag{3.4}$$

In these expressions, ρ is the population correlation coefficient of X and Y, μ_1 and μ_2, respectively, are the means of X and Y, and σ_1 and σ_2, respectively, are the standard deviations of X and Y. Furthermore ρ is defined as in (3.2).

Notice that if $\rho = 0$, (3.3) above reduces to

$$f(X, Y) = \frac{1}{2\pi\sigma_1\sigma_2} e^{-\frac{1}{2}[(X-\mu_1)^2/\sigma_1^2 + (Y-\mu_2)^2/\sigma_2^2]} \tag{3.6}$$

Therefore, if X and Y are uncorrelated, their joint density is factorable in that

$$f(X, Y) = g(X)h(Y) = \frac{1}{\sqrt{2\pi}\sigma_1} e^{-(X-\mu_1)^2/2\sigma_1^2} \frac{1}{\sqrt{2\pi}\sigma_2} e^{-(Y-\mu_2)^2/2\sigma_2^2} \tag{3.7}$$

and it follows that X and Y are independent. See, for instance, Birnbaum (1962, p. 53). That noncorrelation or zero covariance implies independence is a fact of theory related to variables with normal distributions. The implication does not hold true in general for any two random variables. The literature is replete with examples of this statement, for instance, Birnbaum (1962, p. 153) and Wilks (1962, p. 78). To put it differently, there are bivariate distributions in which the covariance of the variables is zero and yet one of the variables is dependent on the other. Of course, by definition, the independence of two variables implies noncorrelation. These comments ought to sufficiently caution the reader against mixing the words noncorrelation and independence freely. They also serve to show that sometimes it is awkward to use a correlation coefficient for checking dependence between two random variables empirically.

3.1.2. *Statistical Inference about* ρ

Given that X and Y have the joint density (3.3), it has been shown that for $\rho = 0$, the sampling distribution of r is

$$f(r \mid n, \rho = 0) = \frac{\Gamma\left(\frac{n-1}{2}\right)}{\sqrt{\pi}\,\Gamma\left(\frac{n-2}{2}\right)} (1 - r^2)^{(n-4)/2} \tag{3.8}$$

and that for $\rho \neq 0$, the sampling distribution of r is

$$f(r \mid n, \rho) = \frac{(1-\rho^2)^{(n-1)/2}}{\pi(n-3)!} (1 - r^2)^{(n-4)/2} \frac{d^{n-2}}{d(r\rho)^{n-2}} \left(\frac{\operatorname{arc\,cos}(-\rho r)}{\sqrt{1-\rho^2 r^2}}\right) \tag{3.9}$$

Here d is the derivative sign. F. N. David (1954) has constructed tables of probability integral and ordinate that are accurate to five decimal places for the densities (3.8) and (3.9). The tables are for each of sample sizes

$n = 3$ through $n = 25$, and also for $n = 50, 100, 200$, and 400. For each size of sample, ten distributions or r have been computed for $\rho = 0.0, 0.1, 0.2, 0.3, 0.4, 0.5, 0.6, 0.7, 0.8$, and 0.9. Thus most questions regarding the construction of confidence intervals and the test of ρ can be answered by the use of these tables.

If David's *Tables* is not readily available and if n is reasonably large, an approximately normal distribution is obtained by the transformation of r and ρ as follows.

$$z = \frac{1}{2} \log \frac{1 + r}{1 - r}: \qquad m = \frac{1}{2} \log \frac{1 + \rho}{1 - \rho}$$

This is called Fisher's z-transformation, and z has an approximately normal distribution with mean m and standard deviation $1/\sqrt{n - 3}$.

If X and Y are not jointly normally distributed, then parametric inferences about ρ are no longer possible. This limits somewhat the ability of the researcher who wants to make a confidence or probability statement about the population correlation coefficient in the light of a random sample.

We now move on to simple regression analysis.

3.2. Simple Linear Regression Analysis—an Introduction

For continuity with the preceding discussion, we first consider the conditional distributions related to the bivariate normal distribution. Although these conditional distributions define linear regression lines of, say, X on Y, or vice versa, it will be explained that this type of regression model is but one of the many in theoretical and empirical usage. Before embarking on this and other tasks, however, let us clear up some terminology questions.

By simple regression, we mean the dependent relation of one variable on another, that is, a two-variable one-way relationship. By linear regression, we mean a regression equation linear in the parameters. Thus a simple linear regression equation may look like

$$Y = \alpha + \beta X + u \tag{3.10}$$

and a nonlinear regression equation may look like

$$Y = \alpha X^{\beta} e^{u} \tag{3.11}$$

In both (3.10) and (3.11), α and β are parameters and u is the disturbance term. In (3.11), e is the base of natural logarithms. A multiple linear regression equation is a generalization of equations of the type (3.10) and has two or more independent variables or regressors. Notice that we are only concerned with linearity in the parameters when we define linear regression. It can be that some regressors are not linear variables (say, X^2, or $\log X$) in a regression equation, for example, (3.11), but the equation will be considered linear as long as the parameters are all linear. For instance, the following is essentially a (multiple) linear regression equation.

$$Y = \alpha + \beta X^2 + \gamma \log Z + u \tag{3.12}$$

3.2.1. Simple Linear Regression Models

REGRESSION LINE OF Y ON X. Anyone familiar with operations with probability distributions can easily show that the conditional distribution of Y on X, given that X and Y are distributed jointly normally according to (3.3) [those who find the following result difficult to derive may wish to consult, for example, P. G. Hoel (1962, pp. 200–201).], is as follows:

$$f(Y \mid X) = \frac{1}{\sqrt{2\pi}\sigma_2\sqrt{1-\rho^2}} \times \exp\left\{-\frac{1}{2\sigma_2^2(1-\rho^2)}\left[(Y \mid X) - \mu_2 - \rho\frac{\sigma_2}{\sigma_1}(X-\mu_1)\right]^2\right\} \tag{3.13}$$

Consider any random variable U normally distributed with mean μ and variance σ^2. Then, U's density is

$$f(U) = \frac{1}{\sqrt{2\pi}\sigma} e^{-(1/2\sigma^2)(U-\mu)^2} \tag{3.14}$$

Allowing the correspondence

$$(Y \mid X) = U; \quad \sigma_2^2(1-\rho^2) = \sigma^2 \quad \text{and} \quad \mu_2 + \rho\frac{\sigma_2}{\sigma_1}(X-\mu_1) = \mu$$

between (3.13) and (3.14), we observe that the conditional distribution of Y on X has mean $\mu_2 + \rho(\sigma_2/\sigma_1)(X - \mu_1)$ and variance $\sigma_2^2(1 - \rho^2)$. Or random variable Y, given X, has the expectation:

$$E(Y \mid X) = \mu_2 + \rho\frac{\sigma_2}{\sigma_1}(X-\mu_1) \tag{3.15}$$

As indicated in Figure 3.1, the graph of the expectation of Y as a function of X is called the regression line of Y on X. Notice that the variability of Y around this line is given by $\sigma_2(1 - \rho^2)$. Also, notice that (3.15) is a linear relation as a direct consequence of the bivariate normal distribution assumed for X and Y. If X and Y have a joint density not of the form (3.3), $E(Y \mid X)$ would not necessarily have a linear form like (3.15). It is only a short step away from (3.15) to write a more familiar linear regression model of Y on X:

$$Y = \alpha + \beta X + u \tag{3.10}$$

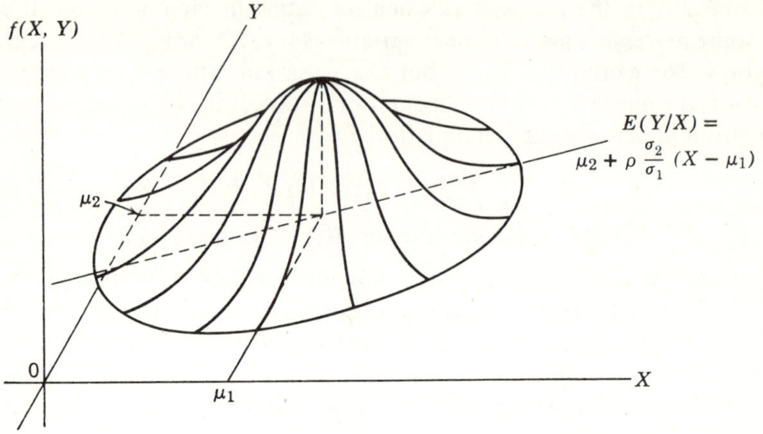

Figure 3.1*a* and *b*

Only one needs to specify, at least, that u is normally distributed with mean zero and a constant variance, while letting

$$\beta = \rho \frac{\sigma_2}{\sigma_1} \qquad \text{and} \qquad \alpha = \mu_2 - \beta\mu_1$$

Now all of these discussions about the conditional expectation of Y on X can be carried out for the reverse case where the conditional distribution of X on Y is in question. The result will be just a simple switching of Y and X in the expression (3.13) through (3.15). We then obtain a linear regression model of X on Y

$$x = \alpha' + \beta' Y + v \tag{3.16}$$

where

$$\alpha' = \mu_1 - \beta'\mu_2, \qquad \beta' = \rho \frac{\sigma_1}{\sigma_2}$$

and v has a normal distribution with mean zero and a constant variance.

Some simple linear regression models. It is clear from the above discussion that, given a bivariate normal distribution, we can derive population regression lines of Y on X and of X on Y. These elegant consequences of the bivariate normal distribution, however, are not always encountered in the theoretical and empirical work in economics and, for that matter, in many other behavioral disciplines. The situations one frequently encounters include the following: one of the variables may have a nonnormal distribution; knowledge about the distributions of the two variables does not exist; it is not known what the form of the conditional distribution of one variable given the other is; and so on. In addition, the interest of a researcher might not lie in the mutual regressions of one variable on the other. He might be more interested in a one-way causation between the two variables he is studying. Furthermore, he might not ask questions about what joint probability distribution governs the variables and might be merely interested in just a descriptive relationship of one variable *as influenced by* another. Given the diversity of interests of researchers in problems where a linear relationship might be applicable, and given the variety of statistical properties that can be specified for the variables in the relationship, various types of simple linear regression models in economics and other behavioral disciplines have been developed.

Recall that in Chapter 2 we defined regression models as models of dependence relations. But correlation models are also models of dependence relations. We distinguish regression from correlation by restricting the dependence to that of one-way, not two-way. Now, given two variables, say Y and X, we can be interested only in how linear dependence of Y on X is measured, of course, assuming that Y does depend on X. We do not necessarily care whether Y given X is normally distributed or otherwise, whether X is random or controllable, whether X is measured with error, and whether any statistical properties should be attached to the portion of Y not explainable by X. Thus, our interest is a purely descriptive linear regression model. One sound method for analyzing a descriptive model is the least-squares method, which we shall discuss extensively in the next section.

Now, if we are not satisfied by description alone and wish to inquire further into statistical properties of estimators in the linear relationship, a variety of models emerges depending on what we care to or can assume about the statistical properties of the variables in the relationship. Depending on assumptions, then, there are (1) the classical linear regression model, (2) the normal linear regression model, (3) the errors-in-the variables linear regression model, and (4) the stochastic linear regression model, among others. It is our task in the subsequent sections and chapters to elaborate on the assumptions made in linear regressions, on the effects of these assumptions on the properties of the estimators, and on the related problems of inference in selected regression models. Before moving on, however, let us refer briefly to some terms which are used for identifying the variables in the regression equation.

TERMINOLOGY. Since regression analysis is used in a wide variety of situations in various sciences, it is natural that many different terms are used for identifying the variables in the left-hand side and in the right-hand side of a regression equation. We cannot enumerate all of the terms, but we can point out a few that are more frequently used along with some comments about them.

Perhaps, the most frequently seen terminology is in the independent-dependent context. This usage probably comes from mathematics where the left-hand-side variable in a function is usually referred to as dependent and those in the right-hand side as independent. Here, the meaning is that the right-hand-side variables change in a known or controllable fashion and give rise to change in the left-hand-side variable according to the form of the function. There is an implication in this usage that the variables in the right-hand side are cause and that in the left-hand side is effect.

Much similar to the usage just discussed is the stimuli-response formulation often seen in medical research, chemical engineering, and agricultural experiment, etc. Here, the right-hand-side variables represent stimuli, such as varying amounts of dosage, and the left-hand-side variable represents the different degrees of response to the stimuli. In this formulation, what is often called independent variables in mathematics is termed stimuli or stimulus variables and dependent variable response or response variable. The curve or surface traced out by the response variable is referred to as response surface, generally.

Somewhat less specific with respect to cause-effect specification than the two usages just mentioned is the explanatory-explained relation. The variables in the right-hand side are called explanatory either because they are a cause for explaining the effect reflected in the left-hand-side variable or because the change in the left-hand-side variable is "explainable" by the change in the right-hand-side variables without implying that the latter are a cause and the former the effect. The last usage is often seen in models where description or prediction is of main interest.

A general way of referring to the variables in a regression, capable of representing different types and forms of the variables, is the regressor-regressand usage. It does not matter whether the right-hand side variables represent cause, or whether they are subject to a feedback from the left-hand-side variable, or whether a right-hand-side variable is decomposed into several categories (or dummy variable system, of which we shall have more to say later), we call the right-hand-side variables regressors. The left-hand side variable is a regressand.

This description covers about all that we shall say regarding the terminology of variables in a regression. It seems necessary to labor the point as different usages connote different models at times. Loose uses of terms can be confusing for clearly seeing the basic nature that the user accords to the variables in his regression model.

3.2.2. The Least-Squares Method

The method of least-squares is the bread and butter of regression analysis. The method is simple to understand and to use. And it is applicable to a wide range of situations. For instance, the least-squares method can be used for a mere description of a linear dependence relation between two variables: the method can be applied to regression models that contain varying assumptions; the method is used frequently in place of theoretically superior methods because of its computational ease, and so on. Our purpose in this section is to expose the least-squares method, leaving questions of research strategy and interpretation to the various respective models that appear in the later sections.

The least-squares method is in the nature of curve fitting. Suppose that there are two variables X and Y whose linear dependence relation is

$$Y = \alpha + \beta X + u \tag{3.17}$$

in the population with u being a random disturbance. And the task is to find the estimates of α and β using a "best" method. Under very general conditions the least-squares method is one of the best methods available for the estimation of α and β. Given that n pairs of observations are collected on (X, Y), the least-squares method determines the parameter estimates, $\hat{\alpha}$ and $\hat{\beta}$, in such a way as to make

$$\sum_{i=1}^{n} \hat{u}_i^2 = \sum_{i=1}^{n} (Y_i - \hat{Y}_i)^2 = \sum_{i=1}^{n} [Y_i - (\hat{\alpha} + \hat{\beta} X_i)]^2 \tag{3.18}$$

minimum, with $\hat{Y}_i = \hat{\alpha} + \hat{\beta} X_i$. The symbols are identified in Figure 3.2 for any i. The rationale here is that, instead of allowing the fitting of the scattered points of Figure 3.2 to be determined arbitrarily by a freehand drawing of a line or to be dictated by some other criterion, the least squares criterion chooses that set of $\hat{\alpha}$ and $\hat{\beta}$ that minimizes the sum of the squares

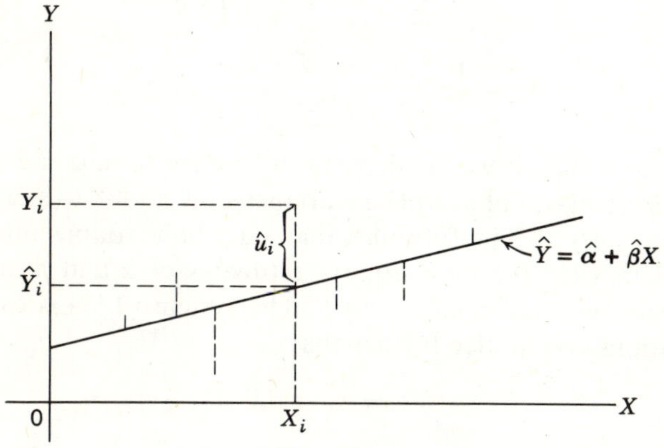

Figure 3.2

of the difference between the actual observation on Y_i and the estimated value $\hat{Y}_i$—a mathematical criterion that renders itself to easy manipulation.

To obtain the parameter estimates that will satisfy the least-squares criterion, we proceed to take partial derivatives of $\sum \hat{u}_i^2$, which is a function of $\hat{\alpha}$ and $\hat{\beta}$ as the latter are allowed to vary, and to solve for the parameter estimates in terms of sample observations. That is, the necessary conditions for $\sum \hat{u}_i^2$ to have a minimum are

$$\frac{\partial}{\partial \hat{\alpha}}(\sum \hat{u}_i^2) = 2 \sum [Y_i - (\hat{\alpha} + \hat{\beta} X_i)](-1) = 0$$

$$\frac{\partial}{\partial \hat{\beta}} \sum (\hat{u}_i^2) = 2 \sum [Y_i - (\hat{\alpha} + \hat{\beta} X_i)](-X_i) = 0^*$$

By rewriting these conditions, we have the system

$$n\hat{\alpha} + (\sum X_i)\hat{\beta} = \sum Y_i \tag{3.19a}$$

$$(\sum X_i)\hat{\alpha} + (\sum X_i^2)\hat{\beta} = \sum X_i Y_i \tag{3.19b}$$

The solution of the system (3.19) yields

$$\hat{\beta} = \frac{n(\sum X_i Y_i) - (\sum X_i)(\sum Y_i)}{n(\sum X_i^2) - (\sum X_i)^2} \tag{3.20a}$$

$$\hat{\alpha} = \frac{(\sum Y_i)(\sum X_i^2) - (\sum X_i Y_i)(\sum X_i)}{n(\sum X_i^2) - (\sum X_i)^2} \tag{3.20b}$$

Alternative equivalent expressions for $\hat{\beta}$ and $\hat{\alpha}$ are

$$\hat{\beta} = \frac{\sum X_i Y_i - [(\sum X_i)(\sum Y_i)]/n}{\sum X_i^2 - [(\sum X_i)^2]/n} \tag{3.20c}$$

$$\hat{\alpha} = \frac{\overline{Y} \sum X_i^2 - \overline{X} \sum X_i Y_i}{\sum X_i^2 - [(\sum X_i)^2]/n} \tag{3.20d}$$

Notice that, in (3.20c), if we divide both the numerator and the denominator by n, we get as the ratio of sample covariance of X and Y to sample variance of X, or $\hat{\beta} = s_{XY}/s_X^2$. The formulas that state how sample values ought to be calculated to yield the least-squares estimates of α and β are known as the least-squares estimators of α and β. The system (3.19) is called a set of normal equations. We notice further that

$$\hat{\alpha} = \overline{Y} - \hat{\beta}\overline{X} \tag{3.20e}$$

* It is advisable to notice that the second-order or sufficient conditions for $\sum \hat{u}_i^2$ to have a minimum are met because this is a quadratic problem.

Table 3.1 Quantities Needed for the Computation of α and β

Column	1	2	3	4	5
Variable	X	Y	XY	X^2	Y^2
Observations and products	X_1 X_2 . . . X_n	Y_1 Y_2 . . . Y_n	X_1Y_1 X_2Y_2 . . . X_nY_n	X_1^2 X_2^2 . . . X_n^2	Y_1^2 Y_2^2 . . . Y_n^2
Sum	$\sum X_i$	$\sum Y_i$	$\sum X_iY_i$	$\sum X_i^2$	$\sum Y_i^2$

The computation of the least-squares estimates might proceed as follows. Given the observations in columns 1 and 2 in Table 3.1, we first obtain the cross product of X and Y for each pair of observations and then the squares of observed X and of Y. Now the sums for columns 1 to 5 provide all the necessary quantities for computing the desired estimates. We illustrate the use of these quantities below.

Example 1. From Table 3.2 a random sample of size 11 is taken and sample observations are set forth in the second and the third columns of Table 3.3. We wish to estimate a linear relation of the form

$$Y_i = \alpha + \beta X_i + u_i \tag{3.17E}$$

according to the least squares criterion. By using (3.20a) and (3.20b), we find, from the quantities in Table 3.3, that

$$\hat{\beta} = \frac{(11)(20{,}756) - (443)(477)}{(11)(19{,}801) - (196{,}249)} = 0.789$$

and, by the use of (3.20e), we obtain

$$\hat{\alpha} = 43.36 - (0.789)(40.27) = 11.59$$

Sometimes for reasons of computational convenience a least-squares line may be computed on the deviations from sample means instead of the raw data. That is, if we denote

$$x_i = X_i - \bar{X} \qquad \text{and} \qquad y_i = Y_i - \bar{Y}$$

then we have, for the ith observation,

$$Y_i = \alpha + \beta X_i + u_i$$

Table 3.2 Fictitious Marks of Students in a Statistics Course (Y) and Scores on a Quantitative Aptitude Test (X) Taken Prior to the Course

Student Number	X	Y	Student Number	X	Y	Student Number	X	Y
1	47	55	36	16	34	71	20	25
2	41	59	37	29	35	72	48	52
3	30	32	38	61	67	73	52	39
4	50	62	39	59	52	74	44	61
5	14	35	40	39	43	75	38	47
6	56	57	41	58	56	76	60	59
7	55	47	42	47	53	77	57	62
8	58	55	43	28	31	78	58	56
9	52	62	44	41	40	79	42	48
10	57	50	45	52	67	80	37	32
11	53	58	46	40	52	81	38	42
12	26	30	47	56	59	82	48	43
13	19	36	48	14	26	83	51	58
14	55	50	49	33	29	84	57	52
15	38	42	50	29	45	85	27	29
16	46	57	51	31	47	86	51	47
17	59	56	52	52	49	87	29	44
18	60	63	53	47	42	88	38	39
19	39	43	54	54	69	89	45	47
20	37	47	55	49	49	90	36	50
21	28	49	56	41	39	91	37	42
22	24	43	57	28	33	92	39	51
23	61	64	58	47	64	93	53	52
24	43	57	59	43	52	94	61	58
25	47	46	60	57	52	95	57	60
26	49	57	61	22	50	96	39	40
27	31	35	62	36	42	97	44	61
28	54	56	63	49	60	98	60	52
29	37	40	64	58	55	99	25	28
30	38	48	65	42	40	100	33	42
31	20	42	66	19	26			
32	52	57	67	29	28			
33	58	56	68	36	41			
34	45	40	69	60	52			
35	54	59	70	48	50			

Table 3.3 Sample Quantities for Calculation of α and β

i	X_i	Y_i	X_iY_i	X_i^2	Y_i^2	$\hat{Y}_i$	$\hat{u}_i$
1	31	35	1085	961	1225	36.1	−1.1
2	19	36	684	361	1296	26.6	9.4
3	37	32	1184	1369	1024	40.8	−8.8
4	57	50	2850	3249	2500	56.6	−6.6
5	54	69	3726	2916	4761	54.2	14.8
6	51	52	2652	2601	2704	51.8	−0.2
7	33	29	957	1089	841	37.6	−8.6
8	58	56	3248	3364	3136	57.3	−1.3
9	19	26	494	361	676	26.6	−0.6
10	41	40	1640	1681	1600	43.9	−3.9
11	43	52	2236	1849	2704	45.5	6.5
Sum	443	477	20,756	19,801	22,467	477.6	−0.6

$\bar{X} = 40.27$; $\bar{Y} = 43.36$
$(\sum X_i)^2 = 196{,}249$
$(\sum X_i)(\sum Y_i) = 211{,}311$

an equivalent expression

$$y_i = \beta x_i + u_i \tag{3.17a}$$

Or, graphically, we are making the following transformation, as shown in Figures 3.3*a* and 3.3*b*.

From the computational point of view, $\hat{\beta}$ for β in (3.17a) is obtained in exactly the same manner as the formula in (3.20a) and is

$$\hat{\beta} = \frac{\sum x_i y_i}{\sum x_i^2} \tag{3.20f}$$

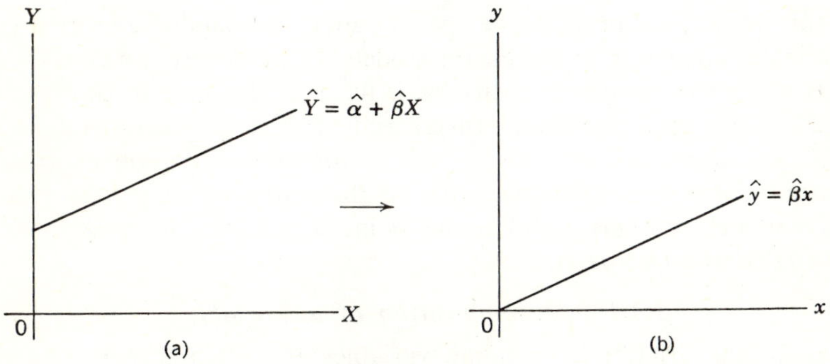

Figure 3.3*a* and *b*

It is easy to verify that (3.20a) is the same as the formula using raw data. Another way of looking at (3.20f), of course, is to regard the transformed data, or the deviations from respective sample means, as raw data and to consider minimization of the sum of squares around the line as shown in Figure 3.2. In the case, then, β is the only parameter to be estimated and the sum of squares is a function only of $\hat{\beta}$ in the sense of (3.18). Interestingly, we notice that a relationship exists between the regression coefficient estimate $\hat{\beta}$ and the sample correlation of X and Y in the way that corresponds to the population situation in (3.15). Multiplying $(\sqrt{\sum y_i^2} / \sqrt{\sum y_i^2})$ to (3.20f), we have

$$\begin{aligned}\hat{\beta} &= \frac{\sum x_i y_i}{\sum x_i^2} \cdot \frac{\sqrt{\sum y_i^2}}{\sqrt{\sum y_i^2}} \\ &= \frac{\sum x_i y_i}{\sqrt{\sum x_i^2 \sum y_i^2}} \cdot \frac{\sqrt{\sum y_i^2}}{\sqrt{\sum x_i^2}} \\ &= r \frac{\sqrt{\sum y_i^2/(n-1)}}{\sqrt{\sum x_i^2/(n-1)}} \\ &= r \frac{s_y}{s_x} \qquad (3.20g)\end{aligned}$$

As mentioned previously and as we shall observe later, the least-squares criterion is but one of the many methods available for analyzing a linear regression model. To determine how powerful the least-squares method is and to consider some of the method's limitations, we shall discuss, in the remainder of this chapter, several types of regression models frequently encountered in economic and other research.

3.3. The Fixed Models

In this section we begin to be specific about probabilistic properties of the variables appearing in regression models. In particular, we consider the models where the independent variable is fixed, in the sense to be explained later, and where the disturbance term is specified to have a general or unknown distribution in one model and to have a normal distribution in another. In order to have a good appreciation of the statistical properties of a regression model, it seems useful at this point to consider the decomposition of a variable into two parts.

3.3.1. Decomposition of a Variable

In analyzing variation in economic variables, it is often proper to imagine and to treat the variables as decomposable into two parts—the systematic part and the random part. This is because the variation of a variable, either by nature or by design, may follow a systematic rule that is identifiable by experience and/or theoretical considerations; but the systematic rule of variation does not constitute all of the variation in the variable. That part of the variation not accounted for by the systematic rule is, then, the random part. For illustration, consider the Keynesian consumption function

$$C = \alpha + \beta Y \tag{3.21}$$

where C = Consumption expenditures, Y = Income, and $0 < \beta < 1$. Now this relation may hold up very well if the averages of consumption and incomes for groups of people are concerned. But, if we look at an individual's consumption expenditures and income, it may be that (3.21) holds true but with some degree of error. That is, it is most likely that the systematic pattern in (3.21) is violated for an individual because of his taste, outlook, etc. To elaborate, two individuals may have the same amount of income and the same behavior parameters α and β but may have different consumption expenditures. Hence, if we consider C as a random variable, the systematic part of C is regularly related to Y through the expression $\alpha + \beta Y$ and the unexplainable part due to taste, etc. is the random part.

In general, variable X may be written

$$X = \mu_x + \epsilon \tag{3.22}$$

where μ_x is systematic and ϵ is random. Furthermore, it is often possible that

$$\mu_x = f(Z) \tag{3.23}$$

where Z is another variable or is a set of variables and f is any function. A special case of interest is when $f(Z)$ is linear in the parameters, say,

$$f(Z) = a + bZ \tag{3.24}$$

By combining the preceding three expressions, we see that we have a linear regression model of X on Z.

3.3.2. Fixed Model A

We now discuss the first of the fixed models using a more familiar notation of a linear regression equation

$$Y = \alpha + \beta X + u \tag{3.25}$$

What we call Fixed Model A is sometimes also referred to as the classical linear regression model and is characterized by three assumptions about the variables in the model. The assumptions are:

The values of the independent variable X are fixed (3.26a)

The expectation of u is zero, or $Eu = 0$ (3.26b)

The conditional variance of Y given X is a constant, or (3.26c)*

$$\text{var}(Y \mid X) = \sigma^2 \qquad \text{or} \qquad E(u^2) = \sigma^2$$

We shall dwell on the meaning of the assumptions (3.26) at some length as a thorough appreciation of the assumptions is indispensible to the proper application of the model.

INDEPENDENT VARIABLE FIXED. The word fixed is used in direct contrast with the term random. When one says that the values of the independent variable are fixed, he could mean any of the following situations: the variable has values that are known or controllable; the values of the variables are fixed from sample to sample, leaving only the dependent variables to change; when a number of populations enter the model, the independent variable has a fixed value (constant) for each population; when a bivariate distribution of a population is in question, the independent variable has values that are known or given and nonstochastic.

EXPECTATION OF THE DISTURBANCE. As we know, if the expectation of u exists, it is defined as

$$E(u) = \int_{-\infty}^{\infty} uf(u)\,du$$

for u continuous. It is a weighted sum of all the values of u. (It is hoped that the discrete version is obvious to the reader.) Intuitively, the weights are probabilities of different values of u. What the assumption of (3.26b) means is that, on the average, Y's systematic part is $(\alpha + \beta X)$. Or, mathematically,

$$E(Y) = \alpha + \beta E(X) = \alpha + \beta X$$

This also means that, in the population, $[E(X), E(Y)]$ is a point on the regression line of Y on X. (See Expression 3.15.)

If, in fact, $E(u)$ were not zero, but its expectation were $E(u) = \gamma Z$, then the systematic part of Y would be incorrectly specified. One should add the term γZ to the original specification, that is,

$$Y = \alpha + \beta X + \gamma Z + u'$$

where, for the new error term u', $E(u') = 0$.

* For an additional assumption about the disturbance u_i, see Section 3.4.2.

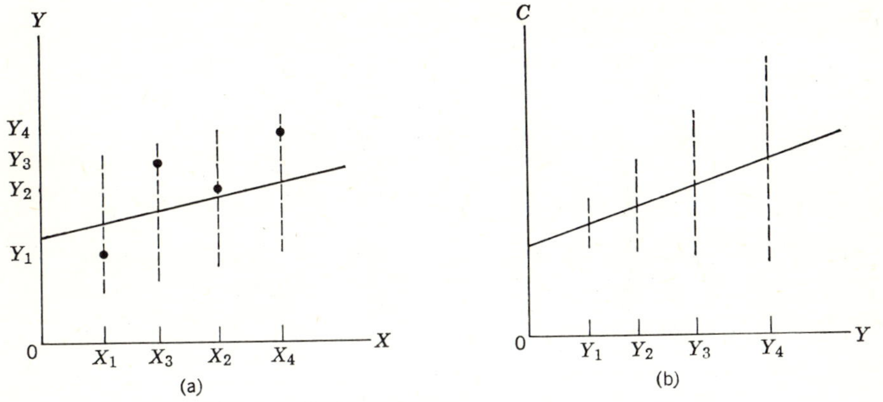

Figure 3.4*a* and *b*

CONSTANT VARIANCE. Since the variance of u is $Eu^2 = E(Y - \alpha - \beta X)^2$ from the model constancy of the (conditional) variance of Y given X means that each Y for a given fixed value of X has the same variance. Graphically, the constancy of conditional variance may be illustrated in Figure 3.4*a* where for each X_i corresponds a Y_i and where var $(Y_i|X_i) = \sigma^2$ for any i. The dotted lines show that the variation in $(Y \mid X)$ is the same, although the value of X may be increasing and at the same time may be causing Y to increase on the average. This case of constant variance is called homoscedasticity. Its meaning is best understood in relation to the phenomenon of heteroscedasticity where the conditional variance of Y given X may vary over the range of values of X. Heteroscedasticity is often observed in cross-section studies where, for instance, variation in consumption expenditures (C) gets larger as level of income (Y) rises. This is illustrated in Figure 3.4*b*. We shall discuss in detail the problem of heteroscedasticity in Chapter 6 where some special problems are treated in some generality. Referring to the second and third assumptions in (3.26), we notice that we are saying that the disturbance has a probability distribution with mean = 0 and variance = σ^2. It is not specified just what distribution (shape of spread not only the size of spread) u has. This is the advantage as well as the disadvantage of Fixed Model A; it is an advantage in the sense that in many situations the distribution of u is not exactly known but yet one wants to proceed with an analysis of the linear dependence relation by the use of the linear regression model (where one assumes only that u has a mean and a constant variance), and a disadvantage in the sense that what one can do in the way of statistical inference in a regression model when u's shape of distribution is unknown may be rather limited. This last point will become clear as we deal with the inferential questions about the fixed models in Section 3.4.

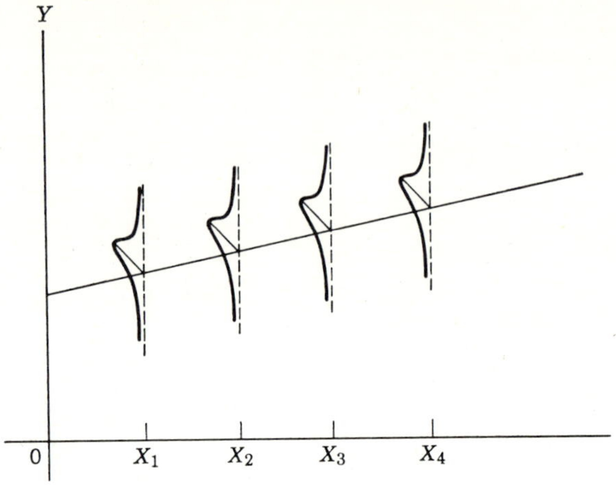

Figure 3.5

3.3.3. *Fixed Model B*

Now, if we assume that the distribution of u is normal in addition to the assumptions in (3.26), we have a linear regression model sometimes referred to as the classical normal linear regression model. Here we call it Fixed Model B. An illustration of the distribution of u in this case is shown in Figure 3.5. It is the intention of Figure 3.5 to show that u is normal for any value of X, and for each value of X the variance of u is a constant (the spread is the same). It should be clear that Model B is a special case of Model A.

We now discuss some questions of inference with regard to the fixed models.

3.4. Statistical Inference in Fixed Models

In this section we are concerned with the estimating and the testing of parameters in the fixed models. First, we review some desirable properties of statistical estimators together with some utilitarian comments. The review is standard, so that some readers may wish to proceed to Subsection 3.4.2 directly. For an excellent discussion of the properties of statistical estimators, see Christ (1966, pp. 251–277).

3.4.1. *Some Properties of a Statistical Estimator*

Let X now be a random variable. Suppose that X has the probability distribution $f(X; \theta_1, \theta_2, \ldots, \theta_p)$ where θ_i are the parameters of the distribution. The problem in estimation is to use sample observations on X and to make guesses as to the true values of $\theta_1, \theta_2, \ldots, \theta_p$. For convenience, we use θ as the only parameter characterizing the density $f(X)$.

An estimator $\hat{\theta}$ of a parameter θ is a function of sample observations, for example,

$$\hat{\theta} = \delta(X_1, X_2, \ldots, X_n) \tag{3.27}$$

where X_i is the ith observation on the random variable X and θ may be the mean of X or the variance of X, or some other parameter. Since, in general, we consider random samples, we shall regard an estimator as a random variable because, for instance, the X_i in (3.27) are random. It follows that an estimator usually has a probability distribution or what is also called a sampling distribution. Do not confuse an estimate with an estimator. The former is a fixed value obtained by plugging values of X into a function, such as the one in (3.27), but the latter is the formula or form of the function through which an estimate is found. An estimator is random if the sample observations are random.

Now, of course, one would like to have a "good" estimator. Several criteria of a good estimator exist, and the choice of a criterion or of a set of criteria jointly depends much on the purposes for which the estimator is to be used. Estimators can be classified according to the following properties, a subset of which may be used as a criterion of a good estimator.

1. The types of functions used, such as linear or quadratic estimator.
2. The value on the average, such as an unbiased estimator.
3. The degree of concentration about the value to be estimated, such as a minimum variance estimator.
4. The mean and variance of the limiting distribution, such as an asymptotically efficient estimator.
5. The type of stochastic limit, such as a consistent estimator.

Most estimators satisfy more than one of these properties. For example, an estimator $\hat{\mu}$ of the mean of X, μ, for X normal,

$$\hat{\mu} = \frac{1}{n}\sum_{i=1}^{n} X_i$$

is a linear estimator of μ, is unbiased, has minimum variance (in the class of unbiased estimators), and is consistent (also asymptotically efficient). We now discuss a few of the more desirable properties of an estimator and comment on the usefulness of these properties. For our purposes, the meaning of the properties of an estimator as used in applied work is of greater interest than the mathematical analysis of these properties.

UNBIASEDNESS. From (3.27) and the comments that follow, the estimator θ, for example, defined by

$$\hat{\theta} = \delta(X_1, X_2, \ldots, X_n)$$

where $X_1, X_2, \ldots, X_n$ from a random sample, can have a probability distribution, say $f(\hat{\theta})$, so that its expectation may be discussed. We say that the estimator $\hat{\theta}$ is unbiased for θ if

$$E(\hat{\theta}) = \int_{-\infty}^{\infty} \hat{\theta} f(\hat{\theta})\, d\hat{\theta} = \theta \tag{3.28}$$

or

$$E(\hat{\theta} - \theta) = 0 \tag{3.28a}$$

Intuitively, mathematical expectation is a weighted average, so that we might say that, if we follow the scheme δ above in estimating θ sample after sample for samples of fixed size n, the estimates thus obtained will have a weighted average equal to θ. The weights are the probabilities of the values of $\hat{\theta}$ generated by the scheme δ. Thus, unbiasedness can be had as long as all the deviations of parameter estimates from the true parameter value cancel out. For instance, for $E\hat{\theta} = \theta = 0$, if the estimator $\hat{\theta}$ yields estimates $\frac{1}{2}$ and $-\frac{1}{2}$ with the same probability, or for that matter, -50 and $+50$ with the same probability, $\hat{\theta}$ will still satisfy the criterion of unbiasedness. That is, the criterion merely depends on the condition (3.28a), which is satisfied as long as the deviations in the form of $(\hat{\theta} - \theta)$ cancel out. Since the deviation of 50 is much larger than that of $\frac{1}{2}$ and since a smaller deviation implies a more precise estimator, naturally, a search is often made to find an estimator that on the average yields estimates with as small deviations as possible from the true parameter. This motivates the discussion of a minimum variance estimator.

MINIMUM VARIANCE. An estimator $\hat{\theta}$ is said to be a minimum variance estimator for θ if, for a given sample size,

$$E[\hat{\theta} - E(\hat{\theta})]^2 \leq E[\tilde{\theta} - E(\tilde{\theta})]^2 \tag{3.32}$$

where $\tilde{\theta}$ is any other estimator of θ, and $E(\hat{\theta}) = E(\tilde{\theta})$. Notice that $E(\hat{\theta})$ or $E(\tilde{\theta})$ does not necessarily equal θ. Thus, if one had chosen an estimator on the criterion of minimum variance, he might obtain $\hat{\theta}$ as compared with $\tilde{\theta}$ as illustrated in Figure 3.6*a*.

BEST UNBIASEDNESS. An estimator $\hat{\theta}$ is said to be best unbiased for θ if $\hat{\theta}$ is unbiased for θ *and* has a minimum variance in the sense of (3.32). Best unbiasedness is a property found in the class of unbiased estimators.

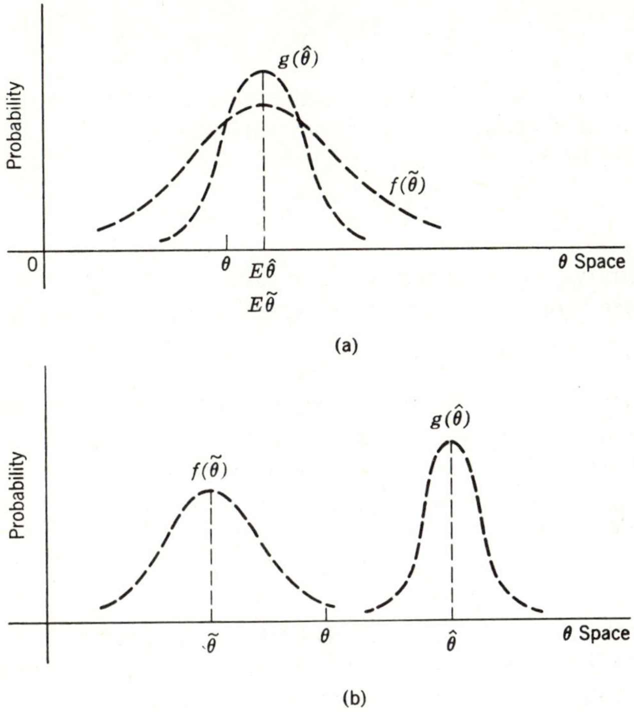

Figure 3.6*a* and *b*

MINIMUM MEAN-SQUARE-ERROR. It is possible that two estimators of θ, say $\hat{\theta}$ and $\tilde{\theta}$, do not have the same expectation and are each biased for θ. Consider Figure 3.6*b* where $\hat{\theta}$ is farther away from θ than is $\tilde{\theta}$ but has much smaller spread than $\tilde{\theta}$. Now the mean-square-error of estimator θ is defined as

$$\begin{aligned} E[\hat{\theta} - \theta]^2 &= E\{[\hat{\theta} - E(\hat{\theta})] + [E(\hat{\theta}) - \theta]\}^2 \\ &= E[\hat{\theta} - E(\hat{\theta})]^2 + E[E(\hat{\theta}) - \theta]^2 \end{aligned} \tag{3.33}$$

the first term in the last expression measuring the sample variability of $\hat{\theta}$ and the second term measuring the average squared bias. Thus (3.33), when used as a criterion for minimization, takes into account both bias and sampling variation of an estimator simultaneously. It is sometimes the case that an estimator with some bias but with a small spread may be preferred to an estimator that has little or no bias but a large spread.

CONSISTENCY. Suppose that we can, physically or conceptually, take larger and larger samples so that instead of a fixed sample estimator

$$\hat{\theta} = \delta(X_1, X_2, \ldots, X_n)$$

we have

$$\hat{\theta}_n = \delta_n(X_1, X_2, \ldots, X_n) \tag{3.29}$$

where n is a variable sample size, although the form of δ is unchanged. We say that if a sequence of random variables $\hat{\theta}_1, \hat{\theta}_2, \ldots, \hat{\theta}_n$ converges in probability to θ or if (in notation)

$$\lim_{n\to\infty} \text{Prob}\,\{|\hat{\theta}_n - \theta| > \epsilon\} = 0 \tag{3.30}$$

for an arbitrary positive ϵ, then $\hat{\theta}_n$ is consistent for θ. When it is clear that $\hat{\theta}_n$ is a sequence of random variables, we write (3.30) equivalently as

$$\text{plim}\,\hat{\theta}_n = \theta \qquad \text{or} \qquad \text{plim}\,\hat{\theta} = \theta \tag{3.31}$$

What this says is that where the sample is large, θ_n becomes close to θ and that the probability that $\hat{\theta}_n = \theta$ is very close to 1. Closely related to this concept of consistency (referred to sometimes as simple consistency) is consistency in mean square. This concept requires that the sequence $\hat{\theta}_n$ defined in (3.30) satisfy

$$\lim_{n\to\infty} E(\hat{\theta} - \theta)^2 = 0 \tag{3.32}$$

for all θ in the parameter space. Thus both the variance of $\hat{\theta}_n$ and the bias vanish in the limit. Generally, when $\hat{\theta}_n$ is unbiased, condition (3.32) ensures that (3.31) hold true, so that these are used to show the consistency of an estimator.

That is, if it can be shown that the estimator $\hat{\theta}$ meets the conditions that

$$E\hat{\theta} = \theta$$

and

$$\lim_{n\to\infty} \text{var}\,(\hat{\theta}_n) = 0 \qquad [(3.32)]$$

then $\hat{\theta}$ is consistent for θ. This means that the probability distribution of a consistent estimator $\hat{\theta}$ is degenerate at parameter θ (or is concentrated on θ) in the limit. Figure 3.7 illustrates one pattern of the densities of $\hat{\theta}_n$ for $n = 10$, $n = 50$, and $n = 100$.

Asymptotic efficiency. In the class of consistent estimators some of them may have smaller asymptotic variance than others. Hence, it stands to reason that the estimator with smaller asymptotic variance would be preferred asymptotically. An estimator $\hat{\theta}_n$ is asymptotically unbiased if

$$\lim_{n\to\infty} E\hat{\theta}_n = 0$$

Thus for $\hat{\theta}_n = \delta_n(X_1, X_2, \ldots, X_n)$ asymptotically unbiased for θ and with

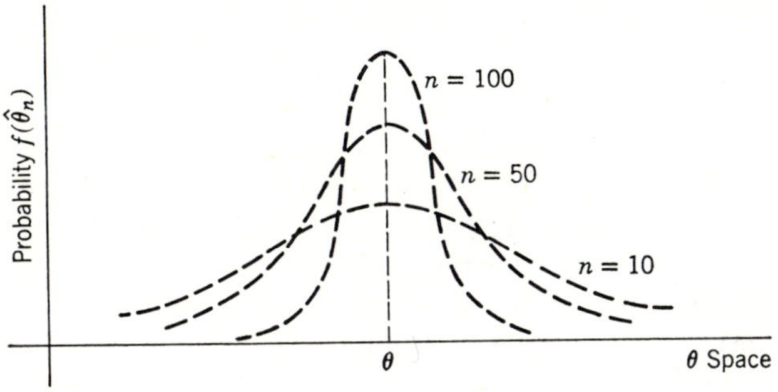

Figure 3.7

the property that

$$E(\hat{\theta}_n - \theta)^2 \le E(\tilde{\theta}_n - \theta)^2 \qquad \text{for } n \to \infty \tag{3.34}$$

where $\tilde{\theta}_n$ is any other asymptotically unbiased estimator of θ, we say that $\hat{\theta}_n$ is asymptotically efficient. Many estimators of practical interest are normally distributed in the limit because of the central limit theorem, so that for large samples the distribution of these estimators depend only on their means and variances. As a consequence, the comparison of the variances becomes a good way to discriminate among these estimators.

It can be shown, under fairly general conditions (see the paragraph below) that if an estimator exists that has the properties of consistency and asymptotic normality and efficiency, the maximum likelihood estimator will give such properties. We review this last estimation criterion next.

Maximum likelihood estimation. Up to this point we have taken the form of an estimator to be given and have discussed the properties of the estimator. Here, we shall consider an estimator the form or derivation of which is guided by a principle called maximum likelihood (ML). This is a very general method, applicable in many problems of practical interest.

Suppose that we have a random sample $X_1, X_2, \ldots, X_n$ from the distribution of a random variable X, say $f(X; \theta)$, which satisfies a set of fairly general conditions. For the conditions, see for example, Cramer (1946, pp. 500–501). Now the joint density of the sample observations is

$$\prod_{i=1}^{n} f(X_i; \theta) \tag{3.35}$$

This is a density with the parameter θ given and X_i variable. Suppose now that we consider this density as a function of θ with X_i given, then the density is called the likelihood function of the random sample (or the probability of the sample) and is, in general, denoted by

$$L(\theta; X_1, X_2, \ldots, X_n) \tag{3.35a}$$

Why consider (3.35a) a function of θ with X_i given? We might think that in our sampling process Nature has given us the most probable sample observations on X, so that following through this assumption we might estimate the parameter θ from this sample in such a way as to maximize the likelihood. The procedure, then, is to take the first and second derivatives of L with respect to θ and to find the solution for θ in terms of sample observations at which L attains a maximum. An illustration follows.

Let random variable X be distributed $N(\mu, 1)$ and assume that interest centers on finding a maximum likelihood estimator of μ. The joint density of a random sample of size n is

$$L(\mu, 1; X_1, X_2, \ldots, X_n) = \left(\frac{1}{\sqrt{2\pi}}\right)^n \prod_{i=1}^{n} e^{-\frac{1}{2}(X_i-\mu)^2} \tag{3.36}$$

We notice that, if logarithmic transformation is applied to (3.36), the transformed function will be easier to differentiate and yet the transformation leaves the point at which L is maximum invariate.

$$\mathscr{L} = \ln L = n \ln \frac{1}{\sqrt{2\pi}} + \left(-\tfrac{1}{2}\sum_{i=1}^{n}(X_i - \mu)^2\right) \tag{3.37}$$

Thus

$$\frac{\partial \mathscr{L}}{\partial \mu} = \sum_{i=1}^{n}(X_i - \mu) = 0$$

Therefore

$$\hat{\mu} = \frac{\sum X_i}{n}$$

is a maximum likelihood estimator of μ. (Verify that the second-order condition for a maximum is satisfied.)

It is also a property of a maximum likelihood estimator that it is invariant with respect to functional transformation. That is, if $\hat{\theta}$ is a maximum likelihood estimator for θ, then $f(\hat{\theta})$ is a maximum likelihood estimator of $f(\theta)$ if f is single-valued.

In concluding our discussion of the properties of a statistical estimator we might ask: What is the use of all these? The answer, in part, may have been given in the course of our discussion, but the main point is that we are interested in the reliability of an estimate in probability terms and sample size and in the comparison of reliability. These notions are useful for research planning and sampling design and for serving as a convention for communication among research workers.

3.4.2. *The Least-Squares (LS) Coefficient Estimators*

In this subsection we discuss some standard properties of the least-squares coefficient estimators. We first consider the implications that the standard assumptions in the fixed models have on the sample observations. We revert to the situation where X is not random.

THE SAMPLE AND THE MODEL. The parameters of interest in the fixed models of the form

$$Y = \alpha + \beta X + u \tag{3.17}$$

are α, β, and $E(u^2) = \sigma^2$. Now, if a random sample for Y is $Y_1, Y_2, \ldots, Y_n$ and their corresponding fixed X's are, respectively, $X_1, X_2, \ldots, X_n$, then (3.17) means that

$$Y_i = \alpha + \beta X_i + u_i \tag{3.39}$$

for $i = 1, 2, \ldots, n$. Furthermore, the last two assumptions in (3.26) imply that

$$E(u_i) = 0 \tag{3.40}$$

and that

$$E(u_i^2) = \sigma^2 \tag{3.41a}$$

for $i = 1, 2, \ldots, n$. In addition, we assume, as a part of the standard assumptions (3.25, a to c) that

$$E(u_i u_j) = 0 \tag{3.41b}$$

for $i \neq j$. This last condition means that the disturbance, say u_1, is at least uncorrelated with any other subscripted u. In other words, given $u_1, u_2, \ldots, u_n$, $E(u_1u_2) = E(u_1u_3) = \cdots = E(u_1u_n) = E(u_2u_3) = E(u_2u_4) = \cdots = 0$, etc. If the subscript denotes time period, the condition (3.41b) means the error this time period is not correlated with the error next period, nor with the error two periods hence, etc.—namely, lack of autocorrelation with any number of lags.

It follows from the randomness of u_i that the latter is at least uncorrelated with X_i, since X_i is fixed.

THE *LS* ESTIMATORS α AND β ARE BLUE. By BLUE we mean best linear unbiased estimator. The estimators as shown in (3.20 a to b) are linear combinations of Y_i's with the coefficients of the combinations determined by the X_i's. Take, for instance,

$$\hat{\beta} = \frac{n \sum X_i Y_i - (\sum X_i)(\sum Y_i)}{n \sum X_i^2 - (\sum X_i)^2} \tag{3.20a}$$

Here if we let

$$w = n \sum X_i^2 - (\sum X_i)^2$$

(3.20a) becomes

$$\hat{\beta} = n \sum \left(\frac{X_i}{w}\right) Y_i - \sum \left(\frac{X_i}{w}\right)\left(\sum Y_i\right)$$

Also, by noticing that w is fixed for a given sample and by taking the expectation of (3.20a), we have

$$\begin{aligned} E\hat{\beta} &= \frac{1}{w} E\left[n \sum X_i Y_i - \sum X_i \sum Y_i\right] \\ &= \frac{1}{w} E\left[n \sum X_i(\alpha + \beta X_i + u_i) - \sum X_i \sum (\alpha + \beta X_i + u_i)\right] \\ &= \frac{1}{w} \left\{n\left(\alpha \sum X_i + \beta \sum X_i^2\right) - \left[n\alpha \sum X_i + \beta\left(\sum X_i\right)^2\right]\right\} \\ &= \frac{1}{w} \beta\left[n \sum X_i^2 - \left(\sum X_i\right)^2\right] \\ &= \beta \end{aligned}$$

Thus, $\hat{\beta}$ is unbiased for β. The student is invited to show, by using (3.20b), that $\hat{\alpha}$ is unbiased. It can also be shown that $\hat{\beta}$ has the minimum-variance property, but we shall let the proof be given in a more general way in Chapter 4. It suffices here to indicate that there are two ways to arrive at the BLUE-ness:* (1) by getting the least-squares estimators and by checking to determine if the criteria of linearity, unbiasedness, and minimum variance are satisfied by the estimators; and (2) by assuming that BLUE-ness is the goal and by obtaining a linear estimator, say, $\sum a_i Y_i = \theta_e$, so that the unbiasedness condition is satisfied [so that a_i's and X_i's are expressed to satisfy $E(\theta_e) = \theta$, say], and then by using these conditions as side conditions for the minimum variance of the linear estimator.

Since $\hat{\alpha}$ and $\hat{\beta}$ are estimators, they will have sampling distributions. The variances and the covariance of the BLUE $\hat{\alpha}$ and $\hat{\beta}$ are†

$$\operatorname{var}(\hat{\beta}) = E(\hat{\beta} - \beta)^2 = \frac{1}{\sum (X_i - \bar{X})^2} \sigma^2 \tag{3.42a}$$

$$\operatorname{var}(\hat{\alpha}) = E(\hat{\alpha} - \alpha)^2 = \frac{\sum X_i^2}{n \sum (X_i - \bar{X})^2} \sigma^2 \tag{3.42b}$$

$$\operatorname{cov}(\hat{\alpha}, \hat{\beta}) = E(\hat{\alpha} - \alpha)(\hat{\beta} - \beta) = -\frac{\sum X_i}{n \sum (X_i - \bar{X})^2} \sigma^2 \tag{3.42c}$$

* An excellent exposition of these two alternative procedures can be found in Johnston (1963, pp. 14–19).

† See, for example, A. M. Mood and F. A. Graybill (1963, pp. 333–343).

As an exercise, let us develop (3.42a) below. First,

$$\begin{aligned}
\hat{\beta} - \beta &= \frac{1}{w}\left(n \sum X_i Y_i - \sum X_i \sum Y_i\right) - \beta \\
&= \frac{1}{w}\left[n \sum X_i(\alpha + \beta X_i + u_i) - \sum X_i \sum (\alpha + \beta X_i + u_i)\right] - \beta \\
&= \frac{1}{w}\Big[n\left(\alpha \sum X_i + \beta \sum X_i^2 + \sum X_i u_i\right) \\
&\qquad - \sum X_i\left(n\alpha + \beta \sum X_i + \sum u_i\right)\Big] - \beta \\
&= \frac{1}{w}\left\{\beta\left[n \sum X_i^2 - \left(\sum X_i\right)^2\right] + n \sum X_i u_i - \sum X_i \sum u_i\right\} - \beta \\
&= \beta + \frac{n}{w}\left(\sum X_i u_i - \bar{X} \sum u_i\right) - \beta \\
&= \frac{\sum X_i u_i - \bar{X} \sum u_i}{\sum X_i^2 - \left(\sum X_i\right)^2/n} \\
&= \frac{\sum X_i u_i - \bar{X} \sum u_i}{\sum (X_i - \bar{X})^2}
\end{aligned}$$

By squaring and taking the expectation of $(\hat{\beta} - \beta)^2$, we have

$$\begin{aligned}
E(\hat{\beta} - \beta)^2 &= \frac{E\left(\sum X_i u_i - \bar{X} \sum u_i\right)^2}{\left[\sum (X_i - \bar{X}^2\right]^2} \\
&= \frac{E\left[\left(\sum X_i u_i\right)^2 - 2\left(\sum X_i u_i\right)\bar{X}\left(\sum u_i\right) + \bar{X}^2\left(\sum u_i\right)^2\right]}{\left[\sum (X_i - \bar{X})^2\right]^2} \\
&= \frac{\sum X_i^2 \sigma^2 - 2n\bar{X}^2\sigma^2 + n\bar{X}^2\sigma^2}{\left[\sum (X_i - \bar{X})^2\right]^2} \\
&= \frac{\sigma^2\left(\sum X_i^2 - n\bar{X}^2\right)}{\left[\sum (X_i - \bar{X})^2\right]^2} \\
&= \frac{\sigma^2}{\sum (X_i - \bar{X})^2}
\end{aligned}$$

since

$$\sum X_i^2 - n\bar{X}^2 = \sum X_i^2 - \frac{\left(\sum X_i\right)^2}{n} = \sum (X_i - \bar{X})^2$$

SOME LEAST-SQUARES PROPERTIES. It is of some interest and significance to observe the relationships that arise between the least-squares estimates of $\hat{Y}_i$ and the actual Y_i themselves. The relationships are of three related results.

1. The sum of deviations of observed Y_i from the least-squares estimated Y_i is zero. That is, the deviations from the least-squares line sum to zero. Or

$$\begin{aligned} \sum \hat{u}_i &= \sum (Y_i - \hat{Y}_i) \\ &= \sum Y_i - \sum \hat{Y}_i \\ &= \sum Y_i - \sum Y_i && \text{[by (3.19a)]} \\ &= 0 && (3.43) \end{aligned}$$

Or,

$$\sum Y_i = \sum \hat{Y}_i \tag{3.44}$$

2. The sum of the cross products $X_i\hat{u}_i$ is zero. Or

$$\begin{aligned} \sum X_i\hat{u}_i &= \sum X_i(Y_i - \hat{Y}_i) \\ &= \sum X_iY_i - \sum X_i\hat{Y}_i \\ &= \sum X_iY_i - [\hat{\alpha} \sum X_i + \hat{\beta} \sum X_i^2] && (3.45) \\ &= \sum X_iY_i - X_i \sum Y_i && \text{[by (3.19b)]} \\ &= 0 \end{aligned}$$

3. $\sum \hat{Y}_i\hat{u}_i = 0$. Or, the sum of cross products of the estimated value of Y_i with its corresponding error of estimate is zero. Notice that this condition is not the same as $\sum Y_i\hat{u}_i = 0$, which does not hold true. The result is shown as follows.

$$\begin{aligned} \sum \hat{Y}_i\hat{u}_i &= \sum \hat{Y}_i(Y_i - \hat{Y}_i) \\ &= \sum \hat{Y}_iY_i - \sum \hat{Y}_i\hat{Y}_i \\ &= \sum Y_i(\hat{\alpha} + \hat{\beta}X_i) - \sum \hat{Y}_i(\hat{\alpha} + \hat{\beta}X_i) && (3.46) \\ &= \hat{\alpha} \sum (Y_i - \hat{Y}_i) + \hat{\beta}X_i \sum (Y_i - \hat{Y}_i) \\ &= \hat{\alpha} \cdot 0 - \hat{\beta} \cdot 0 && \text{[by (3.42) and (3.44)]} \\ &= 0 \end{aligned}$$

These three results have very intuitive correspondence to some theoretical assumptions in regression analysis. We notice that (1) the assumption $E(u) = 0$ has an intuitive expression in (3.43), (2) the assumption of independence between X and u seems directly reflected in (3.45); and (3) the fact that Y is assumed to be a linear function of fixed X leads to the result in (3.46) as the consequence of (3.45).

We advise checking the results in (3.44) against the quantities already calculated in Table 3.2 (notice any round-off errors).

3.4.3. Variation about the Regression Line

Now that we have learned something about the least-squares estimators for the regression coefficients the next question of interest is: How might one evaluate the "goodness" of the fit that the estimated line gives to the sample data? The answer to this question may be given under two related concepts: (1) the standard error of estimate, and (2) the coefficient of (multiple) determination.

STANDARD ERROR OF ESTIMATE. Given that

$$Y_i = \alpha + \beta X_i + u_i \tag{3.39}$$

for $i = 1, 2, \ldots, n$ [Notice that (3.39) says that (3.17) is true of the n sample observations but n can be as large as the size of a finite population], we have observed in the preceding subsection that $\sum \hat{u}_i = 0$. This was a direct consequence of the fact that

$$\hat{u}_i = Y_i - \hat{Y}_i = Y_i - (\hat{\alpha} + \hat{\beta} X_i)$$

where $\hat{\alpha}$ and $\hat{\beta}$ are the least-squares estimates, and $\hat{u}_i$ is the estimate of u_i. Now this knowledge does not help us very much in making the analysis about the spread of Y_i's from the fitted line. One solution is to compute the sum of squares of $\hat{u}_i$, so that for samples taken from a population one can obtain different sums of squares to compare the individual fit of different equations. We represent such a sum of squares by

$$\sum \hat{u}_i^2 = \sum (Y_i - \hat{Y}_i)^2 \tag{3.47}$$

and call it an error sum of squares (ESS). Other solutions are possible. For instance, we can use

$$\sum |\hat{u}_i|$$

as a basis for comparison among equations, but this quantity is not easy to manipulate mathematically. Thus viewed, the so-called mean error sum of squares

$$\frac{1}{n}\left(\sum \hat{u}_i^2\right) \tag{3.48}$$

seems to be a reasonable measure of the variation of Y about the estimated line and, consequently, a "good" estimate of the $E(u^2) = \sigma^2$. The term estimate refers to the calculated value $\hat{Y}_i$ for the corresponding Y_i, and the quantity in (3.48) measures a variability of $(Y \mid X) = (\alpha + \beta X_i)$ as reflected in the distance of Y_i from the estimated line (vertically along the Y-axis). It can be shown that the quantity in (3.48) is a maximum likelihood estimate of $E(u^2)$ if the u_i's are normally and identically distributed so that it has all the nice asymptotic properties that can be imparted by a maximum likelihood estimator. But, with or without the normality assumption, a small sample bias exists in (3.48). Indeed, for small samples, we develop the unbiased estimator for $E(u^2)$ as follows.

From (3.39), we can write

$$\begin{aligned} u_i &= Y_i - (\alpha + \beta X_i) \\ &= (Y_i - \hat{Y}_i) + [\hat{Y}_i - (\alpha + \beta X_i)] \end{aligned}$$

(by adding and subtracting $\hat{Y}_i$). By squaring both sides, summing over n, and taking expectation, we have

$$E(\sum u_i^2) = E\{\sum (Y_i - \hat{Y}_i)^2 + \sum [(\hat{\alpha} - \alpha) + (\hat{\beta} - \beta)X_i]^2\} \tag{3.50}$$

since $E\hat{u}_i[\hat{Y}_i - (\alpha + \beta X_i)] = 0$ for all i. By combining (3.50) and (3.42 a to c), and by collecting terms, (3.50) resolves into

$$n\sigma^2 = E[\sum (Y_i - \hat{Y}_i)^2] + 2\sigma^2$$

Or

$$E\frac{\sum \hat{u}_i^2}{n-2} = \sigma^2 \tag{3.51}$$

It follows then that $\sum \hat{u}_i^2/(n-2)$ is unbiased for σ^2; the standard error of estimate is usually defined to be the positive square root of the unbiased estimate and is denoted S_e.

Now, if the number of observations is small, it is not too difficult to compute S_e^2 but, if n is large, the computation of S_e^2 by the expression shown under the expectation in (3.51) becomes tedious. A convenient computational formula is

$$S_e^2 = \frac{1}{n-2}\left\{\sum Y_i^2 - \frac{1}{n}(\sum Y_i)^2 - \hat{\beta}\left[\sum X_i Y_i - \frac{1}{n}(\sum X_i)(\sum Y_i)\right]\right\} \tag{3.52}$$

we observe that once $\hat{\beta}$ is found the quantities necessary for computation of S_e are available from Table 3.1 in an earlier section.

Coefficient of determination. The significance of the size of S_e from a given sample is not at all obvious unless one is familiar with the nature of the data in the model for which S_e is computed. Thus, for a given problem, we may be able to tell that such and such sizes for S_e represent the small variation of the Y_i's around the regression line, but we might also ask: Small with respect to what? Besides dealing with samples from the same population, there will be occasions where there is interest in the comparison of the degree of variation around the regression line among different linear models which may or may not be concerned with the same problem. Therefore, it seems desirable to have a standard or unit-free measure of "goodness of fit," so that it can be used as a basic measure for relative comparison. It is intuitively clear that division by the variance of Y will serve just this purpose; for instance, a large or a small S_e^2, with respect to the variation

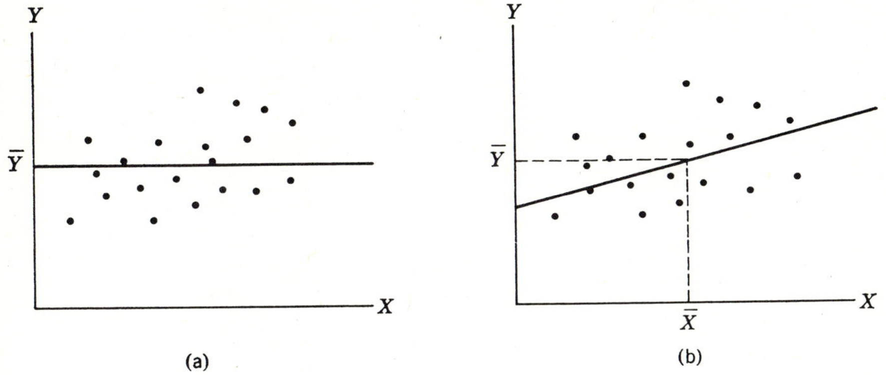

Figure 3.8*a* and *b*

of Y_i's around the sample mean, is a useful point of reference. To discuss this notion and to develop a measure called the coefficient of determination, we must discuss the decomposition of variance.

Given the variable Y and its sample observations Y_i's, it is a standard practice to estimate the variability of Y by finding the sum of squares of the deviations of Y_i's from the sample mean $\bar{Y}$. Or

$$\sum (Y_i - \bar{Y})^2 \tag{3.53}$$

Let us call this the total sum of squares (TSS). Or, in a graph, $\bar{Y}$ goes through the scatter of Y_i's as in Figure 3.8*a*; here (3.53) gives the sample variation of Y around the line $Y = \bar{Y}$ (which, incidentally, is fixed once a sample is chosen). Now the regression model, for example, of the type (3.17), says that Y is dependent on X so that the average level of Y_i moves with the given values of X. That is, $\hat{Y}_i$ is an estimate of the average of the Y values *corresponding to* or *explained by* the particular X_i. Thus, for different values of X, Y_i's form a line, and the measure (3.48), for instance, is another way of measuring the variability of Y *but* it is a variability *conditional on* X. (See Figure 3.8*b* for visual appreciation.) In other words, the variability of Y, without the aid of X to explain it, is $\sum (Y_i - \bar{Y})^2$, but the variability of Y, with X explaining it, is $\sum (Y_i - \hat{Y}_i)^2$. It seems clear that $\sum (Y_i - \bar{Y})^2 \geq \sum (Y_i - \hat{Y}_i)^2$. The equality holds true when X explains nothing of the variation in Y, that is, when the regression coefficients are zero. Indeed, we can write mathematically

$$Y_i - \bar{Y} = (Y_i - \hat{Y}_i) + (\hat{Y}_i - \bar{Y}) \tag{3.54}$$

By squaring and summing over n, we obtain

$$\sum (Y_i - \bar{Y})^2 = \sum (Y_i - \hat{Y}_i)^2 + \sum (\hat{Y}_i - \bar{Y})^2 \tag{3.55}$$

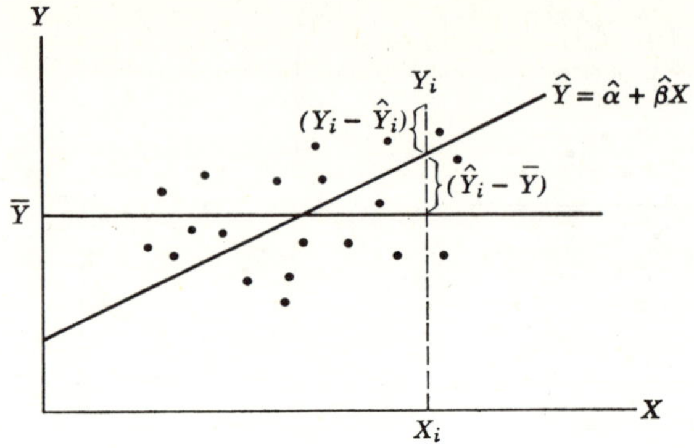

Figure 3.9

That is, the total sum of squares of Y is decomposable into the error sum of squares (ESS), $\sum (Y_i - \hat{Y}_i)^2$, and the regression sum of squares (RSS), $\sum (\hat{Y}_i - \bar{Y})^2$. Sometimes, ESS and RSS are referred to as the sum of squares due to error and the sum of squares due to regression, respectively. We advise verifying (3.55) by referring to (3.43) and (3.46). Graphically, the decomposition (3.54) is shown in Figure 3.9. If, now, we divide through (3.55) by the total sum of squares (TSS), we obtain

$$1 = ESS/TSS + RSS/TSS \tag{3.56}$$

From this, we define

$$\text{Coefficient of determination} = R^2 = 1 - ESS/TSS = RSS/TSS \tag{3.57}$$

Clearly, R^2 measures that portion of the variability of Y_i's explained by the aid of X. To gain an intuitive appreciation of this coefficient, let us imagine two extreme situations. In Figure 3.10*a*, the variation in Y is not related to

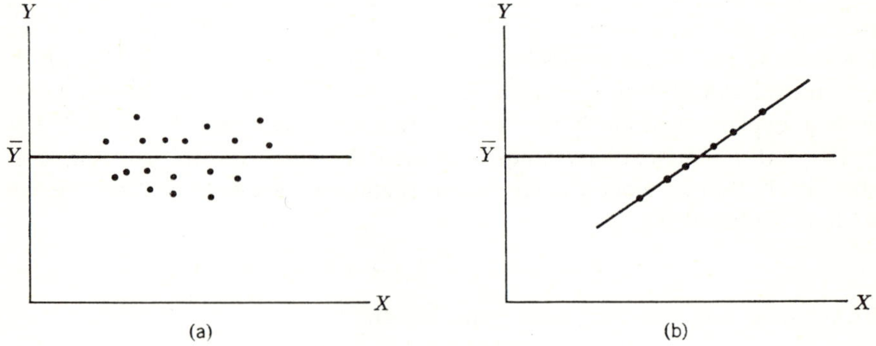

Figure 3.10*a* and *b*

the values of X. Or, we can think of the line $y = \bar{Y}$ as identically a constant for all values of X. In other words, we can say that $\sum (Y_i - \bar{Y})^2 = \sum (Y_i - \hat{Y}_i)^2$. Since $ESS = TSS$, it follows that $R^2 = 0$. By studying Figure 3.10*b*, we observe that the sample observations all fall on the estimated line. Therefore, $Y_i = \hat{Y}_i$ and, as the consequence, $\sum (Y_i - \bar{Y})^2 = \sum (\hat{Y}_i - \bar{Y})^2$. Thus, $R^2 = 1$ (since $ESS = 0$).

Usefully, we notice that in the notation of (3.17a) and (3.20f) at the end of Section 3.2, we can write

$$\sum (\hat{Y}_i - \bar{Y}) = \sum \hat{y}_i$$

so that

$$R^2 = \frac{\sum \hat{y}_i^2}{\sum y_i^2} = \frac{\sum (\hat{\beta} x_i)^2}{\sum y_i^2}$$

$$= \hat{\beta}^2 \frac{\sum x_i^2}{\sum y_i^2} = r^2$$

Thus, we see [here and from (3.20g)] that the coefficient of determination is the same as the square of the correlation coefficient between X and Y in simple regression analysis.

3.4.4. Testing Hypotheses in the Fixed Models

Thus far, we have investigated some of the properties of the least-squares estimators for α, β, and σ^2 in a simple linear regression model. We have found that, whether or not normality of u is assumed, the least-squares estimators for α and β are BLUE and the related $\sum \hat{u}_i^2/(n - 2)$ is unbiased for σ^2. Furthermore, if it is assumed that u is normal, then $\hat{\alpha}$, $\hat{\beta}$, and $\sum \hat{u}_i^2/n$ are maximum likelihood estimators for α, β, and σ^2, respectively; and, as such, these estimators will have the nice asymptotic properties indicated. These are the consequences of the classical estimation theory. With respect to the test of hypotheses about these regression parameters, the classical theory requires that u is normally distributed. Thus, in order to proceed, we shall confine ourselves to Fixed Model B for now. However, later we shall briefly show that recent work indicates that some classical tests are quite good even when the normality assumption about u is violated (robustness).

We review the test of hypothesis very briefly before continuing. A test presupposes that a parameter space exists that contains the one or more parameters of interest to us. Call this parameter θ. A hypothesis, then, is a statement about (or claim about the distribution of) θ. Under this statement or claim we set up a rule that tells us whether or not the hypothesis about θ is reasonable in the light of the sample observations that we make. A test is a device that decomposes the sample space into rejection and acceptance

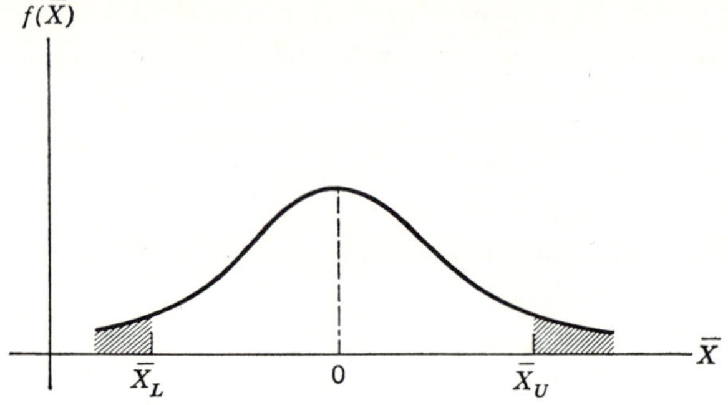

Figure 3.11

areas. For example, say that X is assumed normal with mean unknown and variance equal to σ^2. It then follows that, theoretically, if many samples of size n are obtained and as many X's calculated, the X's will form a normal distribution with the same but unknown mean and the variance equal to σ^2/n. Suppose that someone comes along and tells you that he believes the mean to be zero. Now suppose that you cannot compute the mean from the entire population but that you can observe a sample of size n to check on the accuracy of the claim put to you. The classical rule tells you that within a certain probability consideration, if the observed $\bar{X}$ is less than $\bar{X}_L$ or greater than $\bar{X}_U$, as shown in Figure 3.11, the claim is not accepted. Thus, the test rule here decomposes the possible values of $\bar{X}$ into two classes: (1) the ones falling between $\bar{X}_L$ and $\bar{X}_U$ (acceptance region), and (2) the ones below $\bar{X}_L$ and above $\bar{X}_U$ (rejection region).

It is frequently of interest to hypothesize and to test whether or not the true coefficients of a simple linear regression model are in fact zero. The question is particularly relevant where a researcher looks for or is interested in the extent of the determination of Y by fixed X's. If it develops that one cannot maintain that β in, say

$$Y = \alpha + \beta X + u \tag{3.17}$$

is zero, then he must accept with some confidence that there is some positive or negative effect of X on Y. In this section, we shall discuss (1) tests of α and β separately and (2) tests of α and β by a and b jointly.

SEPARATE TESTS. The ingredients of the tests for α and β are already available in (3.42 a to b) and in Fixed Model B where u is assumed normal, among other things. First, suppose that σ^2 is known in (3.17) and that we wish to test

$$H_0\colon \beta = 0 \qquad \text{against} \qquad H_a\colon \beta \neq 0 \tag{3.58}$$

Then, by using (3.42a), we observe that under H_0

$$z = \frac{b - \beta}{\sqrt{\text{var}\,(\hat{\beta})}} = \frac{b}{\sqrt{\text{var}\,(\hat{\beta})}} \tag{3.59}$$

has a $N(0, 1)$ distribution.* Thus it is possible to use the standardized normal distribution for the test. Suppose that now instead of (3.58) we are interested in testing if β is equal to a constant c. For this case, we have

$$H_0\colon \beta = c \qquad \text{against} \qquad H_a\colon \beta \neq c \tag{3.60}$$

Now $(b - c)$ is still normal and has the same variance as $(b - 0) = b$, so that the same test as that in (3.58) is used for (3.60), namely,

$$z = \frac{b - c}{\sqrt{\text{var}\,(b)}} \tag{3.61}$$

has a $N(0, 1)$ distribution under the null hypothesis.

Easily, we observe that similar procedures are used in the test of

$$H_0\colon \alpha = k \qquad \text{against} \qquad H_a\colon \alpha \neq k \tag{3.62}$$

where k is any constant including $k = 0$.

Suppose now that σ^2 is not known and that it is necessary to estimate σ^2 from the sample data. For most practical purposes the procedures described in the paragraphs immediately preceding are applicable when sample size is fairly large. That is, one needs only to replace σ^2 by a sample estimate of it in (3.59), (3.61), and by similar expressions for test statistics related to α and β. But, if the sample is small, another procedure is more appropriate. This procedure invokes the so-called Student's distribution. Let us discuss this distribution in conjunction with some of the statistics we calculate in the regression analysis.

For convenience in notation, let var $(b) = \sigma_b^2$ and its estimate be denoted by s_b^2. By forming the ratio

$$\frac{s_b^2}{\sigma_b^2} = \frac{[\sum (Y_i - \hat{Y}_i)^2]/[(n-2)\sum (X_i - \bar{X})^2]}{\sigma^2/[\sum (X_i - \bar{X})^2]} = \frac{\sum (Y_i - \hat{Y}_i)^2}{(n-2)\sigma^2} \tag{3.63}$$

we notice that the ratio is a χ^2 distributed variable with $n - 2$ degrees of freedom divided by $n - 2$.† We arrive at this distribution as follows.

* Recall that Y_i's, conditional on X_i, are normally distributed and that b is a linear combination of Y_i's, so that b, conditional on X_i, has a normal distribution.

† The last term of (3.63) has the numerator $\Sigma\,[Y_i - (a + bX_i)]^2$ where a and b estimates are obtained from the sample; hence, the loss of 2 degrees of freedom. For a discussion of degrees of freedom see an excellent article by Helen Walker (1940).

By assumption, u is normally distributed with mean 0 and variance σ^2; hence, u_i^2/σ^2 for any i has a χ^2 distribution with one degree of freedom. Sampling from the distribution of u implies that the statistic $\sum_{i=1}^{n} (\hat{u}_i^2 \mid \sigma^2)$ has a χ^2 distribution with $(n - 2)$ degrees of freedom. This number of degrees of freedom comes from estimating α and β, or two linear restrictions placed on the sample of size n. Or, $(n - 2)s_b^2/\sigma_b^2$ has a χ^2 distribution with $n - 2$ degrees of freedom. By forming another ratio

$$\frac{b - \beta}{\sigma_b} \tag{3.64}$$

we find it to be distributed as $N(0, 1)$. Now a result due to Student (a pen name of W. S. Gosset) states that if (1) W is a $N(0, 1)$ variable, (2) V is a χ^2 variable with m degrees of freedom, and (3) W and V are independent, then

$$t = \frac{W}{\sqrt{V/m}} \tag{3.65}$$

has the Student's t distribution with m degrees of freedom.*

By combining and contrasting (3.63), (3.64), and (3.65), we observe that

$$t = \frac{b - \beta}{s_b} \tag{3.66}$$

is t distributed with $n - 2$ degrees of freedom. To illustrate this test, let us consider the null hypothesis $\beta = 0$ against the alternative hypothesis that $\beta \neq 0$ in model (3.17E). First, we find from the last column of Table 3.3 that $\sum \hat{u}_i^2 = 563.2$. Also, from the same table, $\sum (X_i - \bar{X})^2$ is calculated to be 1960.2. Thus

$$\begin{aligned} t &= \frac{b - \beta_0}{s_b} \\ &= \frac{b}{\sqrt{\dfrac{\sum u_i^2}{n - 2} \cdot \dfrac{1}{\sum (X_i - \bar{X})^2}}} \\ &= \frac{0.789}{\sqrt{\dfrac{563.2}{(9)(1960.2)}}} \\ &= 4.41 \end{aligned}$$

* For derivation of this result, see, for example, A. M. Mood and F. A. Graybill (1963, p. 233).

If the level of significance was selected to be at the 5 percent level, the critical value of t with 9 degrees of freedom is 2.262. Or

$$\Pr\,[t > |2.262|] = 0.05$$

Thus we cannot accept the null hypothesis and conclude that the slope is significantly different from zero.

Again, the test procedure for hypothesis about α is similar to those for β just discussed. It is often the case that σ^2 is unknown and that the t tests are more usually seen.

JOINT TEST. Sometimes it is of interest to ask if the entire linear relationship that involves the two parameters is at all significant. In this case, we consider joint tests of α and β such as

$$H_0\colon \alpha = 0, \beta = 0 \qquad \text{against} \qquad H_a\colon \alpha \neq 0, \beta \neq 0 \tag{3.67}$$

Here, we must consider the so-called F test which involves the ratio of variables having χ^2 distributions with varying degrees of freedom.* For the F distribution, the standard mathematical statistics texts state the following. If (1) U is a χ^2 variable with n degrees of freedom, (2) V is a χ^2 variable with m degrees of freedom, and (3) U and V are independent, then the ratio

$$F = \frac{U/n}{V/m} \tag{3.69}$$

has the F distribution with (n, m) degrees of freedom.

To appreciate the meaning of this distribution as it relates to the joint test we seek, let us return momentarily to the concept of the decomposition of variance touched on in Section 3.4.3. Before we begin, let us first assume that (1) α_0 and β_0 are true but unknown values of the parameters α and β, respectively, and that (2) σ^2 is known [this assumption is necessary only for our development, since in computation σ^2 vanishes when a ratio of the type (3.69) is formed]. Now the sum of squares of the true disturbances in our regression model is (under the null hypothesis)

$$Q_1 = \sum u_i^2 = \sum [Y_i - (\alpha_0 + \beta_0 X_i)]^2 \tag{3.70}$$

which, when divided by σ^2 has a χ^2 distribution with n degrees of freedom. Consider the error sum of squares from the data

$$Q_2 = \sum (Y_i - \hat{Y}_i)^2 = \sum [Y_i - (a + bX_i)]^2 \tag{3.71}$$

As is clear from (3.63), this quantity divided by σ^2 has a χ^2 distribution

* Further discussion of the rationale and procedure of the joint test are found in Sections 5.1 and 5.2.

with $n - 2$ degrees of freedom. Suppose that we form a difference between the estimate of expected Y_i and the expected Y_i, computed from the sample. We have

$$(\hat{Y}_i - EY_i) = [(a + bX_i) - (\alpha_0 + \beta_0 X_i)] \tag{3.72a}$$

By squaring and summing over n, we further obtain

$$Q_3 = \sum [(a - \alpha_0) + (b - \beta_0)X_i]^2 \tag{3.72}$$

This quantity, when divided by σ^2, has a χ^2 distribution with 2 degrees of freedom. And by Cochran's theorem,* Q_2 and Q_3 are independent. The geometric considerations in Figure 3.12 leads to the equality

$$Q_1 = Q_2 + Q_3 \tag{3.73}$$

Here, Q_3 measures the overall discrepancy of the estimated values of α and β from the true values α_0 and β_0 given X_i's. Thus Q_3 gets larger as $|a - \alpha_0|$

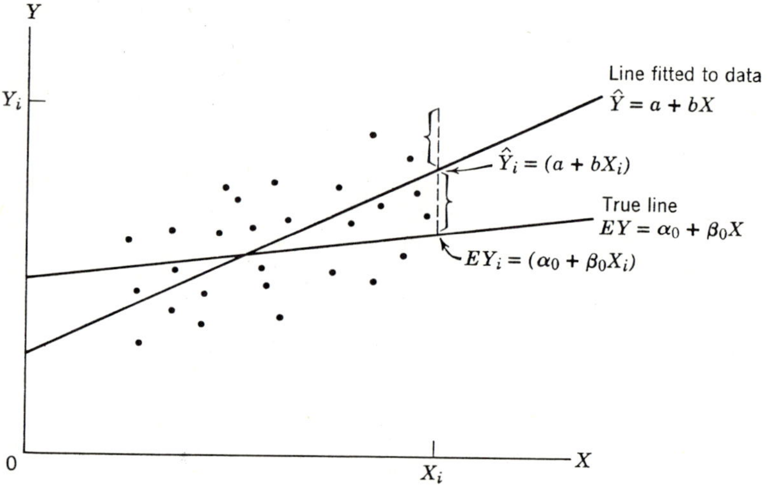

Figure 3.12

* A theorem due to Cochran states: let $x_1, \ldots, x_n$ be normally and independently distributed random variables with 0 mean and unit variance. And, further, let

$$Q_1 + Q_2 + \cdots + Q_k = \sum_{i=1}^{n} x_i^2$$

where Q_i is a quadratic form of rank n_i. The Q_i's are independently χ^2 distributed with n_i degrees of freedom if, and only if,

$$n_1 + n_2 + \cdots + n_k = n$$

and $|b - \beta_0|$ get larger. But Q_2, on the other hand, is a measure of the degree of discrepancy between what is observed in the form of data, Y_i's, and what is determined by the data, $\hat{Y}_i$'s, *irrespective* of the true but unknown values of α_0 and β_0; Q_2, then, is a good estimator of the true *ESS* obtainable from the sample. Thus it seems reasonable to consider the ratio

$$F = \frac{Q_3/2\sigma^2}{Q_2/(n-2)\sigma^2} = \frac{Q_3/2}{Q_2/(n-2)} \tag{3.74}$$

as an appropriate test statistic for the joint test of (3.67). As already suggested by the preceding comments, the large value of F tends to cause the null hypothesis in (3.67) to be rejected.

To illustrate this test consider, for the model (3.17e) $Y_i = \alpha + \beta X_i + u_i$, the test of null hypothesis $\alpha = 0$ and $\beta = 0$. We compute from Table 3.3

$$\begin{aligned} Q_2 &= \sum (Y_i - \hat{Y}_i)^2 = 563.2 \\ Q_3 &= Q_1 - Q_2 \\ &= \sum [Y_i - (0 + 0X_i)]^2 - Q_2 \text{ [under the null hypothesis]} \\ &= 22{,}467 - 563.2 \\ &= 21{,}903.8 \end{aligned}$$

Thus

$$\begin{aligned} F &= \frac{Q_3/2}{Q_2/(n-2)} \\ &= \frac{21{,}903.8/2}{563.2/9} \\ &= 175.0 \end{aligned}$$

At the level of significance of 5 percent, we have

$$\Pr[F > 4.26] = 0.05$$

hence, the null hypothesis is not accepted, and we conclude that a and b are jointly significant.

Since in practice it is often not known whether u is normal, questions may arise as to the appropriateness of the procedure that requires a normality assumption. No definite answer has yet been worked out for Fixed Model A, but Box and Watson (1962) have shown that, even if u is not normal, the joint test described above may still be quite good if the distribution of the independent variable X approximates the normal distribution.

It is possible to test for the significance of a linear combination of α and β, but we postpone the discussion of this test until Section 5.3 where the test is given in a general context.

3.4.5. Prediction and Forecasting

In economics as in other behavioral sciences there often arises the need for or an interest in bringing about certain desired results in behavior at a future point in time or for a coverage in space. For instance, policy makers in Washington might wish to bring the gross national product to $10 trillion by the end of 1980; or, an educational psychologist might be interested in bringing about a certain level of scholastic achievement by students in Class B in a subject as, for example, compared with the students in Class A. One rational way to achieve or nearly to achieve these goals is to study the relevant variables in the behavior of interest and to project the pattern of behavior on the assumption that the pattern that has persisted will hold true in the future or in some other space. For our examples, the economists involved in the policy-making might want to study how the GNP has been determined by key variables in the economy, to project the level of the GNP in 1980, given the levels of the key variables in 1980 or before, and to decide the adjustments in the levels of the key variables in 1980, or prior to it, so that the goal of $10 trillion might be satisfied. The educational psychologist might already have a knowledge of the methods of teaching and the quality of students in Class A and their scholastic achievement in the subject of interest. Thus, to bring about a certain level of achievement in the students in Class B, the teacher might wish to adjust the methods of teaching given the quality of the students in Class B and given that the patterns of response to teaching existing in Class A will possibly carry over to Class B. The reason for this somewhat lengthy introductory comment is to bring out two important features of statistical prediction or forecasting: (1) it is assumed that a theory or a model adequately describes the behavior of interest; and (2) the validity of the model persists into the future or over another space, so that the estimates of the parameters of the model can be used for making projection. This approach is sometimes called scientific prediction or forecasting,* as compared with the more pragmatic (but not necessarily inferior) approach where knowledge or facts not included in a model are used together with the model for prediction or forecasting. We confine ourselves to a discussion of statistical prediction or forecasting within the context of Fixed Model B, or the classical normal linear regression model.

For lack of better terminology we distinguish two situations of prediction by calling one case prediction and another forecasting when these terms are used in connection with regression models. By prediction we mean projection of an expected or average value of a regressand, and by forecasting we mean projection of a specific or single value of a regressand. Whether or

* H. Theil (1966, Chapter 1) describes in detail the meaning of "scientific" forecasting along with a general discussion of verifiability of forecasts and forecasting procedure.

not the values of the regressors are in the sample range of values is not particularly relevant in the general context, although we observe that in economics, more often than not, we are interested in the values of the variables in a model outside of the sample range. Now given an estimated relation, for instance,

$$Y = a + bX \tag{3.75}$$

for the model

$$Y = \alpha + \beta X + u \tag{3.17}$$

and given a specific value of X, for instance X_0, the predicted mean value of Y corresponding to X_0 is

$$\widehat{E(Y_0)} = a + bX_0[= \hat{\mu}_{(Y|X_0)}] \tag{3.76}$$

and a projected single value of Y corresponding to X_0 is

$$\hat{Y}_0 = a + bX_0 \tag{3.77}$$

Notice that the formula used for prediction or forecasting is the same but that the values being projected are quite distinct conceptually. See Figure 3.13 for the concepts of prediction and of forecasting that we try to distinguish. The Figure is also helpful in elucidating the nature of the errors of projection in the two cases.

ERROR OF PREDICTION. The error involved in calculating the mean value of Y corresponding to X_0 comes from two sources: (1) the error in a and (2) the error in b. That is, given that α and β are the true parameters, the error of prediction is

$$e_p = \alpha + \beta X_0 - (a + bX_0) = (\alpha - a) + (\beta - b)X_0 \tag{3.78}$$

By squaring and taking expectation, we have the variability of the predicted

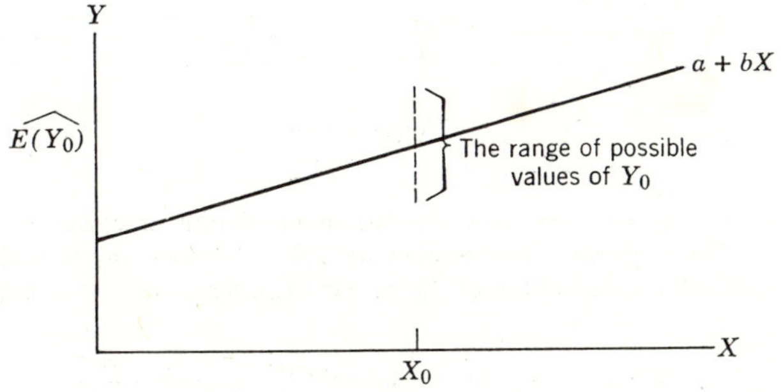

Figure 3.13

Y value as follows*

$$E(e_p)^2 = \text{var}\,(e_p) = \sigma^2\left[\frac{1}{n} + \frac{(X_0 - \bar{X})^2}{\sum (X_i - \bar{X})^2}\right] \tag{3.79}$$

since it is assumed that the conditional variance of Y is σ^2. Thus we can make a confidence statement about the predicted value of Y corresponding to X_0. Notice that var (e_p) [and, consequently, the confidence limits of $\widehat{E(Y_0)}$] gets large as X_0 departs from the sample mean $\bar{X}$. This phenomenon is shown in Figure 3.14. Furthermore, at the sample mean, say $X_0 = \bar{X}$, the prediction error reduces to σ^2/n because at the mean all that is needed is the predicted value of the mean of Y_i's and the values of a and b do not matter.

ERROR OF FORECAST. The error in forecasting a single value of Y corresponding to X_0 is greater than the error of prediction in the related situation because, in addition to the error of prediction mentioned above, the error emanating from the variability of single Y values corresponding to X_0 (as

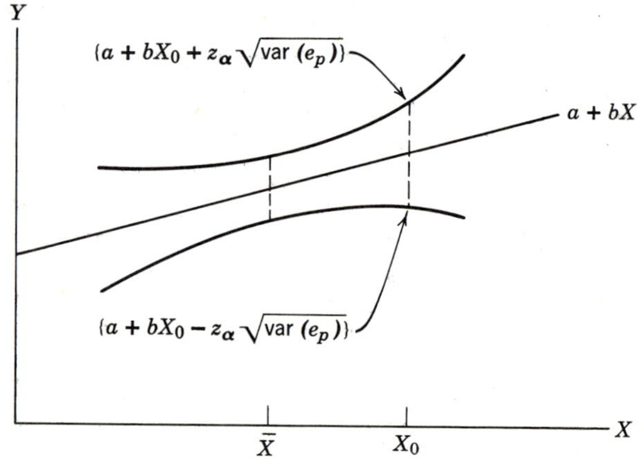

Figure 3.14

* The student may wish to carry out the manipulation necessary to arrive at (3.79) as an exercise. This suggestion also applies to (3.81). Another way to look at the error of prediction is to find the variability of $\hat{Y}_0$, or var $(\hat{Y}_0)$. Since $\hat{Y}_0 = a + bX_0$, we have

$$\text{var}\,(\hat{Y}_0) = \text{var}\,(a + bX_0) = \sigma^2\left[\frac{1}{n} + \frac{(X_0 - \bar{X})^2}{\sum (X_i - \bar{X})^2}\right]$$

from the definition of the variances of a and b.

seen in Figure 3.13) must be taken into consideration. That is

$$e_F = Y_0 - \hat{Y}_0 = (\alpha + \beta X_0 + u) - (a + bX_0)$$
$$= (\alpha - a) + (\beta - b)X_0 + u \tag{3.80}$$

Therefore, the variability in the forecast value of Y corresponding to X_0 is

$$E(e_F)^2 = \text{var}(e_F) = \text{var}(e_p) + \text{var}(u)$$
$$= \sigma^2\left[1 + \frac{1}{n} + \frac{(X_0 - \bar{X})^2}{\sum (X_i - \bar{X})^2}\right] \tag{3.81}$$

Again, based on the knowledge that u is normally distributed, we can make confidence statements about the forecast value of Y given X_0.

Of course, if the conditional variance of Y is not known, then it must be estimated from the sample and the σ^2 in (3.79) and (3.81) must be replaced by the sample estimate, $\sum (Y_i - \hat{Y}_i)^2/(n - 2)$. Consequently, the confidence statements that are made on the predicted or the forecast value of Y will have to be based on the Student's t distribution instead of on the normal distribution.

Again, we refer to model (3.17e) and the estimates thereof to illustrate the concepts discussed in this subsection. Given that the estimates are $a = 11.59$ and $b = 0.789$ and, for instance, $X_0 = 50$. We want to find the variability of the average values of Y associated with the particular X_0 value. Thus

$$\widehat{E(Y|X_0)} = 11.59 + 0.789(50)$$
$$= 51.05$$

Since σ^2 is not known, we estimate it by

$$s^2 = \frac{\sum \hat{u}_i^2}{n - 2} = \frac{563.2}{9} = 62.58$$

It then follows that

$$\text{var}(e_p) = 62.58\left[\frac{1}{11} + \frac{(50 - 40.3)^2}{1960.2}\right]$$
$$= 62.58(0.091 + 0.048)$$
$$= 8.70$$

The confidence interval, by using the level of significance of 5 percent, is

$$51.05 \pm (t_{\alpha=0.05})\sqrt{8.70}$$

or

$$51.05 \pm (2.262)(2.95)$$

or

$$51.05 \pm 6.67$$

Work out the case for projecting the single value of Y for $X_0 = 50$.

CHAPTER 4

MULTIPLE LINEAR REGRESSION

In Chapter 3 we considered in detail the basic elements of the standard linear regression models, simple or multiple. At least from one viewpoint, multiple linear regression is a straightforward extension of simple linear regression. However, computational burden may increase very rapidly in multiple regression as the number of variables and observations go up. Yet this problem becomes less awesome in view of the modern development in computer technology. Thus it becomes incumbent on writers of regression analysis to be fairly certain that the rationale of a method, the assumptions that must be satisfied or nearly satisfied before the method is used, and the interpretation of various statistics that are outcomes of applying the method are understood.

The main task of this and the next three chapters is the generalization and elaboration of the concepts and the results contained in Chapter 3. We shall show that much of what we said in connection with interpreting a simple model and the derived statistics remains intact in a corresponding multiple model—that the nature of the assumptions in a simple model carries over to the multiple model, except that in a multiple model an efficient system of notations must be devised to cope with the conceptual and computational complexities that arise in multivariable situations. On this last question we are fortunate to have available to us the vast body of knowledge accumulated in matrix algebra and the knowledge of its use in mathematical statistics.

In a sense, the present chapter provides a bridge between the materials given in Chapter 3 and the more advanced materials that follow it. We shall introduce matrix notations in this chapter, and they will be used throughout the rest of this book. However, we shall also, at times, refer to or shall use the more conventional summation and its signs in this and the following chapters in order to provide a contrast between the matrix and the "conventional" notations and in order to ease the student into the advanced treatment in the later chapters.

In any case, it is advisable to take pains, at the onset, to learn the matrix notations well, as these notations are one of the basic tools that we shall be using later. Those who are not acquainted with matrix algebra are advised to secure, at least, one standard text, for example, G. Hadley (1961) and Graybill (1969), for use as a constant source of reference in our further work. It is best if the student learns the necessary algebra as we move along, but this can become difficult at times for some. In such an event, it would be advisable for the student to master only the matrix notations so that he would be able to appreciate the various quantities, statistics, and so forth that are given. He would then go through the rest of the book without learning the derivations and proofs that lead to the various results, but nevertheless he would have learned the rationale of a technique, the technique itself, and its applications. In any case, to proceed to the next chapters, the following concepts in matrices are required: a vector, a matrix, the transpose of a matrix, the addition of two or more matrices, the multiplication of matrices, the conformability of two matrices, the linear combination of vectors, the inverse of a matrix, the rank of a matrix, the linear dependence and orthogonality, and quadratic forms, etc. Some of these concepts will be discussed briefly where appropriate and in Appendix B. For others, see Hadley (1961) or Graybill (1969).

4.1. Introduction to Multiple Linear Regression

We begin by comparing a simple linear regression model and a multiple linear regression model with respect to the decomposition of a variable into two parts—systematic and random. That is, if a variable moves in some systematic fashion with respect to another variable, then it is often the case that that variable can be decomposed into (1) the systematic part, which is some function of the latter variable, and (2) the random part, the behavior of which is unpredictable. Thus, by writing a combined variant of (3.22), (3.23), and (3.24), we have a simple linear regression model

$$Y = \alpha + \beta X + u$$

In terms of the decomposition of the variable Y, then, $(\alpha + \beta X)$ is its systematic part and u its random part. Let us now consider a multiple linear regression model

$$Y_1 = \beta_0 + \beta_1 X_1 + \beta_2 X_2 + \cdots + \beta_k X_k + u$$

Then it is generally implied that the variation in Y_1 is systematically explainable by the part of the Y_1 that is represented as $(\beta_0 + \beta_1 X_1 + B_2 X_2 + \cdots + \beta_k X_k)$ and that the part of Y_1 not explained by the X's is represented by u.

Let us suppose, for discussion, that Y and Y_1 each has a normal distribution, both having the same conditional variance, say, σ^2. It follows directly that Y has the normal distribution

$$N(\alpha + \beta X, \sigma^2)$$

and that Y_1 has the normal distribution

$$N(\beta_0 + \beta_1 X_1 + \cdots + \beta_k X_k, \sigma^2)$$

4.1.1. *The Generality of Decomposition*

The spirit of the decomposition described above has, indeed, a very general application. Write, in general, for the two-variable case

$$Y = f(X) + u$$

then

$$f(X) = \alpha + \beta X$$

or

$$f(X) = \delta X^{\gamma}$$

or

$$f(X) = \theta_1 X + \theta_2 X^2$$

and so on. Clearly, similar extensions are directly possible for

$$Y = f(X_1, X_2, \ldots, X_k) + u$$

In general, we might think of the decomposition of a variable as having the following properties.

1. The decomposition is applicable to any random variable (if its expectation exists).
2. The expectation of the variable is the conditional expectation of the variable and is equal to the systematic part in the decomposition.
3. The variance of the random part is the conditional variance of the variable being decomposed, given the systematic part.
4. The systematic part can have any deterministic mathematical functional form.

As is clear, then, the concept of decomposition is applicable in both linear and nonlinear models where the disturbances are or can be made additive.

4.1.2. *Some Notational Agreements*

Thus far, we have defined variables and have denoted them in certain fashion as occasions demanded. This practice may become awkward when there are many variables to consider. Therefore it seems at this point useful to devise a minimum number of standard notations to be used in the subsequent developments.

We shall denote a variable or a random variable by a lower case letter and its values or sample observations by a subscripted capital letter. Thus, if we have a variable x, its n values or n observations form a column vector

$$\mathbf{x} = \begin{bmatrix} X_1 \\ X_2 \\ \cdot \\ \cdot \\ \cdot \\ X_n \end{bmatrix}$$

If we have k variables, we shall write them $x, x_2, \ldots, x_k$. We further agree that sample values or observations of several variables are to be noted by an ordinary capital letter with two subscripts, where the first subscript refers to the number of observations and the second refers to the number of the variable. We shall denote a matrix by a bold-faced capital letter; so that to summarize as well as to illustrate the agreements thus far, if we have k variables $x_1, x_2, \ldots, x_k$, n observations on each of these variables will give rise to a matrix of dimensionality $n \times k$ as follows:

$$\mathbf{X} = [\mathbf{x}_1\ \mathbf{x}_2\ \cdots\ \mathbf{x}_k] = \begin{bmatrix} X_{11} & X_{12} & \cdots & X_{1k} \\ X_{21} & X_{22} & \cdots & X_{2k} \\ \cdot & \cdot & \cdot & \cdot \\ \cdot & \cdot & \cdot & \cdot \\ \cdot & \cdot & \cdot & \cdot \\ X_{n1} & X_{n2} & \cdots & X_{nk} \end{bmatrix} \tag{4.1}$$

Notice that $\mathbf{x}_i$ is the column vector of n observations on the ith variable, x_i.

Generally, a Greek letter denotes a parameter. A summation sign, in general, is for summing over all the sample values of the variable, indicated by its subscript below the sign. Thus, if one is interested in finding the sum of observations on x_1, as shown in (4.1), he writes

$$\sum_i X_{i1}$$

And if we wish to form all the cross products of the observations on variables x_1 and x_k, we write

$$\sum_i \sum_j X_{i1} X_{jk}$$

in each case realizing that $i, j = 1, 2, \ldots, n$.

4.1.3. Representation of a Linear Regression Model

Having agreed on some of the notational arrangements, let us now determine how a multiple linear regression model is written and in what way the model and its sample are related. This effort, although pedestrian in appearance, will be a useful stepping-stone in our future work.

A multiple linear regression model is usually written as follows

$$y = \beta_0 + \beta_1 x_1 + \beta_2 x_2 + \cdots + \beta_k x_k + u \tag{4.2}$$

where there are k variables but effectively $k + 1$ regressors. One can also regard β_0 as a parameter attached to the variable x_0 which assumes a constant value of 1; viewed this way, we could say that (4.2) is a multiple linear regression equation in which y is dependent on $k + 1$ variables and that the manner or extent of the dependence is determined by the $k + 1$ parameters β_i's. This relationship is assumed to hold true for each element of a population under study so that, if the population is finite, say of size N, then (4.2) can be represented by

$$Y_i = \beta_0 + \beta_1 X_{i1} + \beta_2 X_{i2} + \cdots + \beta_k X_{ik} + u_i \qquad i = 1, 2, \ldots, N \tag{4.3}$$

(that is, for i running from 1 to N). This relationship states that, when an observation is taken on the ith element, one will have, as a sample point,

$$(Y_i \quad 1 \quad X_{i1} \quad X_{i2} \cdots X_{ik}) \tag{4.4}$$

and that for any ith arbitrary element the effects of the regressor variables, $x_0, x_1, \ldots, x_k$, on y are measured, respectively, by the parameters $\beta_0, \beta_1, \ldots, \beta_k$. Thus, given the parameters, the level of y is determined by (or conditional on) the values of the regressors in a systematic fashion by virtue of the specified relation; but there exists a part of y that cannot be determined systematically, and that part is written as u. A complete description, then, of what (4.3) says is the system

$$\begin{aligned}
Y_1 &= \beta_0 + \beta_1 X_{11} + \beta_2 X_{12} \cdots + \beta_k X_{1k} + u_1 \\
Y_2 &= \beta_0 + \beta_1 X_{21} + \beta_2 X_{22} \cdots + \beta_k X_{2k} + u_2 \\
&\cdot \quad \cdot \quad \cdot \quad \cdot \quad \cdot \quad \cdot \\
&\cdot \quad \cdot \quad \cdot \quad \cdot \quad \cdot \quad \cdot \\
&\cdot \quad \cdot \quad \cdot \quad \cdot \quad \cdot \quad \cdot \\
Y_N &= \beta_0 + \beta_1 X_{N1} + \beta_2 X_{N2} \cdots + \beta_k X_{Nk} + u_N
\end{aligned} \tag{4.3a}$$

In matrix notation, (4.3a) becomes

$$\begin{bmatrix} Y_1 \\ Y_2 \\ \cdot \\ \cdot \\ \cdot \\ Y_N \end{bmatrix} = \begin{bmatrix} 1 & X_{11} & X_{12} & \cdots & X_{1k} \\ 1 & X_{21} & X_{22} & \cdots & X_{2k} \\ \cdot & \cdot & \cdot & \cdot & \cdot \\ \cdot & \cdot & \cdot & \cdot & \cdot \\ \cdot & \cdot & \cdot & \cdot & \cdot \\ 1 & X_{N1} & X_{N2} & \cdots & X_{Nk} \end{bmatrix} \begin{bmatrix} \beta_0 \\ \beta_1 \\ \cdot \\ \cdot \\ \cdot \\ \beta_k \end{bmatrix} + \begin{bmatrix} u_1 \\ u_2 \\ \cdot \\ \cdot \\ \cdot \\ u_N \end{bmatrix} \tag{4.3b}$$

or

$$\mathbf{y} = \mathbf{X}\boldsymbol{\beta} + \mathbf{u} \tag{4.5}$$

where $\mathbf{y}$ and $\mathbf{u}$ are each a vector of dimension $N \times 1$, $\mathbf{X}$ is a matrix of order $N \times (k + 1)$, and $\boldsymbol{\beta}$ is a vector of dimension $(k + 1) \times 1$. Notice that the ith row of $\mathbf{X}$ is the vector shown in (4.4) less the first element, so that

$$\mathbf{X} = \begin{bmatrix} 1 & X_{11} & X_{12} & \cdots & X_{1k} \\ 1 & X_{21} & X_{22} & \cdots & X_{2k} \\ \cdot & \cdot & \cdot & \cdot & \cdot \\ \cdot & \cdot & \cdot & \cdot & \cdot \\ \cdot & \cdot & \cdot & \cdot & \cdot \\ 1 & X_{N1} & X_{N2} & \cdots & X_{Nk} \end{bmatrix}$$

Now, if a sample of size n is taken to study the relationship (4.2), we can write out the entire sample observations in accordance with the assumed relation (4.2) as in (4.3a), but it is necessary to replace N by n everywhere in that expression. We observe then that (4.5) is, indeed, a very general expression for giving a complete description of how all the values of variables are related in a population *or in a sample*. In the case of the latter, the matrix containing the sample values of the x_i variables will be $n \times (k + 1)$, rather then $N \times (k + 1)$, as in (4.6).

4.1.4. *The Least-Squares Method*

Insofar as either (4.2) or (4.5) is a hypothesized relationship for a population, it may be called a regression model.

Now suppose that we are given a set of data and are interested in making a summary presentation of information about a dependence relation of one variable to others for which the data are collected. In such a case, it is appropriate not to think of using (4.2) or (4.5) as a model, but instead as an equation to describe the data. In this sense, one may think of observations on y as being determined by observations on x_i's, although that part of the y value not accounted for by the relation with the values of x_i's is a residual. Frequent uses of a regression equation in this manner are made, and the least-squares

(LS) method seems appropriate here. This method solves for the coefficients in the equation in such a way as to minimize the sum of squares of the residuals. In notation, for n observations the descriptive device may be

$$Y_i = \beta_0 + \beta_i X_{i1} + \beta_2 X_{i2} + \cdots + \beta_k X_{ik} + r_i \qquad (4.6)$$
$$i = 1, 2, \ldots, n$$

where the LS method is to choose the coefficients $\hat{\beta}_i$'s so that

$$\sum_i (Y_i - \hat{\beta}_0 - \hat{\beta}_1 X_{i1} - \cdots - \hat{\beta}_k X_{ik})^2 = \sum_i \hat{r}_i^2 \qquad (4.7)$$

is a minimum. We refer to $\hat{r}_i$'s as the LS residuals. Notice that r here is used for convenience and differs from the sample correlation coefficient r used in Section 3.1.

Let us use matrix notations to derive the least-squares coefficients. By rewriting (4.6), we have

$$\mathbf{y} = \mathbf{X}\hat{\boldsymbol{\beta}} + \hat{\mathbf{r}} \qquad (4.6a)$$

where $\mathbf{y}$ and $\hat{\mathbf{r}}$ are each $n \times 1$, $\mathbf{X}$ is $n \times (k + 1)$, and $\hat{\boldsymbol{\beta}}$ is $(k + 1) \times 1$. Now $\hat{\boldsymbol{\beta}}$ is the set of coefficients whose values are to be determined by making the following a minimum:

$$(\mathbf{y} - \mathbf{X}\hat{\boldsymbol{\beta}})'(\mathbf{y} - \mathbf{X}\hat{\boldsymbol{\beta}}) = \hat{\mathbf{r}}'\hat{\mathbf{r}} = S \qquad (4.7a)$$

Or, in the summation notation,

$$S = \sum [y_i - (\hat{\beta}_0 + \hat{\beta}_1 X_{i1} + \cdots + \hat{\beta}_k X_{ik})]^2 \qquad (4.7b)$$

By multiplying out the left-hand side after performing the indicated transpose operation in (4.7a), we obtain

$$S = \mathbf{y}'\mathbf{y} - \hat{\boldsymbol{\beta}}'\mathbf{X}'\mathbf{y} - \mathbf{y}'\mathbf{X}\hat{\boldsymbol{\beta}} + \hat{\boldsymbol{\beta}}'\mathbf{X}'\mathbf{X}\hat{\boldsymbol{\beta}}$$
$$= \mathbf{y}'\mathbf{y} - 2\hat{\boldsymbol{\beta}}'\mathbf{X}'\mathbf{y} + \hat{\boldsymbol{\beta}}'\mathbf{X}'\mathbf{X}\hat{\boldsymbol{\beta}}$$

where the second term in the last expression is a linear combination of $\hat{\beta}_i$'s and the third expression is a quadratic form in $\hat{\beta}_i$'s (a linear combination of $\hat{\beta}_i \cdot \hat{\beta}_j$'s, $i, j = 0, 1, 2, \ldots, k$). By taking the matrix derivative of S with respect to $\hat{\boldsymbol{\beta}}$ and setting the result equal to zero, we have

$$\frac{\partial S}{\partial \hat{\boldsymbol{\beta}}} = -2\mathbf{X}'\mathbf{y} + 2\mathbf{X}'\mathbf{X}\hat{\boldsymbol{\beta}} = \mathbf{0} \qquad (4.8)$$

where $\mathbf{0}$ is a zero vector of $(k + 1) \times 1$. To see this, we differentiate (4.7b) with respect to each of the $\hat{\beta}_i$'s and obtain

$$\frac{\partial S}{\partial \hat{\beta}_0} = -2 \sum [y_i - (\hat{\beta}_0 + \hat{\beta}_1 X_{i1} + \cdots + \hat{\beta}_k X_{ik})]$$

$$\frac{\partial S}{\partial \hat{\beta}_1} = -2 \sum [y_i - (\hat{\beta}_0 + \hat{\beta}_1 X_{i1} + \cdots + \hat{\beta}_k X_{ik})]X_{i1}$$

$$\vdots \qquad \qquad \vdots$$

$$\frac{\partial S}{\partial \hat{\beta}_k} = -2 \sum [y_i - (\hat{\beta}_0 + \hat{\beta}_1 X_{i1} + \cdots + \hat{\beta}_k X_{ik})]X_{ik}$$

Notice also the equality by definition

$$\frac{\partial S}{\partial \hat{\boldsymbol{\beta}}} = \begin{bmatrix} \dfrac{\partial S}{\partial \hat{\beta}_0} \\ \dfrac{\partial S}{\partial \hat{\beta}_1} \\ \vdots \\ \dfrac{\partial S}{\partial \hat{\beta}_k} \end{bmatrix} \tag{4.8b}$$

The solution of (4.8) gives

$$\hat{\boldsymbol{\beta}} = (\mathbf{X}'\mathbf{X})^{-1}\mathbf{X}'\mathbf{y} \tag{4.9}$$

provided that the matrix $(\mathbf{X}'\mathbf{X})$ is of full rank. [That is, if the rank of $(\mathbf{X}'\mathbf{X})$, $\rho(\mathbf{X}'\mathbf{X})$ is $k + 1$, so that $(\mathbf{X}'\mathbf{X})^{-1}$ exists; or, expressed differently, if there are no linearly dependent columns in $\mathbf{X}$].* For future references and in the interest of standardized notation, let us denote the $\hat{\boldsymbol{\beta}}$ that satisfies (4.9) under the LS criterion by $\mathbf{b}$. Thus, the LS estimated values of y are given by the vector

$$\hat{\mathbf{y}} = \mathbf{X}\mathbf{b} \tag{4.10}$$

* The requirement that there are no linearly dependent columns in $\mathbf{X}$ ensures that $(\mathbf{X}'\mathbf{X})$ will be nonsingular, or that the determinant $|\mathbf{X}'\mathbf{X}| \neq 0$ and, as the consequence, $(\mathbf{X}'\mathbf{X})^{-1}$ exists. For further elaboration, see the discussions near the Expressions 4.37 and 4.37a.

and the discrepancies between the observed and the estimated values of y by the vector

$$\hat{\mathbf{r}} = \mathbf{y} - \hat{\mathbf{y}} = \mathbf{y} - \mathbf{Xb} \tag{4.11}$$

To obtain another view of the quantities discussed in the preceding paragraphs, let us observe that for a sample of size n

$$\mathbf{X'X} = \begin{bmatrix} n & \sum X_{i1} & \sum X_{i2} & \cdots & \sum X_{ik} \\ \sum X_{i1} & \sum X_{i1}^2 & \sum X_{i1}X_{i2} & \cdots & \sum X_{i1}X_{ik} \\ \sum X_{i2} & \sum X_{i1}X_{i2} & \sum X_{i2}^2 & \cdots & \sum X_{i2}X_{ik} \\ \cdot & \cdot & \cdot & \cdot & \cdot \\ \cdot & \cdot & \cdot & \cdot & \cdot \\ \cdot & \cdot & \cdot & \cdot & \cdot \\ \sum X_{ik} & \sum X_{i1}X_{ik} & \sum X_{i2}X_{ik} & \cdots & \sum X_{ik}^2 \end{bmatrix} \tag{4.10}$$

$$\mathbf{X'y} = \begin{bmatrix} \sum Y_i \\ \sum X_{i1}Y_i \\ \cdot \\ \cdot \\ \cdot \\ \sum X_{ik}Y_i \end{bmatrix} \tag{4.11}$$

so that the LS residual sum of squares is

$$\begin{aligned} \hat{\mathbf{r}}'\hat{\mathbf{r}} &= (\mathbf{y} - \mathbf{Xb})'(\mathbf{y} - \mathbf{Xb}) = (\mathbf{y}' - \mathbf{b}'\mathbf{X}')(\mathbf{y} - \mathbf{Xb}) \\ &= \mathbf{y}'\mathbf{y} - \mathbf{b}'\mathbf{Xy} - \mathbf{y}'\mathbf{Xb} + \mathbf{b}'\mathbf{X}'\mathbf{Xb} \\ &= \mathbf{y}'\mathbf{y} - 2\mathbf{b}'\mathbf{X}'\mathbf{y} + \mathbf{b}'\mathbf{X}'\mathbf{Xb} \\ &= \mathbf{y}'\mathbf{y} - \mathbf{b}'\mathbf{X}'\mathbf{y} \\ &= \mathbf{y}'\mathbf{y} - \mathbf{b}'\mathbf{X}'\mathbf{Xb} \end{aligned} \tag{4.12*}$$

We may also observe that (4.12) gives

$$\mathbf{y}'\mathbf{y} = \hat{\mathbf{r}}'\hat{\mathbf{r}} + \mathbf{b}'\mathbf{X}'\mathbf{Xb}$$

which says that the sum of squares of the dependent variable is composed of the sum of squares of the residuals and the sum of squares of the estimated

* Notice that $\mathbf{b}'\mathbf{X}'\mathbf{y} = \mathbf{y}'\mathbf{Xb}$, that is, the transpose of a scalar is the scalar itself. For the proof, or $\mathbf{X}'(\mathbf{y} - \mathbf{Xb}) = \mathbf{0}$, see Section 4.1.6.

systematic part **Xb**. Now $\mathbf{y}'\mathbf{y} = \sum Y_i^2$, therefore,*

$$\hat{\mathbf{r}}'\hat{\mathbf{r}} = \sum \left\{ Y_i^2 - \sum_{j=0}^{k} [b_j \sum X_{ij} Y_i] \right\} \tag{4.13}$$

Here $X_{i0} = 1$ for $i = 1, 2, \ldots, n$.

4.1.5. An Illustration

We shall now use an actual example to illustrate some of the computations indicated in the preceding discussion. The following is a set of imaginary observations, over a 16-year period, on the total income (x) and the total savings (y) of community A (see Table 4.1). Suppose that we posit an income-

Table 4.1 Data on Income and Savings of Community A (in Millions of Dollars)

Time (t)	Savings (y)	Income (x)
1	0.27	9.3
2	0.26	10.5
3	0.13	11.0
4	0.16	11.9
5	0.53	12.9
6	0.65	13.5
7	0.56	14.3
8	0.77	16.0
9	1.17	16.7
10	1.24	18.2
11	1.07	18.6
12	1.35	19.7
13	1.99	21.9
14	2.54	22.8
15	2.28	23.9
16	2.81	27.1

savings relation as

$$y = \alpha + \beta x + u$$

then in matrices the sample observations corresponding to the model in

* Notice that the $\sum$ sign below sums over i except when otherwise specified.

(4.3b) are

$$\begin{bmatrix} 0.27 \\ 0.26 \\ 0.13 \\ 0.16 \\ 0.53 \\ 0.65 \\ 0.56 \\ 0.77 \\ 1.17 \\ 1.24 \\ 1.07 \\ 1.35 \\ 1.99 \\ 2.54 \\ 2.28 \\ 2.81 \end{bmatrix} = \begin{bmatrix} 1 & 9.3 \\ 1 & 10.5 \\ 1 & 11.0 \\ 1 & 11.9 \\ 1 & 12.9 \\ 1 & 13.5 \\ 1 & 14.3 \\ 1 & 16.0 \\ 1 & 16.7 \\ 1 & 18.2 \\ 1 & 18.6 \\ 1 & 19.7 \\ 1 & 21.9 \\ 1 & 22.8 \\ 1 & 23.9 \\ 1 & 27.1 \end{bmatrix} \begin{bmatrix} \alpha \\ \beta \end{bmatrix} + \begin{bmatrix} u_1 \\ u_2 \\ \cdot \\ \cdot \\ \cdot \\ u_{16} \end{bmatrix} \tag{4.3s}$$

in accordance with

$$\mathbf{y} = \mathbf{X}\boldsymbol{\beta} + \mathbf{u} \tag{4.5}$$

so that, to illustrate further, the first two rows of (4.5) are

$$0.27 = \alpha + \beta(9.3) + u_1$$

and

$$0.26 = \alpha + \beta(10.5) + u_2$$

Clearly, the vector $\boldsymbol{\beta}$ for the problem at hand is $\begin{bmatrix} \alpha \\ \beta \end{bmatrix}$.

Given (4.3s) and (4.5) the quantities needed for LS estimation of α and β are

$$\hat{\boldsymbol{\beta}} = \begin{bmatrix} \hat{\alpha} \\ \hat{\beta} \end{bmatrix} = \begin{bmatrix} a \\ b \end{bmatrix} = (\mathbf{X}'\mathbf{X})^{-1}(\mathbf{X}'\mathbf{y})$$

and

$$(\mathbf{X}'\mathbf{X}) = \begin{pmatrix} 16 & 268.3 \\ 268.3 & 4917.8 \end{pmatrix}$$

$$(\mathbf{X}'\mathbf{X})^{-1} = \frac{1}{6699.1}\begin{bmatrix} 4917.8 & -268.3 \\ -268.3 & 16 \end{bmatrix} = \begin{bmatrix} 0.7341 & -0.04005 \\ -0.04005 & 0.002388 \end{bmatrix}$$

$$(\mathbf{X}'\mathbf{y}) = \begin{bmatrix} 17.78 \\ 365.60 \end{bmatrix}$$

so that

$$\begin{bmatrix} a \\ b \end{bmatrix} = \begin{bmatrix} 0.7341 & -0.04005 \\ -0.04005 & 0.002388 \end{bmatrix} \begin{bmatrix} 17.78 \\ 365.6 \end{bmatrix} = \begin{bmatrix} -1.576 \\ 0.1603 \end{bmatrix}$$

For calculation of residuals, we have

$$\hat{\mathbf{r}} = \mathbf{y} - \mathbf{Xb} = \mathbf{y} - \hat{\mathbf{y}} = \begin{bmatrix} 0.27 \\ 0.26 \\ 0.13 \\ 0.16 \\ 0.53 \\ 0.65 \\ 0.56 \\ 0.77 \\ 1.17 \\ 1.24 \\ 1.07 \\ 1.35 \\ 1.99 \\ 2.54 \\ 2.28 \\ 2.81 \end{bmatrix} - \begin{bmatrix} -0.09 \\ 0.11 \\ 0.19 \\ 0.33 \\ 0.49 \\ 0.59 \\ 0.72 \\ 0.99 \\ 1.10 \\ 1.34 \\ 1.41 \\ 1.58 \\ 1.93 \\ 2.08 \\ 2.25 \\ 2.77 \end{bmatrix} = \begin{bmatrix} 0.36 \\ 0.15 \\ -0.06 \\ -0.17 \\ 0.04 \\ 0.06 \\ -0.16 \\ -0.22 \\ 0.07 \\ -0.10 \\ -0.34 \\ -0.23 \\ 0.06 \\ 0.46 \\ 0.03 \\ 0.04 \end{bmatrix}$$

We verify (4.12) from the data as follows

$$\hat{\mathbf{r}}'\hat{\mathbf{r}} = \sum \hat{\mathbf{r}}_i^2 = 0.658$$

$$\mathbf{y}'\mathbf{y} = \sum Y_i^2 = 31.173$$

$$\mathbf{b}'\mathbf{X}'\mathbf{y} = \begin{bmatrix} -1.576 \\ 0.1603 \end{bmatrix} \begin{bmatrix} 17.78 \\ 365.6 \end{bmatrix} = 30.515$$

so that

$$\hat{\mathbf{r}}'\hat{\mathbf{r}} = \mathbf{y}'\mathbf{y} - \mathbf{b}'\mathbf{X}'\mathbf{y}$$

As an exercise, the student is advised to take the observations for X and Y in Table 3.3, form matrices $(\mathbf{X}'\mathbf{X})$ and $(\mathbf{X}'\mathbf{y})$, to find the inverse of $(\mathbf{X}'\mathbf{X})$, and to estimate the coefficient vector $\hat{\boldsymbol{\beta}}$ by $(\mathbf{X}'\mathbf{X})^{-1}\mathbf{X}'\mathbf{y}$ strictly by matrix operation.

4.1.6. Some Interesting LS Properties

There are several consequences of choosing the coefficient estimates according to the LS criterion. We cite some of the more interesting ones at this time, keeping in mind that these results arise without having to make any assumptions except that the estimates minimize the sum of the squares of the residuals.

b IS A LINEAR TRANSFORM. The estimates are obtained by a linear transformation of the values of y; that is

$$\mathbf{b} = (\mathbf{X}'\mathbf{X})^{-1}\mathbf{X}'\mathbf{y} = \mathbf{A}\mathbf{y} \tag{4.15}$$

where $\mathbf{A} = (\mathbf{X}'\mathbf{X})^{-1}\mathbf{X}'$ is a matrix of constants determined from sample values of the x_i variables.

MATRICES **M** AND **I** − **M**. The matrix that transforms the observed values of y into the LS estimated values of y, or $\hat{y}$, is of considerable significance. From (4.15), the LS estimate of **y** is

$$\hat{\mathbf{y}} = \mathbf{X}\mathbf{b} = \mathbf{X}\mathbf{A}\mathbf{y} \tag{4.16}$$

By letting

$$\mathbf{M} = \mathbf{I} - \mathbf{X}\mathbf{A} = \mathbf{I} - \mathbf{X}(\mathbf{X}'\mathbf{X})^{-1}\mathbf{X}'$$

we observe that

$$\mathbf{y} - \hat{\mathbf{y}} = \mathbf{y} - (\mathbf{I} - \mathbf{M})\mathbf{y} = \mathbf{M}\mathbf{y} \tag{4.17}$$

That is, the LS residuals are obtained by transformation of **y** by **M**, although the estimated or calculated values of **y** are obtained by premultiplication of **y** by **I** − **M**. We add some very important properties of **M** below.

M is symmetric. That is,

$$\mathbf{M}' = [\mathbf{I} - \mathbf{X}(\mathbf{X}'\mathbf{X})^{-1}\mathbf{X}']' = \mathbf{I} - \mathbf{X}(\mathbf{X}'\mathbf{X})^{-1}\mathbf{X}' = \mathbf{M} \tag{4.18}$$

It follows that **I** − **M** is symmetric (show this), or that

$$(\mathbf{I} - \mathbf{M})' = (\mathbf{I} - \mathbf{M}) \tag{4.19}$$

Furthermore

$$(\mathbf{I} - \mathbf{M})\mathbf{X} = [\mathbf{X}(\mathbf{X}'\mathbf{X})^{-1}\mathbf{X}']\mathbf{X} = \mathbf{X}[(\mathbf{X}'\mathbf{X})^{-1}\mathbf{X}'\mathbf{X}] = \mathbf{X} \tag{4.20}$$

Therefore

$$\mathbf{M}\mathbf{X} = \mathbf{0} \tag{4.21}$$

M is idempotent, or*

$$\mathbf{M}^2 = \mathbf{M} \tag{4.22}$$

Consequently

$$(\mathbf{I} - \mathbf{M})^2 = (\mathbf{I} - \mathbf{M}) \tag{4.23}$$

$$\mathbf{M}(\mathbf{I} - \mathbf{M}) = \mathbf{M} - \mathbf{M}^2 = \mathbf{0} \tag{4.24}$$

* Show (4.22) and (4.23) as an exercise in matrices. Write $\mathbf{M}^2 = \mathbf{M} \cdot \mathbf{M} = [\mathbf{I} - \mathbf{X}(\mathbf{X}'\mathbf{X})^{-1}\mathbf{X}'][\mathbf{I} - \mathbf{X}(\mathbf{X}'\mathbf{X})^{-1}\mathbf{X}']$, etc.

LEAST-SQUARES HYPERPLANE. The LS coefficient estimates can be used to generate a hyperplane in the $(k+1)$ dimensional space by feeding all possible values of the x_i variables into the scheme

$$\hat{\mathbf{y}} = \mathbf{X}\mathbf{b}$$

When we confine the generation of points to just the sample observations (from which $\mathbf{b}$ was obtained), interesting things happen. First, the sum $\sum y_i = \sum \hat{y}_i$. This is true because from (4.8) we have a set of normal equations

$$\mathbf{X}'\mathbf{X}\mathbf{b} = \mathbf{X}'\mathbf{y} \tag{4.25}$$

whose first row is

$$\boldsymbol{\iota}'\hat{\mathbf{y}} = \boldsymbol{\iota}'\mathbf{X}\mathbf{b} = \boldsymbol{\iota}'\mathbf{y} \tag{4.26}$$

where $\boldsymbol{\iota}$ is a column vector of 1's. Hence, by (4.26),

$$\boldsymbol{\iota}'\hat{\mathbf{r}} = \boldsymbol{\iota}'(\mathbf{y} - \hat{\mathbf{y}}) = \boldsymbol{\iota}'\hat{\mathbf{y}} - \boldsymbol{\iota}'\hat{\mathbf{y}} = \mathbf{0} \tag{4.27}$$

The reader is advised to contrast this and the following results with the ones in (3.43) through (3.46).

The sum of cross products between the estimated values of y and the respective residuals is zero. Or

$$\hat{\mathbf{y}}'\hat{\mathbf{r}} = \mathbf{0} \tag{4.28}$$

We show this by rewriting it according to (4.16) and (4.17). Thus,

$$\begin{aligned} \hat{\mathbf{y}}'\hat{\mathbf{r}} &= [(\mathbf{I} - \mathbf{M})\mathbf{y}]'(\mathbf{M}\mathbf{y}) \\ &= \mathbf{y}'(\mathbf{I} - \mathbf{M})\mathbf{M}\mathbf{y} \\ &= \mathbf{y}'\mathbf{0}\mathbf{y} = \mathbf{0} \qquad \text{by (4.24).} \end{aligned}$$

Also, the sum of cross products between the observations on any one of the x_i variables and the respective residuals is zero. That is, from (4.17),

$$\begin{aligned} \mathbf{X}'(\mathbf{y} - \hat{\mathbf{y}}) &= \mathbf{X}'\hat{\mathbf{r}} = \mathbf{X}'\mathbf{M}\mathbf{y} \\ &= \mathbf{0} \end{aligned} \tag{4.29}$$

[by (4.21)].

Notice that, since $\mathbf{y}$ is linear in $\mathbf{X}$, (4.28) follows from (4.29). Also, (4.27) is a special case of (4.29).

4.2. Fixed Models

In this section we discuss the classical regression models where the regressors are fixed or accurately controllable. As we shall soon observe, most of our discussion will be an extension of the concepts and results studied in the sections related to the fixed models in Chapter 3. Since we now face multivariable situations, we shall have a notational problem to overcome as certain statistical concepts peculiar to multivariate analysis must be understood. Thus this section will be mainly a review of most of Sections 3.3 and 3.4 in a multivariate context and a clarification of some further statistical concepts. Although we separated the discussion of Models A and B in Chapter 3, we shall find it more convenient to treat the fixed models together.

4.2.1. Assumptions of the Models

The regression models of interest are of the type

$$y = \beta_0 + \beta_1 x_1 + \beta_2 x_2 + \cdots + \beta_k x_k + u \tag{4.2}$$

or, in matrix notation,

$$\mathbf{y} = \mathbf{X}\boldsymbol{\beta} + \mathbf{u} \tag{4.5}$$

where the x variables are in some sense fixed. In this subsection, we attempt to list and to elaborate the assumptions that go with the analysis of fixed models.

Assumptions of model A

1. The x_i variables are fixed or nonstochastic.
2. No two or more than two of the x_i variables are perfect linear combinations of one another.
3. The disturbances u have an identical distribution with mean 0, variance σ^2 and, among them, zero covariances.

Assumptions of model B

1. Assumptions 1 and 2 for Model A hold true.
2. The disturbances u are identically normally distributed with mean 0, variance σ^2 and, among them, zero covariances. (4.33)

Notice that these assumptions are identical with the respective models given in the preceding chapter except that here we consider the systematic part of y as being accounted for by the x_i variables rather than by a single variable, say x. This point of view is significant for purposes of interpretation and for some analytical aspects of the regression model. We now shift our point of view somewhat and adopt a sampling theoretic construction of analysis to bring into place certain knowledge in multivariate statistical analysis. This shift is not a change in the substance of the problem and will prove useful for two reasons: (1) we can cast the regression models in a multivariate-analytic context; and (2) the attempt in (1) will enable us to attack further topics with greater efficiency than otherwise possible.

Consider, then, a model of the form (4.5) where **y** and **u** are each a vector of $N \times 1$, **X** is a matrix of $N \times (k + 1)$, and $\boldsymbol{\beta}$ is a vector of $(k + 1) \times 1$. Presumably, (4.5) is a representation of a relationship among the indicated variables about a population of size N. (For convenience, we assume finite population, but in this book a population is usually considered infinite in size.) Suppose that we now divide the population into n subgroups and that from each subgroup one observation is randomly chosen. [Then for the population we shall have

$$N_1 + N_2 + \cdots + N_n = N \tag{4.34}$$

where N_i is the size of the ith subgroup.] For the sample, we have

$$\mathbf{y} = \mathbf{X}\boldsymbol{\beta} + \mathbf{u} \tag{4.35}$$

where **y** and **u** are each $n \times 1$, **X** is $n \times (k + 1)$, and $\boldsymbol{\beta}$ is $(k + 1) \times 1$. Another way to write (4.35) is, of course,

$$Y_i = \beta_0 + \beta_1 X_{i1} + \beta_2 X_{i2} + \cdots + \beta_k X_{ik} + u_i \qquad i = 1, 2, \ldots n. \tag{4.35n}$$

With these in mind, we can now proceed to elaborate on the assumptions stated at the beginning of this subsection and to represent them in matrix notations.

With respect to Assumption 1 in Model A, no further discussion is necessary. We simply extend the concept of fixedness discussed in Subsection 3.3.2 to all the x_i variables. As for the second assumption, the essential idea is to eliminate the redundance of information in the matrix **X** in, say (4.5.) By redundance we mean that any one of the x_i variables can be written as a linear combination of the other variables. To illustrate this point, consider, without losing generality, the possibility that

$$x_k = \sum_{i=0}^{k-1} c_i x_i \tag{4.37}$$

where not all c_i's are zero. A special case of this is that one variable is perfectly collinear with another, say, for $i \neq j$,

$$x_i = c x_j \tag{4.37a}$$

On this assumption, if c is known, there is no point in having observations on the particular x_j, since these observations can be had by dividing c into the observations on the particular x_i variable. Indeed, matrix theory tells us that if (4.37) holds true, the $(k + 1)$ x_i variables are linearly dependent and the rank of **X** will not be $(k + 1)$ and, hence, the rank of $(\mathbf{X}'\mathbf{X})$ also will not be $(k + 1)$. As will be seen later, such a situation will cause difficulty in the estimation of the parameters.

Assumption 3 is discussed in two parts: (3a) that the mean of u is zero and (3b) the variances of u_i's are identically σ^2, and the covariances are 0. Under (3a) we have by Assumption 1 and by the definition of the expectation operator that, from (3.35),

$$E(\mathbf{y}) = E\mathbf{X}\boldsymbol{\beta} = \mathbf{X}\boldsymbol{\beta} \tag{4.38}$$

since $E(u_i) = 0$ for $i = 1, 2, \ldots n$, or $E(\mathbf{u}) = \mathbf{0}$. Here the expected value of y is a function of the x_i variables so that it is a conditional expectation. Thus one might write, for example,

$$E(\mathbf{y}) = E(\mathbf{y} \mid \mathbf{X}) \tag{4.38b}$$

The assumption that the mean of u is zero is really an assumption that the mean of u is a constant provided that there is a free constant term in the equation (since zero and any other constant are going to be indistinguishable, as can be observed by redefining the constant term). Now Assumption 3b states that $E(u_i u_j) = \sigma^2$ for $i = j$ but that $E(u_i u_j) = 0$ for $i \neq j$. In other words, the ith u has the constant variance σ^2 for all the n elements (the constant variance or homogeneity assumption), but the covariances among the u_i's are all zero. In matrix, notation, we can write these statements about the variances and covariances of u_i's in the following form:

$$E\mathbf{uu}' = E\begin{bmatrix} u_1u_1 & u_1u_2 & \cdots & u_1u_n \\ u_2u_1 & u_2u_2 & \cdots & u_2u_n \\ \cdot & \cdot & \cdot & \cdot \\ \cdot & \cdot & \cdot & \cdot \\ \cdot & \cdot & \cdot & \cdot \\ u_nu_1 & u_nu_2 & \cdots & u_nu_n \end{bmatrix} = \begin{bmatrix} \sigma^2 & 0 & \cdots & 0 \\ 0 & \sigma^2 & \cdots & 0 \\ \cdot & \cdot & \cdot & \cdot \\ \cdot & \cdot & \cdot & \cdot \\ \cdot & \cdot & \cdot & \cdot \\ 0 & 0 & \cdots & \sigma^2 \end{bmatrix} = \sigma^2\mathbf{I} \tag{4.39}$$

Notice that σ^2 is a scalar (a real number) and that $\mathbf{I}$ is an identity matrix of order n.

Summarizing, the statistical properties of the model (4.35) in matrix notation are the following.

1. $\mathbf{X}$ is a matrix of fixed values of $x_0, x_1, \ldots, x_k$ where $x_0 = 1$ at all times.
2. The x_i variables $(i = 0, 1, \ldots, k)$ are linearly independent, or $\rho(\mathbf{X}) = k + 1$. (4.32m)
3. $E(\mathbf{u}) = \mathbf{0}$.
4. $E(uu') = \sigma^2\mathbf{I}$ where $\mathbf{I}$ is the identity matrix of order n; $\sigma^2\mathbf{I}$ is referred to as the variance-covariance matrix of $\mathbf{u}$, or simply covariance of $\mathbf{u}$.*

* The covariance matrix of random variables $v_1, v_2, \ldots, v_n$ is defined as $E(\mathbf{v} - E\mathbf{v})(\mathbf{v} - E\mathbf{v})'$ where $\mathbf{v} = (v_1v_2 \cdots v_n)'$ and E is the expectation sign.

If, in addition, the u_i's have an identical normal distribution, we have the Fixed Model B which contains these assumptions:

1. Assumptions 1 through 3 in (4.32m).
2. $\mathbf{u}$ is normally distributed with mean $\mathbf{0}$ and variance-covariance $\sigma^2\mathbf{I}$. (4.33m)

4.2.2. The LS Estimator for $\boldsymbol{\beta}$

Now that the statistical properties of the variables in the fixed models are set forth, we can evaluate the properties of an estimator for $\boldsymbol{\beta}$. Among the possible estimators, we pick the LS estimator $\mathbf{b}$ for this discussion and show that this estimator has many desirable properties. We first consider the properties of $\mathbf{b}$ with the flexible assumptions regarding the shape of the distribution of $\mathbf{u}$. Later, we extend the results to the case where $\mathbf{u}$ is normally distributed. In both cases (really, Fixed Models A and B), the LS estimator turns out to be optimal in the sense of the minimum sum of squares of the residual errors but, in Fixed Model B it has the additional desirable property of being a maximum likelihood estimator. We now proceed to consider the so-called BLUE (best linear unbiased estimate) property of the LS estimator for $\boldsymbol{\beta}$ in (4.35), $\mathbf{y} = \mathbf{X}\boldsymbol{\beta} + \mathbf{u}$ for Fixed Model A.

b *is* BLUE. As has been discussed in Subsection 4.1.4, the LS estimator for $\boldsymbol{\beta}$ in (4.35) is

$$\mathbf{b} = (\mathbf{X}'\mathbf{X})^{-1}\mathbf{X}'\mathbf{y} \tag{4.40}$$

Contrasting this with (4.15) in Subsection 4.1.6, we observe that $\mathbf{b}$ is a linear combination of the observed values of y, or $\mathbf{b}$ is a linear transform of the vector $\mathbf{y}$, the transforming matrix being $(\mathbf{X}'\mathbf{X})^{-1}\mathbf{X}'$. One need only consider the matrix as a matrix of fixed constants once values of the x_i variables are given (or observed).

Before showing that $\mathbf{b}$ is best unbiased, we must observe that the variance-covariance matrix of the estimator $\mathbf{b}$ is

$$\mathbf{V}(\mathbf{b}) = E(\mathbf{b} - \boldsymbol{\beta})(\mathbf{b} - \boldsymbol{\beta})' = \sigma^2(\mathbf{X}'\mathbf{X})^{-1} \tag{4.41}$$

This follows since, from (4.40),

$$\begin{aligned}\mathbf{b} &= (\mathbf{X}'\mathbf{X})^{-1}\mathbf{X}'(\mathbf{X}\boldsymbol{\beta} + \mathbf{u}) = (\mathbf{X}'\mathbf{X})^{-1}\mathbf{X}'\mathbf{X}\boldsymbol{\beta} + (\mathbf{X}'\mathbf{X})^{-1}\mathbf{X}'\mathbf{u} \\ &= \boldsymbol{\beta} + (\mathbf{X}'\mathbf{X})^{-1}\mathbf{X}'\mathbf{u}\end{aligned}$$

so that

$$\mathbf{b} - \boldsymbol{\beta} = (\mathbf{X}'\mathbf{X})^{-1}\mathbf{X}'\mathbf{u} \tag{4.41a}$$

Then the covariance of $(\mathbf{b} - \boldsymbol{\beta})$ is

$$E(\mathbf{b} - \boldsymbol{\beta})(\mathbf{b} - \boldsymbol{\beta})' = E(\mathbf{X}'\mathbf{X})^{-1}\mathbf{X}'\mathbf{u}\mathbf{u}'\mathbf{X}(\mathbf{X}'\mathbf{X})^{-1}$$

which reduces to (4.41) for the following reasons: (1) for two conformable matrices $\mathbf{A}$ and $\mathbf{B}$, $(\mathbf{AB})' = \mathbf{B}'\mathbf{A}'$; (2) $\mathbf{A}^{-1}\mathbf{A} = \mathbf{I}$; (3) the expectation of a matrix is the matrix of expectations of the elements in the matrix; (4) $E(\mathbf{AZ}) = \mathbf{A}E(\mathbf{X})$ where $\mathbf{Z}$ is a random vector or matrix, provided that $\mathbf{A}$ is a matrix of fixed coefficients; and, finally, (5) the assumption from (4.32m) that $E(\mathbf{uu}') = \sigma^2\mathbf{I}$.

To show that $\mathbf{b}$ is best unbiased, we first show, as a preliminary exercise, that $\mathbf{b}$ is unbiased for $\boldsymbol{\beta}$. From (4.40),

$$\mathbf{b} = (\mathbf{X}'\mathbf{X})^{-1}\mathbf{X}'(\mathbf{X}\boldsymbol{\beta} + \mathbf{u})$$

By taking the expectation, we have

$$\begin{aligned} E\mathbf{b} &= E(\mathbf{X}'\mathbf{X})^{-1}(\mathbf{X}'\mathbf{X})\boldsymbol{\beta} + E(\mathbf{X}'\mathbf{X})^{-1}\mathbf{X}'\mathbf{u} \\ &= \boldsymbol{\beta} + (\mathbf{X}'\mathbf{X})^{-1}\mathbf{X}'E\mathbf{u} \\ &= \boldsymbol{\beta} \end{aligned}$$

Therefore, $\mathbf{b}$ is unbiased for $\boldsymbol{\beta}$. It is worthwhile noticing that the fact that we can write $E(\mathbf{X}'\mathbf{X})^{-1}\mathbf{X}'\mathbf{u}$ as $(\mathbf{X}'\mathbf{X})^{-1}\mathbf{X}'E\mathbf{u}$ is because of the assumptions that $\mathbf{X}$ is fixed and that u_i is independent of $\mathbf{X}_i$ for $i = 1, 2, \ldots, n$.

Now the general proof that $\mathbf{b}$ is best unbiased goes as follows. We first choose an arbitrary linear transform, say $\mathbf{T}$, a $(k + 1) \times n$ matrix, to be multiplied to $\mathbf{y}$ to form an arbitrary estimator of $\boldsymbol{\beta}$, say,

$$\mathbf{b}^* = \mathbf{T}\mathbf{y} \tag{4.42}$$

The strategy here is to choose first a class of linear estimators from which a subclass of unbiased estimators will be selected and then to find a minimum variance estimator from the class of linear unbiased estimators. Hence, for unbiasedness, we require that

$$E(\mathbf{b}^*) = E(\mathbf{T}\mathbf{y}) = \boldsymbol{\beta} \tag{4.43}$$

This condition makes it necessary and sufficient to choose $\mathbf{T}$ so that

$$\mathbf{TX}\boldsymbol{\beta} = \boldsymbol{\beta} \tag{4.44}$$

since

$$E(\mathbf{Ty}) = E[\mathbf{T}(\mathbf{X}\boldsymbol{\beta} + \mathbf{u})] = E(\mathbf{TX}\boldsymbol{\beta}) + \mathbf{0}$$

for $E(\mathbf{Tu}) = \mathbf{T}(E\mathbf{u}) = \mathbf{0}$ by assumption and we want unbiasedness for all true values of $\boldsymbol{\beta}$. If it so happens that we choose

$$\mathbf{T} = (\mathbf{X}'\mathbf{X})^{-1}\mathbf{X}' \tag{4.45}$$

we observe that (4.44) is satisfied and that $\mathbf{b}^* = \mathbf{Ty}$ will be unbiased for $\boldsymbol{\beta}$. Since the estimator $\mathbf{Ty}$ that satisfies (4.45) is identical with the least-squares estimator, we conclude that the least-squares estimator is linear and unbiased. (Compare this with the proof of the unbiasedness of $\mathbf{b}$ given earlier.) We notice further that for $\mathbf{T}$ in (4.45), the variance-covariance of $\mathbf{b}^* = \mathbf{Ty}$ is

$$\mathbf{V}(\mathbf{b}^*) = E(\mathbf{b}^* - \boldsymbol{\beta})(\mathbf{b}^* - \boldsymbol{\beta})' = \sigma^2(\mathbf{X}'\mathbf{X})^{-1} = \mathbf{V}(\mathbf{b}) \tag{4.46}$$

To show the minimum variance property, let us take yet another arbitrary matrix, say $\mathbf{C}$, of order $(k + 1) \times n$ and define it as

$$\mathbf{C} = \mathbf{T} + \mathbf{T}_1 \tag{4.47}$$

where $\mathbf{T}$ is the matrix discussed in the preceding paragraph. By letting $\mathbf{Cy} = \mathbf{b}^{**}$ and by limiting our attention to the class of unbiased estimators, we observe that we must have, for unbiasedness,

$$(\mathbf{T} + \mathbf{T}_1)\mathbf{X}\boldsymbol{\beta} = \boldsymbol{\beta} \tag{4.48}$$

and

$$\mathbf{T}_1\mathbf{X} = \mathbf{0} \tag{4.49}$$

since it has been required that $\mathbf{TX}\boldsymbol{\beta} = \boldsymbol{\beta}$. With these results in mind, we now are in a good position to examine the variance-covariance matrix of $\mathbf{b}^{**}$. Thus,

$$\begin{aligned} E(\mathbf{b}^{**} - \boldsymbol{\beta})(\mathbf{b}^{**} - \boldsymbol{\beta})' &= (\mathbf{T} + \mathbf{T}_1)E(\mathbf{uu}')(\mathbf{T} + \mathbf{T}_1)' \\ &= \sigma^2(\mathbf{T} + \mathbf{T}_1)(\mathbf{T} + \mathbf{T}_1)' \\ &= \sigma^2(\mathbf{TT}' + \mathbf{T}_1\mathbf{T}' + \mathbf{TT}_1' + \mathbf{T}_1\mathbf{T}_1') \\ &= \sigma^2[(\mathbf{X}'\mathbf{X})^{-1}\mathbf{X}'\mathbf{X}(\mathbf{X}'\mathbf{X})^{-1} + \mathbf{T}_1\mathbf{T}_1'] \qquad (4.50) \\ &= \sigma^2[(\mathbf{X}'\mathbf{X})^{-1} + \mathbf{T}_1\mathbf{T}_1'] \end{aligned}$$

Therefore, the covariance matrix of $\mathbf{b}^{**}$ obtained under the $\mathbf{C}$ transformation decomposes into two covariance matrices. Notice that the first of the two matrices is the variance-covariance matrix of the LS estimator $\mathbf{b}$. Now it is known from matrix theory that $\mathbf{T}_1\mathbf{T}_1'$ is positive semidefinite, meaning, among other things, that the diagonal elements of the matrix are each greater than or equal to zero. Or, we can claim that

$$E(b_i^{**} - \beta_i)^2 \geq E(b_i - \beta_i)^2 \qquad i = 0, 1, 2, \ldots, k \tag{4.51}$$

where b_i's are the elements of the LS estimator $\mathbf{b}$. That is, the condition that the diagonal elements of (4.50) be each greater than or equal to zero implies that either $\mathbf{T}_1 = \mathbf{0}$ or else the strict inequality in (4.51) holds true. This is all we need to establish, generally, that the LS estimator $\mathbf{b}$ is BLUE.

4.2.3. The Maximum Likelihood Estimator for $\boldsymbol{\beta}$

We have just observed that under very weak assumptions about the shape of the distribution of $\mathbf{u}$ the LS estimator for $\boldsymbol{\beta}$ has the desirable properties, BLUE. Certainly, the BLUEness is the common property of the LS estimator, regardless of whether $\mathbf{u}$ is normally distributed or has some other distribution as long as **X** is fixed. Thus, for our Fixed Models, we, in effect, have been discussing and proving the so-called Gauss-Markov theorem on the least-squares method. Formally, this theorem is as follows.

> In the linear regression models of the type (4.5), having the assumptions of our Fixed Models, the best linear unbiased estimator of $\boldsymbol{\beta}$ is the least-squares estimator $\mathbf{b} = (\mathbf{X}'\mathbf{X})^{-1}\mathbf{X}'\mathbf{y}$ which has the covariance matrix $\mathbf{V}(\mathbf{b}) = \sigma^2(\mathbf{X}'\mathbf{X})^{-1}$. (4.51)

This is a significant general result and is a strong theoretical justification for the use of the least-squares procedure in various research problems. In fact, the least-squares method of estimation is popular to the extent that it is often used interchangeably with the phrase "regression analysis." Certainly, one cannot overemphasize the usefulness of the least-squares method, but we must also point out that the LS method is but one of the several methods available for the estimation of a regression model. One of these other methods is the maximum likelihood (ML) procedure. This procedure, in principle, is applicable to regression models with error terms that have a probability distribution. However, if the error term is normal, the ML estimator is identical to the LS estimator.

In our Fixed Model B the disturbances are normally distributed so that for a sample of size n the u_i's are normally and identically distributed with mean 0 and variance σ^2. Assuming that we have a random sample, then, the discussion of the ML method in Subsection 3.4.1 tells us that the joint density of the sample is

$$L = (2\pi\sigma^2)^{-(n/2)} \exp\left(-\frac{1}{2\sigma^2}\sum_{i=1}^{n} u_i^2\right)$$

which is equivalent to

$$L = (2\pi\sigma^2)^{-(n/2)} \exp\left[-\frac{1}{2\sigma^2}(\mathbf{y} - \mathbf{X}\boldsymbol{\beta})'(\mathbf{y} - \mathbf{X}\boldsymbol{\beta})\right] \tag{4.52}$$

since

$$\mathbf{u} = \mathbf{y} - \mathbf{X}\boldsymbol{\beta} \quad \text{and} \quad \mathbf{u}'\mathbf{u} = \sum_{i=1}^{n} u_i^2$$

Thus, as is familiar to us from the earlier discussion, the expression in (4.52) is the likelihood function of interest, a function of parameter vector $\boldsymbol{\beta}$ given the observations on **y** and **X**. To find $\hat{\boldsymbol{\beta}}$ in terms of observations so that the likelihood L is maximum is the same as minimizing the exponent of e. Since the logarithmic transformation of (4.52) will not affect the values of $\hat{\boldsymbol{\beta}}$ at which the maximum of L is attained, we take the natural logarithm of both sides of (4.52) and proceed to obtain solutions that will make

$\mathscr{L} = \ln L$ in the following maximum:

$$\mathscr{L} = \ln L = \frac{-n}{2} \ln (2\pi\sigma^2) - \frac{1}{2\sigma^2} (\mathbf{y} - \mathbf{X}\boldsymbol{\beta})'(\mathbf{y} - \mathbf{X}\boldsymbol{\beta}) \tag{4.53}$$

Now the first term in the righthand side of (4.53) is a constant (not involving $\boldsymbol{\beta}$), and $(-1/2\sigma^2)$ in the second term is also a constant. Thus, if we find $\boldsymbol{\beta}$ so that $(\mathbf{y} - \mathbf{X}\boldsymbol{\beta})'(\mathbf{y} - \mathbf{X}\boldsymbol{\beta})$ is minimum, we shall have $\mathscr{L}$ maximum. This is exactly the same as the least-squares procedure described in Subsection 4.1.4, and without further ado we write the maximum likelihood estimator of $\boldsymbol{\beta}$ as

$$\hat{\boldsymbol{\beta}} = (\mathbf{X}'\mathbf{X})^{-1}\mathbf{X}'\mathbf{y}$$

Thus, if $\mathbf{u}$ is normally distributed, the least-squares estimator of $\boldsymbol{\beta}$ is also a maximum likelihood estimator. That is, the least-squares estimator is not only BLUE but also consistent and efficient (both in the large sample sense discussed earlier).* A seemingly more formal way to obtain in (4.52) a maximum likelihood estimator of $\boldsymbol{\beta}$ is to consider, first-order and second-order conditions for $\mathscr{L}$ there to be maximum. This procedure has the advantage of giving us the maximum likelihood estimator for the variance of u, σ^2, also. Accordingly, by setting the first partials of $\mathscr{L}$ with respect to $\boldsymbol{\beta}$ and σ^2 in (4.53) equal to zero, we have

$$\frac{\partial \mathscr{L}}{\partial \boldsymbol{\beta}} = \frac{1}{2\sigma^2} \mathbf{X}'(\mathbf{y} - \mathbf{X}\boldsymbol{\beta}) = \mathbf{0} \tag{4.54}$$

$$\frac{\partial \mathscr{L}}{\partial \sigma^2} = -\frac{n}{2\sigma^2} + \frac{1}{2\sigma^4} (\mathbf{y} - \mathbf{X}\boldsymbol{\beta})'(\mathbf{y} - \mathbf{X}\boldsymbol{\beta}) = \mathbf{0} \tag{4.55}$$

By solving (4.54) for $\boldsymbol{\beta}$ and, in turn, (4.55) for σ^2, we easily find that the maximum likelihood estimators for $\boldsymbol{\beta}$ and σ^2 are, respectively,

$$\hat{\boldsymbol{\beta}} = (\mathbf{X}'\mathbf{X})^{-1}\mathbf{X}'\mathbf{y} \tag{4.56}$$

$$\hat{\sigma}^2 = \frac{1}{n} (\mathbf{y} - \mathbf{X}\hat{\boldsymbol{\beta}})'(\mathbf{y} - \mathbf{X}\hat{\boldsymbol{\beta}}) \tag{4.57}$$

It can be seen easily that the matrix of second partials of $\mathscr{L}$ with respect to $\boldsymbol{\beta}$ and to σ^2 is negative definite, confirming that a maximum of $\mathscr{L}$ can be had. Notice that the estimator for σ^2 in (4.57) is equivalent to the residual sum of squares given in (3.48) in Subsection 3.4.3, where we discussed the variation of y values around the estimated regression line. This estimator turns out to be biased for σ^2, as was shown in that subsection. In the next subsection we discuss this and related problems in the multivariate context.

* Notice that the normality of the disturbances is not required for the LS estimator of $\boldsymbol{\beta}$ to be consistent and asymptotically efficient. Here, we are merely saying that these properties are the consequence of the ML estimator based on normality.

4.2.4. Variation Around the LS Hyperplane

In principle and techniques the material in this section is mainly a restatement of the concepts described in Subsection 3.4.3. We shall discuss: (1) variance estimators, (2) the coefficient of multiple determination, and (3) the uses of the coefficient of multiple determination.

VARIANCE ESTIMATORS. Given that we want to find an unbiased estimator for σ^2, reasonably we find the LS residuals after data are fitted to a fixed model. Hence, for sample size n we recall from (4.17) that, by (4.21),

$$\begin{aligned}\hat{\mathbf{u}} &= \mathbf{y} - \hat{\mathbf{y}} = \mathbf{My} = \mathbf{M}(\mathbf{X}\boldsymbol{\beta} + \mathbf{u}) \\ &= \mathbf{Mu}\end{aligned} \tag{4.58}$$

The reason for writing this expression is that we want to express the residuals (what we obtain from the sample) in terms of the population $\mathbf{u}$ (for which we made assumptions of statistical properties), so that we can evaluate the statistical properties of our estimator (which is a function of the sample observations). It follows then that the expectation of the residual sum of squares is

$$\begin{aligned}E(\hat{\mathbf{u}}'\hat{\mathbf{u}}) &= E[\mathbf{u}'\mathbf{Mu}] && \text{(using symmetry and idempotency of } \mathbf{M}\text{)} \\ &= E\,\mathrm{tr}\,[\mathbf{u}'\mathbf{Mu}] && \text{(all covariances of } u_i\text{'s are zero)} \\ &= E\,\mathrm{tr}\,[\mathbf{Muu}'] && [\mathrm{tr}\,(\mathbf{AB}) = \mathrm{tr}\,(\mathbf{BA}) \text{ if } \mathbf{B} \text{ and } \mathbf{A} \text{ are conformable}] \\ &= \mathrm{tr}\,[\mathbf{M}E(\mathbf{uu}')] \\ &= \sigma^2\,\mathrm{tr}\,\mathbf{M} \\ &= \sigma^2(n - k - 1) && \{\mathrm{tr}\,(I) = n \text{ and } \mathrm{tr}\,[\mathbf{X}(\mathbf{X}'\mathbf{X})^{-1}\mathbf{X}] = k + 1\}\end{aligned} \tag{4.59}$$

Thus we conclude that

$$s^2 = \frac{\hat{\mathbf{u}}'\hat{\mathbf{u}}}{n - k - 1} \tag{4.60}$$

is unbiased for σ^2.

The second variance estimator of interest here is the covariance matrix estimator for $\mathbf{V}(\mathbf{b}) = E(\mathbf{b} - \boldsymbol{\beta})(\mathbf{b} - \boldsymbol{\beta})' = \sigma^2(\mathbf{X}'\mathbf{X})^{-1}$. Since the inverse of the moment matrix involved is constant and since s^2 is unbiased for σ^2, we conclude that

$$\hat{\mathbf{V}}(\mathbf{b}) = s^2(\mathbf{X}'\mathbf{X})^{-1} \tag{4.61}$$

is unbiased for $\mathbf{V}(\mathbf{b})$. Let this particular estimator $\hat{\mathbf{V}}(\mathbf{b})$ be designated by $\mathbf{S}(\mathbf{b})$ henceforth.

COEFFICIENT OF MULTIPLE DETERMINATION R^2. The concept underlying the coefficient of multiple determination has been presented in some detail under the heading "coefficient of determination" in Subsection 3.4.3. In the multiple regression case we obtain the coefficient R^2 by first finding the residual sum of squares. Computationally, we observe that

$$\begin{aligned}\hat{\mathbf{u}}'\hat{\mathbf{u}} &= (\mathbf{y} - \hat{\mathbf{y}})'(\mathbf{y} - \hat{\mathbf{y}}) \\ &= (\mathbf{y} - \mathbf{Xb})'(\mathbf{y} - \mathbf{Xb}) \\ &= \mathbf{y}'\mathbf{y} - \mathbf{b}'\mathbf{X}'\mathbf{y} - \mathbf{y}'\mathbf{Xb} + \mathbf{b}'\mathbf{X}'\mathbf{Xb} \\ &= \mathbf{y}'\mathbf{y} - \mathbf{b}'\mathbf{X}'\mathbf{y} \\ &= \mathbf{y}'\mathbf{y} - \mathbf{b}'\mathbf{X}'\mathbf{Xb}\end{aligned} \tag{4.62}$$

since in the third line of this expression $-\mathbf{b}'\mathbf{X}'\mathbf{y}$ and $\mathbf{b}'\mathbf{X}'\mathbf{Xb}$ cancel out (why?). That is, we break up the sum of squares of the dependent variable into the explained ($\mathbf{b}'\mathbf{X}'\mathbf{Xb}$) and the unexplained ($\hat{\mathbf{u}}'\hat{\mathbf{u}}$) sums of squares. Of course, the reason for finding this error sum of squares is to use it for forming the expression $R^2 = 1 - ESS/TSS$. Now the total sum of squares (TSS) is

$$\sum (y_i - \bar{y})^2 = \mathbf{y}'\mathbf{y} - \frac{\left(\sum_{i=1}^{n} y_i\right)^2}{n} \tag{4.63}$$

and the regression sum of squares (RSS) is, from (4.62) and (4.63),

$$\begin{aligned}RSS = TSS - ESS &= \mathbf{y}'\mathbf{y} - \frac{\left(\sum_{i=1}^{n} y_i\right)^2}{n} - \mathbf{y}'\mathbf{y} + \mathbf{b}'\mathbf{X}'\mathbf{Xb} \\ &= \mathbf{b}'\mathbf{X}'\mathbf{Xb} - \frac{\left(\sum_{i=1}^{n} y_i\right)^2}{n}\end{aligned} \tag{4.64}$$

And, as previously, we have

$$R^2 = \frac{\mathbf{b}'\mathbf{X}'\mathbf{Xb} - \dfrac{\left(\sum_{i=1}^{n} y_i\right)^2}{n}}{\mathbf{y}'\mathbf{y} - \dfrac{\left(\sum_{i=1}^{n} y_i\right)^2}{n}} \tag{4.65}$$

Hence, the quantities necessary for computing R^2 are $\sum_{i=1}^{n} y_i$, $\mathbf{y}'\mathbf{y} = \sum_{i=1}^{n} y_i^2$, the coefficient estimates b_i's, and

$$\mathbf{X}'\mathbf{y} = \begin{bmatrix} \sum X_{0i}y_i \\ \sum X_{1i}y_i \\ \sum X_{2i}y_i \\ \cdot \\ \cdot \\ \cdot \\ \sum X_{ki}y_i \end{bmatrix}$$

since the first column of $\mathbf{X}$ consists of all 1's and the first element of $\mathbf{X}'\mathbf{y}$ is really $\sum_{i=1}^{n} y_i$. And note always that, from the normal equations,

$$\mathbf{b}'\mathbf{X}'\mathbf{X}\mathbf{b} = \mathbf{b}'\mathbf{X}'\mathbf{y}.$$

4.2.5. *A Geometric View of the LS Estimation*

A this point it is useful to gain a geometric appreciation of some of the basic concepts discussed in the preceding sections. This appreciation will help us in understanding our future work and will aid independent study of the statistical literature.

First we proceed generally and consider the model

$$\mathbf{y} = \mathbf{X}\boldsymbol{\beta} + \mathbf{u} \tag{4.35}$$

where $\mathbf{y}$ and $\mathbf{u}$ are each $n \times 1$, $\mathbf{X}$ is $n \times (k + 1)$, and $\boldsymbol{\beta}$ is $(k + 1) \times 1$. The column vector $\mathbf{y}$ is a point in the n-dimensional space which we call sample space. Presumably many of these points can be drawn from the sample space R. Belonging in this space are also the $k + 1$ column vectors of matrix $\mathbf{X}$, each vector being $n \times 1$. The vector $\boldsymbol{\beta}$, similarly, is a point in the $(k + 1)$-dimensional parameter space Ω, and again one can think of an infinite number of these points in the parameter space. Now by the assumption of independence of the $(k + 1)$ column vectors in $\mathbf{X}$, which is another way of saying that the rank of $\mathbf{X}$ is $(k + 1)$, or that a linear combination of the $(k + 1)$ vectors belongs in the space generated or spanned by these vectors. Assume that an arbitrary element, say $\boldsymbol{\theta}$, is taken from Ω to form a point $\mathbf{X}\boldsymbol{\theta}$, then this point belongs in the space spanned by the $(k + 1)$ $\mathbf{x}_i$ vectors; this space is a subspace of the sample space and is called the estimation space. Before proceeding, let us illustrate these concepts with a manageable example. In Figure 4.1, we draw a 3-dimensional sample space where any point in the R_1-R_2-R_3 space is a candidate for sample drawing of $\mathbf{y}_1$. Suppose that the parameter space (which is not necessarily a subspace of the sample space)

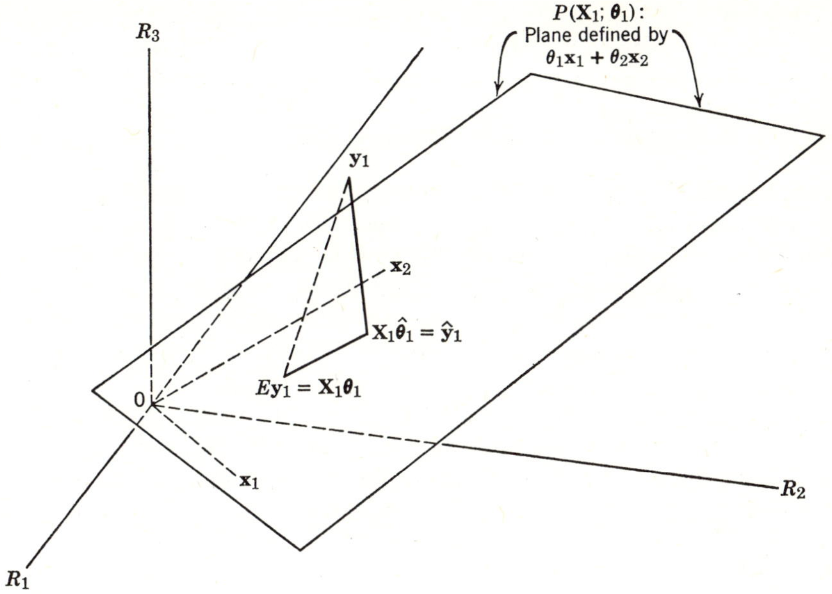

Figure 4.1

is 2-dimensional, so that the model (4.35) will be

$$\begin{pmatrix} Y_1 \\ Y_2 \\ Y_3 \end{pmatrix} = \begin{pmatrix} X_{11} & X_{12} \\ X_{21} & X_{22} \\ X_{31} & X_{32} \end{pmatrix} \begin{pmatrix} \theta_1 \\ \\ \theta_2 \end{pmatrix} + \begin{pmatrix} u_1 \\ u_2 \\ u_3 \end{pmatrix} \tag{4.35e}$$

Now, since $\begin{pmatrix} \theta_1 \\ \theta_2 \end{pmatrix}$ is an arbitrary element of the 2-dimensional parameter space, we can think of allowing θ_1 and θ_2 to vary in value and can form linear combinations of the vectors

$$\mathbf{x}_1 = \begin{pmatrix} X_{11} \\ X_{12} \\ X_{13} \end{pmatrix} \quad \text{and} \quad \mathbf{x}_2 = \begin{pmatrix} X_{21} \\ X_{22} \\ X_{23} \end{pmatrix}$$

by θ_1 and θ_2. These linear combinations are themselves points in the R_1-R_2-R_3 space, but they have the special property that they all rest on the plane generated by the vectors $\mathbf{x}_1$ and $\mathbf{x}_2$. This plane represents the estimation space and is designated as $P(\mathbf{X}_1; \boldsymbol{\theta}_1)$ in Figure 4.1. Notice that point $\mathbf{X}_1\boldsymbol{\theta}_1$ on the plane is obtained from picking a particular $\boldsymbol{\theta}_1$ from the parameter space.

LS ESTIMATION. Suppose now, in general terms, that the $\boldsymbol{\theta}$ picked from Ω is a variable vector and that, given the observation of $\mathbf{y}$, we want to find a specific set of elements for $\boldsymbol{\theta}$ so as to minimize the quantity*

$$(\mathbf{y} - \mathbf{X}\boldsymbol{\theta})'(\mathbf{y} - \mathbf{X}\boldsymbol{\theta}) \tag{4.7e}$$

Since $\mathbf{X}\boldsymbol{\theta}$ defines a point in the estimation space, the quantity (4.7e) is the distance between the point $\mathbf{y}$ and the point $\mathbf{X}\boldsymbol{\theta}$. Thus, minimizing (4.7e) is tantamount to finding a particular $\boldsymbol{\theta}$, call it $\hat{\boldsymbol{\theta}}$, which is the point at the foot of projecting $\mathbf{y}$ perpendicularly to the estimation space, or to finding a point on the estimation space that is closest to the point $\mathbf{y}$. Notice that (4.7e) is the usual sum of squares and corresponds directly to the quantity (4.7a) in Section 4.1.4.

In terms of Figure 4.1, the point $\mathbf{X}_1\hat{\boldsymbol{\theta}}_1$, is on the plane $P(\mathbf{X}_1; \boldsymbol{\theta}_1)$, the estimation space, and is at the projection of $\mathbf{y}_1$ on $P(\mathbf{X}_1; \boldsymbol{\theta}_1)$. Assume now that the point $\mathbf{X}_1\boldsymbol{\theta}_1$ in Figure 4.1 represents the true population expectation of $\mathbf{y}$, then by Pythagorean theorem we determine that

$$(\mathbf{y}_1 - \mathbf{X}_1\hat{\boldsymbol{\theta}}_1)'(\mathbf{y}_1 - \mathbf{X}_1\hat{\boldsymbol{\theta}}_1)$$
$$= (\mathbf{y}_1 - \mathbf{X}_1\boldsymbol{\theta}_1)'(\mathbf{y}_1 - \mathbf{X}_1\boldsymbol{\theta}_1) - (\mathbf{X}_1\hat{\boldsymbol{\theta}}_1 - \mathbf{X}_1\boldsymbol{\theta}_1)'(\mathbf{X}_1\hat{\boldsymbol{\theta}}_1 - \mathbf{X}_1\theta_1)$$

or

$$(\mathbf{y}_1 - \mathbf{X}_1\boldsymbol{\theta}_1)'(\mathbf{y}_1 - \mathbf{X}_1\boldsymbol{\theta}_1)$$
$$= (\mathbf{y}_1 - \mathbf{X}_1\hat{\boldsymbol{\theta}}_1)'(\mathbf{y}_1 - \mathbf{X}_1\hat{\boldsymbol{\theta}}_1) + (\mathbf{X}_1\hat{\boldsymbol{\theta}}_1 - \mathbf{X}_1\boldsymbol{\theta}_1)'(\mathbf{X}_1\hat{\boldsymbol{\theta}}_1 - \mathbf{X}_1\boldsymbol{\theta}_1)$$

In other words, the distance between the $\mathbf{y}_1$ and its expectation (all in terms of points in the 3-dimensional space) can be decomposed into two parts: (1) the distance between $\mathbf{y}_1$ and the estimate $\hat{\mathbf{y}}_1$ and (2) the distance between $\hat{\mathbf{y}}_1$ and $E\mathbf{y}_1$.

In the general case, $\hat{\boldsymbol{\theta}}_1$, just discussed, corresponds to the least-squares estimate of $\boldsymbol{\beta}$ in (4.35), or $\mathbf{b}$. Thus, for (4.35),

$$(\mathbf{y} - \mathbf{X}\boldsymbol{\beta})'(\mathbf{y} - \mathbf{X}\boldsymbol{\beta}) = (\mathbf{y} - \mathbf{X}\mathbf{b})'(\mathbf{y} - \mathbf{X}\mathbf{b}) + (\mathbf{X}\mathbf{b} - \mathbf{X}\boldsymbol{\beta})'(\mathbf{X}\mathbf{b} - \mathbf{X}\boldsymbol{\beta})$$

a decomposition similar to the case illustrated at the end of the preceding paragraph. Furthermore, the vector of residuals $\mathbf{r} = (\mathbf{y} - \hat{\mathbf{y}})$ is, by the LS choice, orthogonal to $\hat{\mathbf{y}}$, that is, $\hat{\mathbf{y}}'\mathbf{r} = \mathbf{0}$, a result shown in (4.28) in Section 4.1.6.

For further geometric treatment of the LS procedures, see N. R. Draper and H. Smith (1966, pp. 285–295) and E. Malinvaud (1966, Chapters 5 and 6).

* We can express this in another way by stating that we want to search for a specific $\boldsymbol{\theta}_0$ so that $\mathbf{X}\boldsymbol{\theta}_0$ "best" estimates $\mathbf{y}$.

4.2.6. *Prediction and Forecasting*

The basic concepts related to prediction and forecasting have been stated in some detail in Subsection 3.4.5. Here, we shall discuss the same concepts in matrix notation.

PREDICTION. Let the observations on the independent variables for the prediction "period" be $\mathbf{X}_p$. Clearly, $\mathbf{X}_p$ is $1 \times (k + 1)$. Since $\mathbf{X}_p$ is picked from the same population as was the sample from which **b** and other statistics are computed, we immediately notice that y in the prediction period Y_p is

$$Y_p = \mathbf{X}_p\boldsymbol{\beta} + u_p \tag{4.66}$$

Now, prediction, in the sense we use it, means calculation of the expected value of Y_p on the basis of X_p and the estimate of $\boldsymbol{\beta}$, **b**. Thus

$$\widehat{E(Y_p)} = \widehat{E(Y_p \mid \mathbf{X}_p)} = \mathbf{X}_p\mathbf{b} \tag{4.67}$$

The variability in $\widehat{E(Y_p)}$ comes from the variability in the **b** vector for which we already have a variance-covariance matrix (4.41). Therefore, we have

$$\text{var}\,(\mathbf{X}_p\mathbf{b}) = \mathbf{X}_p\mathbf{V}(\mathbf{b})\mathbf{X}_p' = \sigma^2\mathbf{X}_p(\mathbf{X}'\mathbf{X})^{-1}\mathbf{X}_p' \tag{4.68}$$

This is a matrix counterpart, so to speak, of the result that, if σ^2 is the variance of random variable x, then the variance of kx is $k^2\sigma^2$. Now, if $\mathbf{V}(\mathbf{b})$ is unknown, it can be estimated by $\mathbf{S}(\mathbf{b})$, or we can replace σ^2 by s^2 in the last term of (4.68).

Under Fixed Model B, then, it is possible to make probability statements about the calculated values in (4.67) and (4.68), since **b** is a linear combination of the normally distributed variables.

It will be instructive for the student to show that the predictor (4.67) is BLUE, following the method used for establishing the same properties for **b**.

FORECAST. For an observation on the independent variables in the forecast period, say $\mathbf{X}_f$ (we use different notations for the extra-sample observations on the independent variables simply for distinguishing the concepts of prediction and forecast; it is quite possible that for the problem at hand $X_p = X_f$) the value of y is

$$Y_f = \mathbf{X}_f\boldsymbol{\beta} + u_f \tag{4.69}$$

and the forecast of such a value can be obtained by the following procedure:

$$\hat{Y}_f = \mathbf{X}_f\mathbf{b} \tag{4.70}$$

The error of the forecast thus is

$$\begin{aligned} e_f = Y_f - \hat{Y}_f &= (\mathbf{X}_f\boldsymbol{\beta} + u_f) - \mathbf{X}_f\mathbf{b} \\ &= \mathbf{X}_f(\boldsymbol{\beta} - \mathbf{b}) + u_f \end{aligned} \tag{4.71}$$

It follows that the variability in the forecast value comes from two sources: (1) the estimated b coefficients, and (2) the disturbance in the forecast period. Thus the variance of the error is

$$\begin{aligned} \text{var}(e_f) &= \text{var}(\mathbf{X}_f\mathbf{b}) + \text{var}(u_f) \\ &= \mathbf{X}_f\mathbf{V}(\mathbf{b})\mathbf{X}_f' + \sigma^2 \\ &= \sigma^2[1 + \mathbf{X}_f(\mathbf{X}'\mathbf{X})^{-1}\mathbf{X}_f'] \end{aligned} \tag{4.72}$$

Again, if σ^2 is not known, it is estimated by s^2. It will be interesting to compare the last expression of (4.72) with (3.81).

We consider tests of the significance of the coefficient estimates in Chapter 5.

4.3. Some Useful Topics

In the preceding two sections our attention has been largely confined to the theoretical and intuitive understanding of the least-squares procedure as applied to linear multiple regression models. It is the task of the present section to supplement the preceding discussion by considering some of the more frequently encountered problems in empirical work. We shall deal with the uses of the coefficient of multiple determination, the interpretation of regression coefficients, and some computational problems.

4.3.1. The Uses of R^2

Continuing the discussion in Section 4.2.4, we notice that in general R^2 measures the percentages of the variation in the dependent variable explained *jointly* by the independent variables. Thus, a high R^2 indicates a good fit of a posited relation, and a low R^2 a poor fit. The use of R^2, in this role, is not as straightforward as it sounds, and we elaborate on this in the following material. Later, we shall comment briefly on the so-called R^2 corrected for degrees of freedom, or $\bar{R}^2$.

Frequently, in economic research one is faced with a situation where theory is only indicative of a number of variables to explain or to determine a variable, but it is not clear as to the order of magnitude of the influence of the respective variables on the dependent variable and as to the appropriate form of the functional relation (whether linear or otherwise). Thus it sometimes becomes necessary to determine, on the basis of the sample information, a preferable subset of all of the theoretically suggested variables for inclusion in the relation as regressors for a given functional form. Since R^2 provides a standard measure of how good a specified relation fits the sample data, the measure can be used as a criterion for the comparison of (1) the goodness of fit between any two alternative (but not necessarily mutually exclusive) sets of independent variables, and (2) the goodness of fit of two functional forms given the same set of independent variables.*

Cutting across these two uses of R^2 is the notion that the simpler the relation or the smaller the number of variables in a relation, the more preferable such a relation is (virtue of simplicity). Particularly with respect to the first use of R^2, noted above, we observe that R^2 is a nondecreasing function of the number of regressors. In a general framework, suppose that we use m variables to estimate a regression equation and find that the coefficient of determination is R_1^2. Now, if p more variables are added to the reestimation as the result of which the coefficient of determination is R_2^2. Since $R_2^2 \geq R_1^2$ as $p \geq 0$, the temptation would be to obtain as high an R^2 as possible because a high value represents good fit. This kind of effort should be attempted with a great deal of caution as it is done at the peril of theory and simplicity. In any case, a measure called the coefficient of multiple determination, corrected for degrees of freedom or denoted by $\bar{R}^2$ and defined below, may be used to compare the strength of the additions of variables. In a regression involving $k + 1$ independent variables

$$\bar{R}^2 = 1 - (1 - R^2)\frac{n - 1}{n - k - 1} \tag{4.73}$$

where n is the usual sample size. This makes it possible for $\bar{R}_2^2 < \bar{R}_1^2$ in the general case just mentioned. This last situation can occur if the increase in R^2 is not large after p variables are added.

There is a theoretical justification for wanting the R^2 corrected for degrees of freedom. It has been shown that if the "true" coefficient of multiple determination is P^2, then its sample estimator R^2 has the expectation

$$E(R^2) = P^2 - \frac{1 - P^2}{n}[(k + 1) - (1 - P^2)(1 + 2P^2)] + o(n^{-1}) \tag{4.74}$$

where $o(n^{-1})$ converges to zero as $n \rightarrow \infty$. The second term on the right of (4.74) indicates that the estimator R^2 has a negative bias even when sample size is relatively large. One way to correct this problem is to obtain unbiased estimators for the population equivalents of the numerator and the denominator forming the ratio R^2 in (4.65). Thus, keeping in mind that an *approximate* unbiased version of

$$1 - R^2 = \frac{\hat{\mathbf{u}}'\hat{\mathbf{u}}}{\mathbf{y}'\mathbf{y} - (\sum y_i)^2/n} \tag{4.75}$$

* Notice that the levels of significance in statistical tests will be affected as the same sample undergoes repeated uses. This point should not worry us greatly here (although appropriate caution is recommended), as we are interested in the question of empirical fit.

for $(1 - P^2)$ is

$$1 - \bar{R}^2 = \frac{\hat{\mathbf{u}}'\hat{\mathbf{u}}/(n - k - 1)}{\left[\mathbf{y}'\mathbf{y} - \frac{(\sum y_i)^2}{n}\right] \Big/ (n - 1)}$$

$$= (1 - R^2)\frac{n - 1}{n - k - 1} \tag{4.76}$$

we observe that an approximate unbiased version of R^2 is as shown in (4.73). Notice that, in general, the expectation of a ratio of functions of random variables is not equal to the ratio of the expectation of the functions. It can be shown further that

$$E(\bar{R}^2) = P^2 - \frac{P^2(1-P^2)(1 + 2P^2)}{n} + o(n^{-1}) \tag{4.77}$$

so that $\bar{R}^2$ has a smaller bias than R^2, although both are consistent. For more detail concerning this or a related discussion, refer to A. P. Barten (1962).

4.3.2. Interpreting the Regression Coefficients

As alluded to briefly in Subsection 4.1.3, the parameters β_i's in a regression model of the form

$$y = \beta_0 + \beta_1 x_1 + \beta_2 x_2 + \cdots + \beta_k x_k + u \tag{4.2}$$

specify the manner in and the extent to which the x_i's determine, affect, or explain the level of y. Of course, from a purely descriptive point of view, the β_i's are just a systematic way of representing the dependence of y on the x_i's as far as the data are concerned. When a theoretical behavioral assumption is contained in (4.2), then β_i's can be thought of as having true values in the space of the parameters that define the way y responds to the changes in the values of the x_i variables or that represent the conditional dependence of y on the x_i variables. A regression model, in its most general context, can have fixed x_i variables, random x_i variables, or x_i variables measured with error; or it can even have a mixture of any number of the three different types of variables in the right-hand side. Regardless of the nature of the variables, the interpretation of the β_i coefficients is (1) its role in the one-way determination of y by x_i variables, or (2) in exhibiting the degree to which y values are conditionally dependent on the values of the x_i variables.

Problems arise in the interpretation of *estimated* coefficients, since it is possible that data can be taken from a cross section, over a time span, and from the combination of the two. Although these problems are basically data problems, it must be observed that theorizing about the behavior of a population is usually carried out with a clear assumption as to the time and/or space dimensions in which the behavior takes place. Economists have too often in the past hypothesized about behavior (particularly, short run behavior) over time but have used cross-section data to estimate and to test the hypotheses.

When a relationship of the form (4.2) is estimated from a set of cross-section data, the coefficients, in effect, measure the interelement (or inter-person if people are involved) difference in behavior. Take a simple case of consumption expenditures E as dependent on disposable income Y. Then, for the relationship,

$$E_i = \alpha + \beta Y_i + u_i \tag{4.79}$$

the $\hat{\alpha}$ and $\hat{\beta}$ obtained from a cross section (say, for a given year) of consumers are basically a measure of how individuals' E_i's will change as we let disposable incomes vary over individuals. If the model (4.79) is set up for a study of a cross section of consumers' behavior, then use of the cross-section data is appropriate. If, however, the model is for a study of behavior over time, then the intertemporal interpretation of $\hat{\alpha}$ and $\hat{\beta}$ tends to be misleading unless there is reason to believe that the "response" coefficients are the same for all individuals. Further elaboration of this problem at this time would lead us into the general problem of aggregation, and it will not be considered here. For now it is sufficient to point out that some questions of proper interpretation exist with respect to the time-space dimensions involved in the models and the data.

4.3.3. Computational Problems

Since multiple regression analysis often involves a large number of variables and observations, a reliance on some rapid methods of computation is necessary. Relatively small problems can be handled adequately by rotary- or electronic-type desk calculators, but large problems depend almost entirely on high-speed electronic computers. In fact, the capabilities of large electronic computers have opened up the possibilities of analyzing some of the problems that a decade ago were considered impracticable. A case in point is the estimation of behavioral relationships using several thousand observations. We first discuss some procedures and problems common to all methods of calculating regression statistics and then consider problems applicable to the uses of desk calculators and large-scale computers. We shall discuss only the LS method of estimation.

FORMATION OF SAMPLE MOMENT MATRICES. Given the relationship of the form (4.35n), the first step that an analyst must take is to obtain the LS estimators of the parameters in the model. They are $\boldsymbol{\beta}$, σ^2, $E(\boldsymbol{\beta} - \mathbf{b})(\boldsymbol{\beta} - \mathbf{b})' = \mathbf{V}(\mathbf{b})$. Concretely, then, he proceeds to form the matrices

$$\mathbf{X}'\mathbf{X} \qquad \mathbf{X}'\mathbf{y} \qquad \mathbf{y}'\mathbf{y} \qquad (\mathbf{X}'\mathbf{X})^{-1}$$

so that the following statistics can be calculated.

$$\mathbf{b} = (\mathbf{X}'\mathbf{X})^{-1}\mathbf{X}'\mathbf{y} \tag{4.80a}$$

$$\hat{\mathbf{u}}'\hat{\mathbf{u}} = \mathbf{y}'\mathbf{y} - \mathbf{b}'\mathbf{X}'\mathbf{y} \qquad (= ESS) \tag{4.80b}$$

$$s^2 = \hat{\mathbf{u}}'\hat{\mathbf{u}}/(n - k - 1) \tag{4.80c}$$

$$\mathbf{S}(\mathbf{b}) = s^2(\mathbf{X}'\mathbf{X})^{-1} \tag{4.80d}$$

$$R^2 = \left[\mathbf{b}'\mathbf{X}'\mathbf{y} - \frac{(\sum y_i)^2}{n}\right] \Big/ \left[\mathbf{y}'\mathbf{y} - \frac{(\sum y_i)^2}{n}\right] = RSS/TSS \tag{4.80e}$$

or

$$R^2 = 1 - \frac{ESS}{TSS} = 1 - \hat{\mathbf{u}}'\hat{\mathbf{u}} \Big/ \left[\mathbf{y}'\mathbf{y} - \frac{(\sum y_i)^2}{n}\right]$$

Two things that require some care in obtaining these quantities are (1) the accuracy of the inverse matrix $(\mathbf{X}'\mathbf{X})^{-1}$ because of round-off errors, and (2) the computation of the residual sum of squares $\hat{\mathbf{u}}'\hat{\mathbf{u}}$.

ROUND-OFF ERRORS. Because the inverse of a matrix is formed by a number of multiplication operations, the accuracies of the elements of the inverse can be affected by round-off errors in the elements of the original matrix. For instance, the following two matrices $\mathbf{A}_1$ and $\mathbf{A}_2$, where the former is the rounded version of the latter, show how rounding can affect the accuracy of the elements of the inverse of $\mathbf{A}_1$.

$$\mathbf{A}_1 = \begin{pmatrix} 1.1 & 0.3 & 1.2 \\ 0.3 & 2.5 & 0.8 \\ 1.2 & 0.8 & 0.9 \end{pmatrix} \qquad \mathbf{A}_2 = \begin{pmatrix} 1.114 & 0.329 & 1.234 \\ 0.329 & 2.529 & 0.774 \\ 1.234 & 0.774 & 0.862 \end{pmatrix}$$

$$\mathbf{A}_1^{-1} = \begin{pmatrix} -1.207 & -0.517 & 2.069 \\ -0.517 & 0.337 & 0.390 \\ 2.069 & 0.390 & -1.994 \end{pmatrix} \quad \mathbf{A}_2^{-1} = \begin{pmatrix} -1.017 & -0.432 & 1.843 \\ -0.432 & 0.362 & 0.293 \\ 1.843 & 0.293 & -1.742 \end{pmatrix}$$

Notice that in many instances the size of the errors are rather considerable. Rounding off, of course, can be done by the researcher or by the computer. In the former case, one should try to retain as many significant digits as possible in the raw moment matrix $\mathbf{X}'\mathbf{X}$ insofar as such digits are meaningful. In the latter case, the difficulty arises because quite often the number of significant digits retained in a computer arithmetic are not large enough to carry a sufficiently large number of significant digits in the products of observations to ensure accuracy. In this case, not much can be done by the researcher except to be aware of the degree of accuracy with which the estimates come off the computer. In most cases, however, computers do carry enough significant digits to meet most accuracy requirements. We observe in passing that, if the researcher wishes to have a very accurate arithmetic, the so-called double-precision arithmetic can be had in most computers.

The problem of round-off error becomes rather serious when the moment matrix is "ill conditioned" (another way of saying that the x_i variables are intercorrelated in the data or in the population). We need to worry about this problem because sample moment matrices obtained for economic variables are usually ill conditioned because of the phenomenon that many economic variables move upward and downward together. R. J. Freund (1963) performed an experiment of estimating the same regression equation by the use of four different computer programs. Among the results shown was the finding that for the same data and the same computer program widely divergent coefficient estimates obtain depending on whether or not double-precision arithmetic is used. For comparison, part of the result is shown in Table 4.2. One should realize that double-precision arithmetic is a bit more time consuming than the single-precision and also requires greater computer storage; but the net result of all of them is a recommendation for the use of the double-precision arithmetic.

COMPUTING *ESS*. In calculating the residual sum of squares, it is tempting first to compute the residuals for the *i*th observation, say, $\mathbf{Y}_i - \mathbf{X}_i\mathbf{b}$, where $\mathbf{X}_i$ is the *i*th row of $\mathbf{X}$, for all *i* and then sum over *i* after each residual is squared. This is time consuming, at best, and the direct use of the Expression 4.80b is recommended. Again, note that $\mathbf{b}'\mathbf{X}'\mathbf{X}\mathbf{b} = \mathbf{b}'\mathbf{X}'\mathbf{y}$.

We also notice in passing that the LS method forces $\sum \hat{u}_i = 0$, so that this equality can be used as a partial test of whether or not one has a set of LS coefficient estimates.

USING CALCULATORS AND COMPUTERS. Before the advent of large-scale computers, rotary-type desk calculators were about the only type of machines that could be used for handling multiple regression analysis. Today, because direct or indirect access to some type of electronic computer is commonplace, one uses desk calculators only in a dire situation or on a very small problem.

If one is dealing with a problem involving up to 4 or more variables and 20 or 30 observations, hand calculation by desk calculator may still be feasible. It is true that desk calculators are readily available and that the cost involved is low, although one must bear in mind that the chance of computing errors increases as the number of variables and of observations become larger. To

Table 4.2 Results of Regression Analysis Reported by R. J. Freund[a]

Variable	Coefficients from IBM 709	Coefficients from IBM 709 Double-Precision
Intercept	−9886.24	−139491.63
c	−15434.14	151719.23
c^2	16215.68	−54456.83
c^3	2881.29	6945.35
p	1882.99	3675.72*
p^2	−23.24*[b]	−21.89*
p^3	−0.28*	−0.35*
t	204.82	290.63
t^2	72.14*	73.73*
t^3	−1.56*	−1.56*
cp	−1066.67	−2815.84*
cp^2	19.68*	21.04*
ct	156.63	60.15
ct^2	−7.75	−8.31
pt	−30.69*	−30.15*
pt^2	−0.27	−0.29
c^2p	−3.90	400.50
c^2t	−41.32	−16.67
p^2t	0.65*	0.64*
$\hat{u}'\hat{u}/n$	1541.25	1518.08

[a] For definition of variables, see the article in *The American Statistician*, December 1963.

[b] The symbol indicates significance at the 0.05 level by conventional test.

minimize errors in hand calculation while obtaining results that are similar to those in (4.80a to e), various methods of Doolittle variety have been used. This type of method requires that all the sums of squares and cross products be laid out on a work sheet in matrix form and that the operations on the rows are carried out in such a way as to invert effectively the moment matrix **X′X** while obtaining various other statistics. The student is advised to refer to the work of J. Friedman and R. T. Foote (1957) and of L. R. Klein (1956) for the exposition of the Doolittle and related techniques.

Whatever advantages the desk calculator might have over electronic computers, those advantages quickly vanish as soon as a regression problem approaches even a modest size, say 3 regressors and 30 observations. Although the use of a computer usually entails (1) having available a program, (2) the preparation of data in punch-card form according to the requirement of the computer program, and (3) submitting the program and data cards to the computer, the accuracy of the computed output far exceeds that which can be achieved by hand calculation. Thus, on balance, for most calculations the use of a computer is advised; nowadays, the least-squares analysis programs are usually a part of standard library programs that must accompany a computer system. Study the write-up of these programs carefully, check to see if the programs perform the computations desired, and spend time to learn how to use these programs.

4.3.4. Presentation of Computed Results

Given a regression model, it is difficult to say what is the minimum number of statistics that should be computed and presented for discussion and evaluation. Much will depend on the purposes of the investigator. In some cases, only the coefficient estimates are desired in order to see what "effects" the regressors have on the dependent variable. In other cases, one might be interested in testing the statistical significance of the coefficients and in making various inferences about the parameters in the regression model. However, usually, the statistics in (4.80a to e) are calculated to give an idea of (1) the "effects" of the regressors on the dependent variable (the $\mathbf{b}$ coefficients), (2) the extent of the variability of the estimated $\mathbf{y}$ along the sample regression plane (s^2, or its square root, referred to as the standard error of estimate), (3) the degree of sample "goodness of fit" (the R^2 or that corrected for degrees of freedom), and (4) the statistical significance of the estimated coefficients (depending on the diagonal elements of the variance-covariance matrix $s^2(\mathbf{X}'\mathbf{X})^{-1}$ or their square roots). We shall discuss the last point in the next chapter. Also, if one uses a canned computer program for calculation, quite often he gets more statistics than we have just listed—statistics such as Durbin-Watson's d and the F ratio for the analysis of variance test of the significance of regression. These topics and the related problems will be discussed in the subsequent development. At this time it is useful to observe that an estimated regression model might be presented as follows.

$$\begin{aligned} y = &\ \underset{(s_{b_0})}{b_0} + \underset{(s_{b_1})}{b_1x_1} + \underset{(s_{b_2})}{b_2x_2} + \cdots + \underset{(s_{b_k})}{b_kx_k} \qquad (4.81) \\ S_e &= ? \\ R^2 &= ? \quad \text{or} \quad (\bar{R}^2 = ?) \\ n &= ? \end{aligned}$$

That is, we write y as a linear combination of the estimated coefficients and the x_i variables; below estimated coefficients we write in the parentheses the estimated standard errors (s_{b_i}); and then the computed mean error sum of squares (S_e^2) or its square root, the coefficient of determination (R^2) and the sample size (n) follow.

CHAPTER 5

TESTING THE FIXED MODELS

A main function of econometric research is to formulate operational economic hypotheses, to estimate the models that may be indicated by the hypotheses, and also to test, in statistical and other senses, the models in the light of the estimated relationships. Sometimes theory is only suggestive of a relationship that uses a number of variables, but not of the specific form of the relation. Thus it is possible that one has several alternative functional forms for a given hypothesis. In this case, the testing problems are not confined to the statistical test of a model (among the many possible) but also are extended to questions of which functional form is appropriate. As indicated in Subsection 4.3.1, one may wish to use R^2 as a guide for choosing a functional form, but such a test does not always give clear-cut answers. Also, one should be aware that the use of a sample repeatedly for testing functional forms affects the levels of significance in successive tests. Another approach is to use extrapolation—taking some observations outside the sample and comparing the closeness of the predicted values with actually observed values. If one functional form gives a better prediction than another, it will be preferred to the other on the ground that the functional form, so the argument goes, must describe the underlying (but not necessarily ascertainable) structure better.

A clarification of the terms, hypothesis, model, and structure, is desirable at this point. We consider a hypothesis to be synonymous with a model. A hypothesis is a set of propositions (they may consist of just one) about the behavioral relationship that involves a number of variables; it purports to describe the true state of things and, as such, its content can be deterministic or stochastic, since there is no requirement that a hypothesized relationship be observable. However, to be testable, a model or hypothesis must embody statements about the general and possibly unknown way in which observations that are related to the variables being studied are generated. Thus an economic hypothesis might be that the quantity (q) of a commodity demanded is a decreasing function of the price (p), but the hypothesis with sharper empirical implication might state that

$$q = \alpha - \beta p + v \tag{5.0}$$

where α and β are positive and v has a certain probability distribution. A structure corresponding to the model in (5.0) may be that

$$q = 7.8 - 12.5p + v \tag{5.0a}$$

where, as can be observed, the coefficients have specified values and the assumed distribution (possibly unobservable) of v remains as a part of the structure. Hence, it is possible that many structures can emerge from one model, since we might have another structure for the model as, for instance,

$$q = 1.2 - 3.4p + v \tag{5.0b}$$

It is well to notice that, in the statistical nomenclature, a test of hypothesis could mean a test of a proposition about a specified aspect of a structure or a model. For example, in the model in (5.0) one may test the hypothesis that $\beta = 0.25$, or that v is normally distributed.

The testing of a regression model implied by an economic hypothesis can mean a number of things. A test can be formal as well as informal. Given a functional form, a test for the significance of the estimated parameters can be performed, and a test for the significance of the entire structural relation is also available. The reference to the significance of a given relation raises the questions regarding the appropriateness of the variables included in the relation and of the form of the relation. As indicated earlier, it is frequently possible to use R^2 as a guide (not a test) in choosing among different functional forms. As for investigating the appropriateness of the inclusion of certain variables (as against the exclusion of certain other variables), no general test is available, although research workers sometimes use R^2, or change in it, as a guide; and sometimes predictive tests are used. Predictive test may also be used for investigating the desirability of various functional forms and for choosing among different estimation procedures.

Given the complicated nature of hypothesis testing in economics, we first confine ourselves to some test procedures about the fixed model. In particular, we consider statistical tests concerning: (1) hypotheses about the regression coefficients, and (2) the stability over the time or the space of a relationship. Later we shall learn that these considerations answer many of the problems referred to in the preceding paragraphs. Although we discuss only the fixed models, we notice that many of the results obtained here will apply to models that violate the assumption of fixed regressors.

5.1. Testing Hypotheses about the Regression Coefficients

The discussion in this section draws on the material given in Section 3.4.4. Mainly because we now deal with multiple regression models, we divide our presentation into (1) the separate test, (2) the joint test, and (3) the partial joint test. We use matrix notations in much of what is developed here, but notice that the substance of the test procedures remains the same as the one in Section 3.4.4. Since the test procedures depend on the normality assumption of the disturbance, this assumption is made throughout the chapter. For pedagogical reasons, we defer the test of general linear hypothesis to Section 5.3.

5.1.1. Separate Test

Suppose that we want to test the null hypothesis that

$$\beta_i = 0 \tag{5.1}$$

for a particular i against the alternative hypothesis that

$$\beta_i \neq 0$$

in the model

$$y = \beta_0 + \beta_1 x_1 + \beta_2 x_2 + \cdots + \beta_k x_k + u \tag{5.2}$$

or, for a sample of size n

$$\mathbf{y} = \mathbf{X}\boldsymbol{\beta} + \mathbf{u} \tag{5.2a}$$

It is obvious that the way to approach this problem is to consider first the sample estimate of the particular β_i. Intuitively, we first have some hypothesized value of β_i, then we assume that we get an estimate of β_i from a given sample. Now, if the sample estimate of the parameter and the hypothesized value of the parameter are far apart, we tend not to believe in the null hypothesis. To decide how far apart the two values must be before one decides to reject the null hypothesis, we must consider the sampling distribution of the estimator of β_i. Now (4.41) states that if σ^2 is known, then for $\mathbf{X}'\mathbf{X} = \mathbf{C}$

$$\mathbf{V}(\mathbf{b}) = \sigma^2(\mathbf{X}'\mathbf{X})^{-1} = \sigma^2\mathbf{C}^{-1} = \sigma^2\{c^{ij}\} \tag{5.3}$$

where c^{ij} is the ijth element of $\mathbf{C}^{-1}$ and, if we are testing for nullness of β_i, the appropriate test statistic is

$$z = \frac{b_i - 0}{\sigma\sqrt{c^{ii}}} \tag{5.4}$$

Recall that b_i is a linear combination of the observed Y_i's that are normally distributed. Therefore, b_i is normally distributed with mean 0 (under null hypothesis) and variance $\sigma^2 c^{ii}$ (under the assumption of the model). Thus the

z variable in (5.4) is a standard normal variable with mean 0 and variance 1. We then look up the critical value of z corresponding to a desired level of significance to determine if the estimated b_i is statistically significant.

If, as will generally be the case, σ^2 is not known, then resort must be made to the familiar Student's t distribution. By looking at (4.61), (4.60), and (4.58) in that order, we notice that the covariance matrix of $\mathbf{b}$ is a function of the vector $\mathbf{u}$. Given that $\mathbf{u}$ is a normal vector, the calculated error sum of squares

$$\hat{\mathbf{u}}'\hat{\mathbf{u}} = \mathbf{u}'\mathbf{M}\mathbf{u} \tag{5.5}$$

is a quadratic form in $\mathbf{u}$ with the matrix of the form being $\mathbf{M}$ which turns out to be an idempotent matrix of rank $(n - k - 1)$. It can be shown that $\hat{\mathbf{u}}'\hat{\mathbf{u}}$ has a $\sigma^2\chi^2$ distribution with $(n - k - 1)$ degrees of freedom; or $\hat{\mathbf{u}}'\hat{\mathbf{u}}/\sigma^2$ has a χ^2 distribution with $(n - k - 1)$ degrees of freedom. Keeping this in mind, we form the ratio

$$t = \frac{(b_i - 0)/\sigma_{b_i}}{s_{b_i}/\sigma_{b_i}} = \frac{b_i}{\sqrt{\hat{\mathbf{u}}'\hat{\mathbf{u}}c^{ii}/(n - k - 1)}} = \frac{b_i}{\sqrt{s^2c^{ii}}} \tag{5.6}$$

which has the t distribution with $(n - k - 1)$ degrees of freedom as indicated in (3.65). Notice that in (5.6), s_{b_i} is the estimated standard error of the coefficient b_i. Heuristically, one can think of t distribution as an approximation to the normal distribution when σ_{b_i} is not known, so that the ratio b_i/s_{b_i}, under the null hypothesis that $\beta_i = 0$, has an approximate $N(0, 1)$ distribution. But the justification of the use of the t as defined in (5.6) comes from the fact that the t statistic is a ratio of a $N(0, 1)$ variable to a χ^2 variable with given degrees of freedom. Of course, if the sample size is 40 or greater one can safely use the z-test.

The tests we have just discussed are really not confined to the hypotheses that individual coefficients in the population are zero. In fact, we observe that the variance-covariance matrix of $\mathbf{b}$ is one that is applicable to any specified set of β_i's having nonzero values. Thus, with normality and the sizes of the standard errors intact, we can state that under the null hypothesis that $\beta_i = \beta_i^*$ where β_i^* is some hypothesized value and, with known σ^2, we shall have

$$z = \frac{b_i - \beta_i^*}{\sigma\sqrt{c^{ii}}} \tag{5.4g}$$

and, with σ^2 unknown, we shall have

$$t = \frac{b_i - \beta_i^*}{s\sqrt{c^{ii}}} \tag{5.6g}$$

as appropriate test statistics. See Table A-1 in the Appendix A for the critical values of t at different significance levels.

5.1.2. *Joint Test*

Suppose that the separate tests for the significance of the individual coefficients prompt us to accept the null hypotheses about the coefficients. Then, should we conclude that the regression relationship must be rejected? Not necessarily. There are reasons for saying this. The data used in the analysis may be plagued with multicollinearity. In this case, as will be detailed later, the individual tests of the significance of the coefficients may not be helpful, as we would get unnecessarily large standard errors for the coefficient estimators. The result might be that the fit (as measured by R^2) is good, but the coefficients are not significant. This points out the serious shortcoming of the separate tests, since in the separate test we are concerned only with the significance of one coefficient *allowing* the other coefficients to have any values. A test taking into account the hypothesized values of the other coefficients can eliminate situations where we might accept the hypothesis of the nonsignificance of the entire regression relation. This is explained as follows. In Figure 5.1, we have drawn a confidence region for the two parameter β_0 and β_1 in the simple regression model

$$y = \beta_0 + \beta_1 x + u \tag{5.7}$$

for which a sample of size n presumably yields the least-squares estimates

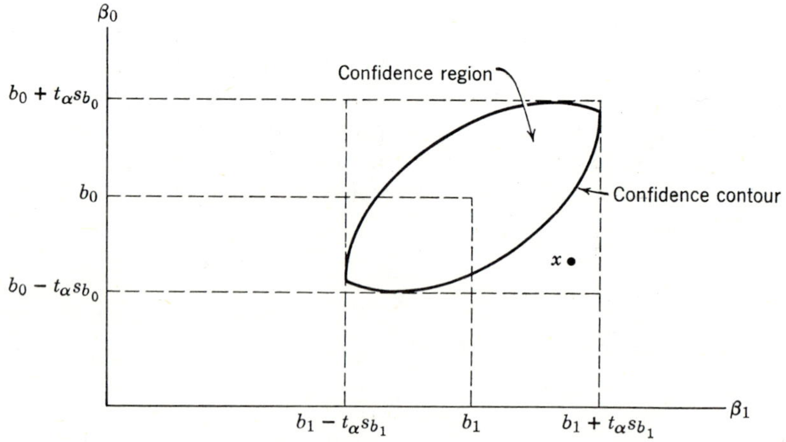

Figure 5.1

b_0 and b_1 and the individual confidence intervals

$$b_0 \pm t_\alpha s_{b_0} \tag{5.8}$$

$$b_1 \pm t_\alpha s_{b_1} \tag{5.9}$$

where t_α is the t statistic corresponding to the level of significance α. The ellipse in the (β_0, β_1) plane, or the joint confidence region, is obtained by defining the F ratio

$$F = \frac{Q_3/2}{Q_2/(n-2)} \tag{5.10}$$

where

$$Q_3 = \sum [(b_0 - \beta_0) + (b_1 - \beta_1)X_i]^2 \tag{5.11}$$

$$Q_2 = \sum u_i^2 - Q_3 \tag{5.12}$$

[Notice that $u_i = y_i - \beta_0 - \beta_1 X_i$ follows from (5.7)] and writing

$$\Pr [F \leq F_\alpha] = 1 - \alpha \tag{5.13}$$

where F_α is a fixed F ratio corresponding to the level of significance α. From (5.10) and (5.13) follows the relation

$$F = F_\alpha$$

which appears as the contour line depicted by the ellipse as shown.

Suppose that we now substitute $\beta_0{}^*$ and $\beta_1{}^*$ for β_0 and β_1 in Q_3 in (5.11) and consider

$$Q_3{}^* = \sum [(b_0 - \beta_0{}^*) + (b_1 - \beta_1{}^*)X_i]^2 \tag{5.14}$$

and

$$\begin{aligned} Q_2{}^* &= \sum u_i^2 - Q_3{}^* \\ &= \sum (Y_i - \beta_0{}^* - \beta_1{}^* X_i)^2 - Q_3{}^* \end{aligned} \tag{5.15}$$

Then for the F ratio corresponding to (5.10) and equivalent to (3.74) we can draw an ellipse representing a contour of the joint distribution of β_0 and β_1 under the assumed values of $\beta_0{}^*$ and $\beta_1{}^*$. Indeed, if the latter values are the ones assumed for the null hypothesis, then we are talking about the joint test that $(\beta_0 - \beta_0{}^*)$ and $(\beta_1 - \beta_1{}^*)$ are each zero, or that $\beta_0 = \beta_0{}^*$ and $\beta_1 = \beta_1{}^*$ respectively. We draw the ellipse under the null hypothesis in Figure 5.2.

Now consider Figure 5.1. The point x represents a pair of values of β_0 and β_1 that fall outside of the confidence region. Such values of coefficients may be quite consistent with the data when viewed individually since, for example, β_0 falls within the interval $(b_0 - t_\alpha s_{b_0}, b_0 + t_\alpha s_{b_0})$ with $(1 - \alpha)$ percent confidence. In contrast, consider point y in Figure 5.2 and let, say, $y = (b_1{}^*, b_0{}^*)$, the sample estimates of β^* and $\beta_0{}^*$. Then it is clear that under separate tests we would accept the separate null hypotheses that $\beta_0 = \beta_0{}^*$

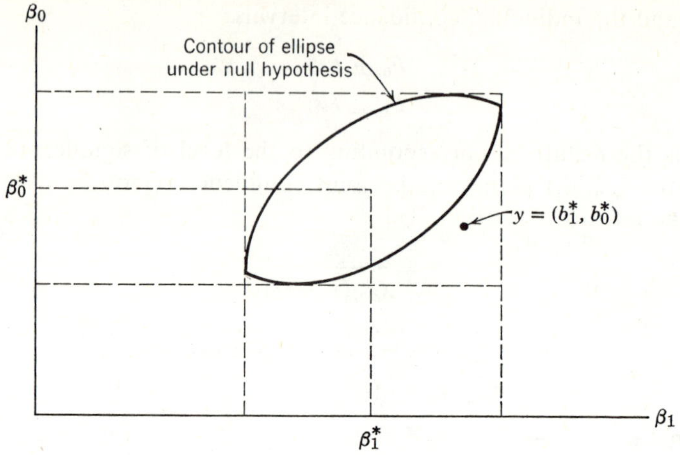

Figure 5.2

and $\beta_1 = \beta_1{}^*$. However, since the point y falls outside of the joint acceptance region, the null hypothesis that

$$\begin{pmatrix}\beta_0\\ \beta_1\end{pmatrix} = \begin{pmatrix}\beta_0{}^*\\ \beta_1{}^*\end{pmatrix} \tag{5.16}$$

will be rejected.

These comments are probably sufficient to show the advantage of testing the hypotheses about the individual coefficients *jointly*. In particular, the joint test is a sharper test, given the hypothesized values of the coefficients, since under separate tests we can come up with the acceptance of the hypothesis that ($\beta_0 = \beta_0{}^*$) and/or $(\beta_1 = \beta_1{}^*)_0$ but under the joint test we would reject the null hypothesis, for instance, the one in (5.16).

We now give the derivation of the test statistic needed for the joint test. Notice that the statistics are derived in the same way as in Section 3.4.4, particularly the part that involves Expressions 3.67 to 3.74. Only the degrees of freedom involved will be different.

We consider for the joint test two types of null hypothesis. The first type is the null hypothesis that the coefficient vector is zero, or that the hypotheses involved are

$$\begin{aligned} H_0&: \boldsymbol{\beta} = \mathbf{0}\\ H_a&: \boldsymbol{\beta} \neq \mathbf{0} \end{aligned} \tag{5.17}$$

say, for the model in (5.2a). And the second type is that the coefficient vector is equal to some nonzero vector, or that the hypotheses involved are

$$\begin{aligned} H_0&: \boldsymbol{\beta} = \boldsymbol{\beta}^* \\ H_a&: \boldsymbol{\beta} \neq \boldsymbol{\beta}^* \end{aligned} \tag{5.18}$$

Although it is true that (5.17) is really a special case of (5.18), we discuss these two cases one after the other because (5.17) is so frequently used that it deserves to be singled out.

Null hypothesis that $\boldsymbol{\beta} = \mathbf{0}$. Consider, first, the true population regression relationship if the null hypothesis in (5.17) holds true. That is, for $\boldsymbol{\beta} = \mathbf{0}$

$$\mathbf{y} = \mathbf{X0} + \mathbf{u} = \mathbf{u} \tag{5.19}$$

and form (4.41a)

$$\mathbf{b} = (\mathbf{X}'\mathbf{X})^{-1}\mathbf{X}'\mathbf{u} \tag{5.20}$$

Therefore, the quadratic form, corresponding to (3.72), is

$$\begin{aligned} Q_3 &= (\mathbf{b} - \mathbf{0})'(\mathbf{X}'\mathbf{X})(\mathbf{b} - \mathbf{0}) \\ &= \mathbf{b}'(\mathbf{X}'\mathbf{X})\mathbf{b} \\ &= \mathbf{u}'\mathbf{X}(\mathbf{X}'\mathbf{X})^{-1}(\mathbf{X}'\mathbf{X})(\mathbf{X}'\mathbf{X})^{-1}\mathbf{X}'\mathbf{u} \\ &= \mathbf{u}'\mathbf{X}(\mathbf{X}'\mathbf{X})^{-1}\mathbf{X}'\mathbf{u} \\ &= \mathbf{u}'(\mathbf{I} - \mathbf{M})\mathbf{u} \end{aligned} \tag{5.21}$$

where $\mathbf{I} - \mathbf{M} = \mathbf{X}'(\mathbf{X}'\mathbf{X})^{-1}\mathbf{X}$ as defined in (4.18) and has various properties described in (4.19) to (4.24). Since the rank of $\mathbf{I} - \mathbf{M}$ is $k + 1$ and since Q_3 is a quadratic form in the normal vector $\mathbf{u}$, we can show that Q_3 divided by σ^2 is a χ^2 variable with $k + 1$ degrees of freedom. Now consider Q_2, the sum of the squares of the residuals that are given by the sample, regardless of what is hypothesized about the true values of $\boldsymbol{\beta}$. It is obvious, from (4.17), that the calculated error sum of squares

$$\begin{aligned} Q_2 &= \hat{\mathbf{u}}'\hat{\mathbf{u}} \\ &= \mathbf{y}'\mathbf{M}\mathbf{y} \\ &= \mathbf{u}'\mathbf{M}\mathbf{u} \end{aligned} \tag{5.22}$$

This quadratic form is identical to the one in (5.5), and Q_2 divided by σ^2 has a χ^2 distribution with $n - k - 1$ degrees of freedom as indicated there. Furthermore, we notice that under the null hypothesis the error sum of squares

$$\begin{aligned} Q_1 &= \mathbf{u}'\mathbf{u} \\ &= Q_2 + Q_3 \end{aligned}$$

and that by Cochran's theorem Q_2 and Q_3 are independent.

Intuitively, Q_2 tells us the variability of the y values around the regression plane on the basis of the sample information. Q_3, on the other hand, is the difference between the sample regression sum of squares and the hypothesized regression sum of squares. Hence, if Q_3 is large in comparison with Q_2, we would be led to believe that the regression relationship is not trivial or that the sample b_i's are so different from zero that the null hypothesis is not acceptable.

Therefore, the ratio

$$F = \frac{Q_3/(k + 1)}{Q_2/(n - k - 1)} \tag{5.23}$$

is an appropriate statistic for testing the null hypothesis against the alternative hypothesis as shown in (5.17). As is clear from the preceding discussion, a large value of F would lead to the rejection of the null hypothesis. Notice that Q_2 can be calculated from the expression (4.62) but that Q_3 can be found by using the second line of (5.21).

It is important to observe that all of the F tests discussed here or later take the form of a ratio in which the numerator is a restricted sum of squares $[(\mathbf{y} - \mathbf{X}\boldsymbol{\beta}^*)'(\mathbf{y} - \mathbf{X}\boldsymbol{\beta}^*)]$ less an unrestricted sum of squares $[(\mathbf{y} - \mathbf{X}\mathbf{b})'(\mathbf{y} - \mathbf{X}\mathbf{b})]$ and in which the denominator is the unrestricted sum of squares (both the numerator and the denominator to be divided by appropriate degrees of freedom). We want to determine if the restrictions (hypotheses to be tested) make any difference and, of course, in the sample, the restrictions will make some difference, even if they are true. We want to determine if that difference is big, so we divide it by the fundamental measure of how big the "noise" is in this system, namely, an unbiased estimate of the variance of the disturbances. Notice that the estimate of the variance of the disturbances is unbiased whether or not the restrictions are true.

Example 1. In Table 5.1, we furnish a hypothetical set of data on quarterly trading stamp redemption (R), total issues (S), and seasonal dummy variables (E_i). The regression model to be estimated is

$$R = \beta_0 + \beta_1 S + \beta_2 E_2 + \beta_3 E_3 + \beta_4 E_4 + u \tag{5.23a}$$

where E_1 has been dropped for reasons to be discussed in Section 7.3. For now (5.23a), hypothesizes that the redemption is a linear function of the total issues and the seasonal factors. The LS estimate of (5.23a) is

$$\begin{aligned} R = -924.3 &+ 0.9145S + 683.5E_2 + 496.9E_3 + 2226.0E_4 \\ (54.76) &\quad (0.09446) \quad (153.1) \quad (155.1) \quad (158.0) \\ R^2 &= 0.923 \end{aligned} \tag{5.23b}$$

Here, as it will be elsewhere, the numbers in the parentheses below the

Table 5.1 Hypothetical Data on Issues and Redemption of Trading Stamps (in Hundreds of Books)

Year and Quarter	Issues (*S*)	Redemption (*R*)	Quarterly E_1	Seasonal E_2	E_3	Dummy E_4
1910						
I	2223	1086	1	0	0	0
II	2294	1760	0	1	0	0
III	2434	1725	0	0	1	0
IV	2566	2890	0	0	0	1
1911						
I	2398	1411	1	0	0	0
II	2905	2092	0	1	0	0
III	2902	2370	0	0	1	0
IV	2973	3809	0	0	0	1
1912						
I	3000	1886	1	0	0	0
II	2798	2463	0	1	0	0
III	3248	2511	0	0	1	0
IV	3156	4156	0	0	0	1
1913						
I	3157	1944	1	0	0	0
II	2936	2843	0	1	0	0
III	3316	2522	0	0	1	0
IV	3209	4225	0	0	0	1
1914						
I	3448	2160	1	0	0	0
II	3002	3093	0	1	0	0
III	3631	2687	0	0	1	0
IV	3563	4412	0	0	0	1
1915						
I	3600	2369	1	0	0	0
II	4210	3219	0	1	0	0
III	4219	3438	0	0	1	0
IV	3508	5201	0	0	0	1
1916						
I	3655	2430	1	0	0	0
II	4416	3462	0	1	0	0
III	4009	3264	0	0	1	0
IV	3772	5223	0	0	0	1
1917						
I	3893	2525	1	0	0	0
II	4028	3458	0	1	0	0
III	4017	3466	0	0	1	0

estimated coefficients are the respective standard deviation estimates. To test the hypothesis that the independent variables have no influence over the dependent ones, we find the sum of the squares as follows:

$$\begin{aligned} TSS &= Q_1 = 31{,}432{,}554 \\ ESS &= Q_2 = 2{,}417{,}280 \\ RSS &= Q_3 = 29{,}015{,}274 \end{aligned} \tag{5.23c}$$

Thus the test statistic

$$F = \frac{(Q_1 - Q_2)/5}{Q_2/(31 - 5)} = 78.021 \tag{5.23d}$$

has an F distribution with (5, 26) degrees of freedom. It follows then that at the 5 percent level of significance we reject the null hypothesis since, as can be seen from Table A-2 in Appendix A that the critical F value is 2.59.

NULL HYPOTHESIS THAT $\boldsymbol{\beta} = \boldsymbol{\beta}^*$. Suppose now that $\boldsymbol{\beta} = \boldsymbol{\beta}^*$, then the population relationship will be

$$\mathbf{y} = \mathbf{X}\boldsymbol{\beta}^* + \mathbf{u} \tag{5.19g}$$

and the least-squares estimator for $\boldsymbol{\beta}^*$ will be

$$\mathbf{b} = \boldsymbol{\beta}^* + (\mathbf{X}'\mathbf{X})^{-1}\mathbf{X}'\mathbf{u}$$

Or, in effect,

$$\mathbf{b} - \boldsymbol{\beta}^* = (\mathbf{X}'\mathbf{X})^{-1}\mathbf{X}'\mathbf{u} \tag{5.20g}$$

And in the quadratic form, corresponding to (3.72), it will be

$$\begin{aligned} Q_3 &= (\mathbf{b} - \boldsymbol{\beta}^*)'(\mathbf{X}'\mathbf{X})(\mathbf{b} - \boldsymbol{\beta}^*) \\ &= \mathbf{u}'(\mathbf{I} - \mathbf{M})\mathbf{u} \end{aligned} \tag{5.21g}$$

which is the same as the form in (5.21). This equality is forced by the strength of the null hypothesis. Therefore, we can conclude that Q_3/σ^2 is χ^2 distributed with $k + 1$ degrees of freedom. Now Q_2 in this case remains the same as that in (5.22) as it is derived from the sample alone, so that Q_2/σ^2 still is a χ^2 variable with $n - k - 1$ degrees of freedom. Thus, on forming the F ratio, we see that the test procedure for the general case is the same as the one in the special case where the null hypothesis states that the coefficient vector is zero.

Another way to determine the equivalence of the test statistics for $\boldsymbol{\beta} = \boldsymbol{\beta}^*$ and $\boldsymbol{\beta} = \mathbf{0}$ is the following. When $\boldsymbol{\beta}^*$ is known, we can always write

$$\begin{aligned} \mathbf{y} - \mathbf{X}\boldsymbol{\beta}^* &= \mathbf{X}\boldsymbol{\beta} + \mathbf{u} - \mathbf{X}\boldsymbol{\beta}^* \\ &= \mathbf{X}(\boldsymbol{\beta} - \boldsymbol{\beta}^*) + \mathbf{u} \end{aligned}$$

can treat the left-hand side as a new variable, and can test the hypothesis that $\boldsymbol{\beta} - \boldsymbol{\beta}^* = \mathbf{0}$; thus, when $\boldsymbol{\beta}^* = \mathbf{0}$, we in effect perform the test that $\boldsymbol{\beta} = \mathbf{0}$ and, if $\boldsymbol{\beta}^*$ is a vector of constants, we test $\boldsymbol{\beta} = \boldsymbol{\beta}^*$. Note, however, that in the case of the test for $\boldsymbol{\beta} = \boldsymbol{\beta}^*$ the calculation of

$$Q_3 = (\mathbf{b} - \boldsymbol{\beta}^*)'\mathbf{X}'\mathbf{X}(\mathbf{b} - \boldsymbol{\beta}^*)$$

is a task a little more complicated than the computation of $Q_3 = \mathbf{b}'\mathbf{X}'\mathbf{X}\mathbf{b}$ (or $\mathbf{b}'\mathbf{X}'\mathbf{y}$) in the case involving the hypothesis $\boldsymbol{\beta} = \mathbf{0}$. For the latter sum of squares are often a part of the standard printout of most least-squares computing programs where they are referred to as the sum of squares of the estimates (if the observations used are the raw data) or the regression sum of squares (if the observations are in the form of deviations from the sample means). In the case involving the hypothesis $\boldsymbol{\beta} = \boldsymbol{\beta}^*$, Q_3 is obtained by first forming the difference of the vectors $\mathbf{b}$ and $\boldsymbol{\beta}^*$, second premultiplying the transpose of the difference vector $(\mathbf{b} - \boldsymbol{\beta}^*)$ to $\mathbf{X}'\mathbf{X}$, and third postmultiplying the result by the difference vector.

5.1.3. Partial Joint Test

Because sometimes interest may center on a subset of the elements in the coefficient vector $\boldsymbol{\beta}$, we might consider testing jointly a number of coefficients, but not all of the ones in the equation. For instance, in a cross-sectional study of a demand relationship for durable goods, containing both economic variables and demographic variables (such as age and sex of head of household), we may test that the demographic variables have no effect, assuming that the coefficients for the economic variables can have any values. In general, then, we can select the coefficients of interest from the $\boldsymbol{\beta}$ vector and can partition the vector into two subvectors as

$$\boldsymbol{\beta} = \begin{pmatrix} \boldsymbol{\beta}_1 \\ \boldsymbol{\beta}_2 \end{pmatrix} \tag{5.24}$$

Suppose that we test the hypothesis that

$$\boldsymbol{\beta}_2 = \boldsymbol{\beta}_2^* \tag{5.25a}$$

against the alternative hypothesis that

$$\boldsymbol{\beta}_2 \neq \boldsymbol{\beta}_2^* \tag{5.25b}$$

Then we would proceed to obtain the appropriate quadratic forms for the test. For the derivation of the test statistics for the partial joint test, see Section 5.3, or Goldberger (1964, pp. 174–175). We shall consider only the test procedure in the following discussion.

For the partition indicated in (5.24), we can write (5.2a) as follows

$$\mathbf{y} = (\mathbf{X}_1\ \mathbf{X}_2)\begin{pmatrix}\boldsymbol{\beta}_1\\ \boldsymbol{\beta}_2\end{pmatrix} + \mathbf{u} \tag{5.26}$$

where $\boldsymbol{\beta}_1$ and $\boldsymbol{\beta}_2$ are, say, respectively, $l \times 1$ and $m \times 1$, so that $l + m = k + 1$ and $\mathbf{X}_1$ and $\mathbf{X}_2$ are, respectively, $n \times l$ and $n \times m$ but $\mathbf{y}$ and $\mathbf{u}$ are each $n \times 1$ as before. In essence, because of the interest in testing either $\boldsymbol{\beta}_1$ or $\boldsymbol{\beta}_2$ (from now on we shall assume that we are interested in testing about $\boldsymbol{\beta}_2$), the entire observation matrix $\mathbf{X}$ is partitioned into $\mathbf{X}_1$ and $\mathbf{X}_2$. Thus, the LS estimator will be

$$\begin{aligned}\mathbf{b} = \begin{pmatrix}\mathbf{b}_1\\ \mathbf{b}_2\end{pmatrix} &= [(\mathbf{X}_1\ \mathbf{X}_2)'(\mathbf{X}_1\ \mathbf{X}_2)]^{-1}(\mathbf{X}_1\ \mathbf{X}_2)'\mathbf{y}\\ &= \begin{pmatrix}\mathbf{X}_1'\mathbf{X}_1 & \mathbf{X}_1'\mathbf{X}_2\\ \mathbf{X}_2'\mathbf{X}_1 & \mathbf{X}_2'\mathbf{X}_2\end{pmatrix}^{-1}\begin{pmatrix}\mathbf{X}_1'\mathbf{y}\\ \mathbf{X}_2'\mathbf{y}\end{pmatrix}\\ &= \begin{pmatrix}(\mathbf{X}_1'\mathbf{X}_1)^{-1}\mathbf{X}_1'\mathbf{y} - (\mathbf{X}_1'\mathbf{X}_1)^{-1}\mathbf{X}_1'\mathbf{X}_2\mathbf{D}^{-1}\mathbf{X}_2'\mathbf{M}_1\mathbf{y}\\ \mathbf{D}^{-1}\mathbf{X}_2'\mathbf{M}_1\mathbf{y}\end{pmatrix}\end{aligned} \tag{5.27}$$

where

$$\mathbf{M}_1 = \mathbf{I} - \mathbf{X}_1(\mathbf{X}_1'\mathbf{X}_1)^{-1}\mathbf{X}_1'$$

and

$$\mathbf{D} = \mathbf{X}_2'\mathbf{X}_2 - \mathbf{X}_2'\mathbf{X}_1(\mathbf{X}_1'\mathbf{X}_1)^{-1}\mathbf{X}_1'\mathbf{X}_2 = \mathbf{X}_2'\mathbf{M}_1\mathbf{X}_2$$

Furthermore, $\mathbf{b}_2$ is reducible into

$$\mathbf{b}_2 = \boldsymbol{\beta}_2 + \mathbf{D}^{-1}\mathbf{X}_2'\mathbf{M}_1\mathbf{u}$$

so that

$$\mathbf{b}_2 - \boldsymbol{\beta}_2 = \mathbf{D}^{-1}\mathbf{X}_2'\mathbf{M}_1\mathbf{u} \tag{5.28}$$

Suppose that for (5.26) we set our null hypothesis to be $\boldsymbol{\beta}_2 = \boldsymbol{\beta}_2^*$, then the relevant sum of squares to be computed is

$$Q_3 = (\mathbf{b}_2 - \boldsymbol{\beta}_2^*)'\mathbf{D}(\mathbf{b}_2 - \boldsymbol{\beta}_2^*) \tag{5.29}$$

which can be shown to have a χ^2 distribution with m degrees of freedom when divided by σ^2. On forming the ratio

$$F = \frac{(Q_3/\sigma^2)/m}{(Q_2/\sigma^2)/(n-k-1)} = \frac{Q_3/m}{Q_2/(n-k-1)} \tag{5.30}$$

we observe that a large value of F will lead to the rejection of the null hypothesis that $\boldsymbol{\beta}_2 = \boldsymbol{\beta}_2^*$. Notice that our discussion has been for the general case, so that the test procedure just mentioned is certainly applicable to the test that $\boldsymbol{\beta}_2 = \mathbf{0}$; we are saying here that $\boldsymbol{\beta}_2^* = \mathbf{0}$. Since this last case is frequently of interest to researchers, we exhibit another, but equivalent, form through which the F test statistic can be obtained.

Suppose that we have access to a computer program for computing a regression of y on $k + 1$, or $l + m$, independent variables (including the constant term). Suppose, further, that we wish to test that the last m variables make no contribution to the explanation of the variation in y. Then we can instruct the computer to find the regression of y on the first l variables. Call the regression sum of squares obtained as the result RSS_1. Furthermore, instruct the machine to find the regression of y on all the $l + m$ variables and call the resulting regression sum of squares RSS and the related error sum of squares ESS. Then it can be shown that*

$$F = \frac{(RSS - RSS_1)/m}{ESS/(n - k - 1)} = \frac{\Delta RSS/m}{ESS/(l + m)} \tag{5.30a}$$

has an F distribution with $(m, n - k - 1)$ degrees of freedom. Also, the denominator statistic in (5.30a) is the same as its counterparts in (5.30) and (5.23).

Example 2. The partial joint tests described in the subsection immediately above can become rather burdensome computationally, especially for the test of the statistic which is given in (5.30). Notice, in (5.30), that to compute Q_3 we first must obtain the matrix $\mathbf{D}$, as defined below the Expression 5.27. Since $\mathbf{M}_1$ is involved in the calculation and since the dimensionality of $\mathbf{M}_1$ is of the order of the sample size, we can become involved in a considerable computational problem if we do not use a high-speed electronic computer. Here we illustrate the F test for the special case that $\boldsymbol{\beta}_2 = \mathbf{0}$, utilizing the expression (5.30a). From the data in Table 5.1 we estimate the relation

$$R = \beta_0 + \beta_1 S + w \tag{5.30i}$$

and by contrasting the results of the estimation with the ones of (5.23a), we test for the null hypothesis that

$$\beta_2 = \beta_3 = \beta_4 = 0 \tag{5.30j}$$

against the alternative that any of these coefficients can be nonzero. The estimate of (5.30i) is

$$\begin{aligned} R &= -79.37 + 0.9032S \\ &\qquad (158.6) \qquad (0.02686) \\ R^2 &= 0.281 \end{aligned} \tag{5.30k}$$

and the related regression sum of squares is

$$RSS_1 = 8{,}817{,}105$$

Thus the appropriate F statistic for (5.30j) is calculated by reference to

* See, for example, Goldberger (1964, p. 177).

the quantities in (5.23c); namely, $RSS = 29{,}015{,}274$ and $ESS = 2{,}417{,}280$ as follows:

$$F = \frac{\Delta RSS/m}{ESS/(n - l - m)} = \frac{(29{,}015{,}274 - 8{,}817{,}105)/3}{(2{,}417{,}280)/(26)} = 72.42 \quad (5.30\text{m})$$

Since the value of F with (3, 26) degrees of freedom at the 0.05 significance level is 2.98, the statistic just calculated above shows that the null hypothesis is rejected.

5.1.4. The Use of R^2 in the Partial Joint Tests

It is apparent by now that in the joint test, discussed in the preceding two subsections, the denominator variable of the F ratios used in the tests is the same for both tests. In fact, if we square the t statistic, we obtain an F ratio so that, in effect, the error sum of squares from the sample provides a basis for a comparison with the difference between the sample regression sum of squares and the hypothesized regression sum of squares for all tests of significance. Moreover, the separate tests and the partial joint test are special cases of the general joint test. Now, R^2 is a measure of goodness of fit of the sample and is directly calculated from the sample regression sum of squares or, from the calculated error sum of squares given the total sum of squares. Let us determine first how these sums of squares are related to R^2 through the various test statistics that are obtainable under the null hypothesis that $\boldsymbol{\beta}_2 = \mathbf{0}$.

First notice that for $k + 1$, or $l + m$, variables

$$R^2 = \frac{RSS}{TSS} = \frac{\mathbf{b}'\mathbf{X}'\mathbf{X}\mathbf{b} - \bar{y}^2}{\mathbf{y}'\mathbf{y} - \bar{y}^2} \tag{5.31}$$

where $\bar{y}^2 = (\sum y)^2/n$. Now the RSS after regressing y on the first l variables is

$$R_1^{\,2} = \frac{RSS_1}{TSS} = \frac{\mathbf{b}_1'\mathbf{X}_1'\mathbf{X}_1\mathbf{b}_1 - \bar{y}^2}{\mathbf{y}'\mathbf{y} - \bar{y}^2} \tag{5.32}$$

Thus, in terms of (5.30a) and from (4.80e),

$$\begin{aligned} F &= \frac{(RSS - RSS_1)/m}{ESS/(n - k - 1)} \\ &= \frac{(R^2 - R_1^{\,2}) \cdot TSS/m}{(1 - R^2)TSS/(n - k - 1)} \\ &= \frac{(R^2 - R_1^{\,2})/m}{(1 - R^2)/(n - k - 1)} \end{aligned}$$

Then, designating the increment in R^2 as the result of adding the m variables to the regression by ΔR^2, we in effect have a test statistic based on R^2's or making any partial joint test about the *nullness* of the coefficients of the additional regressors. Therefore, the statistic

$$F = \frac{\Delta R^2/m}{(1 - R^2)/(n - k - 1)} \tag{5.30b}$$

is appropriate for the test of the significance of the increment in R^2 as the result of m added variables.

Example 3. By reference to the coefficients of determination for (5.30i) and (5.23a), it is possible to test the null hypothesis in (5.30j) by the statistic (5.30b). Observing that $R^2 = 0.923$ and $R_1^2 = 0.281$, we find that*

$$F = \frac{(0.923 - 0.281)/3}{(1 - 0.923)/26} = 72.30$$

The extension of the result in (5.30b) is of interest, particularly in the case of the test of the hypothesis that all the independent variables do not contribute to the explanation of the variation in y. In this case, the test statistic is

$$F = \frac{R^2/k}{(1 - R^2)/(n - k - 1)} \tag{5.30c}$$

as l is equal to 1 and m is equal to k in the general framework of (5.30b) The R^2 in the numerator in (5.30c) is the one that is obtained after regressing y on the $k + 1$ variables. It is divided by k because

$$R^2 = R^2 - R_1^2 = R^2 - 0$$

and the number of parameters corresponding to R^2 is $k + 1$ while one variable corresponds to R_1^2. As an exercise, show that $R_1^2 = 0$. (*Hint.* The LS estimator for the coefficient of the constant term when the k variables are left out is $\bar{y}$.)

5.2. Testing for Structural Stability

The tests discussed in Section 5.1 might be called internal tests, since they use the information available within the sample to test the specific hypothesized values of the individual regression coefficients. Testing for structural stability calls for the use of "external" or additional observations besides the sample that is used to estimate a given model. The need for this type of tests is obvious, since a wide gap may lie between the true state of things and the

* The F value here at 72.30 is a little different from the F value given earlier in (5.30m) because of the rounding-off errors in the R^2 figures. In theory, the values should be the same.

model that purports to explain them, and since it is frequently of interest to know if an economic relationship remains stable over two or more points in time or over space. For instance, we might want to ask if a certain type of consumption function was the same for the prewar and postwar periods, or if a certain behavior relation (in the form of a structure) relating to one community might hold true in some other community. It may be that interest centers on just a subset of coefficients and their intertemporal or spatial stability. For example, in a consumption function involving income and asset variables, it may be desirable to know if the coefficients of the asset variables remained constant through the prewar and postwar periods. In this section, we are primarily concerned with the test procedures that are useful in answering questions about structural stability and the related questions. In particular, we discuss (1) the test of constancy of an entire set of regression coefficients, and (2) the test of constancy of a subset of regression coefficients.

5.2.1. Constancy of Entire Set of Regression Coefficients

In this and the next subsections we discuss the formal procedures for testing the constancy of the entire set of coefficients in a regression equation and the constancy of a subset of the coefficients. Suppose that the model under study is of the form

$$\mathbf{y} = \mathbf{X}\boldsymbol{\beta} + \mathbf{u} \tag{5.35}$$

where $\boldsymbol{\beta}$ is $(k + 1) \times 1$ as before. Now, $\boldsymbol{\beta}$ can further be partitioned into two vectors, $\boldsymbol{\beta}_1$ and $\boldsymbol{\beta}_2$, say $l \times 1$ and $m \times 1$, respectively. Thus, (5.35) can be written in another form

$$\mathbf{y} = (\mathbf{X}_1\ \mathbf{X}_2)\begin{pmatrix}\boldsymbol{\beta}_1\\ \boldsymbol{\beta}_2\end{pmatrix} + \mathbf{u} \tag{5.36}$$

Then we are interested in asking if $\boldsymbol{\beta}$ in (5.35) remains constant over two or more periods of time or over two or more areas in space. A similar question is asked about, for instance, $\boldsymbol{\beta}_1$ in (5.36) in the subsection that follows. In both cases, we assume that two or more samples that correspond to two or more periods in time or two or more points in space are drawn and that the number of observations in all samples are greater than $(k + 1)$. These procedures are treated in a systematic way in a work by Chow (1960) and in Rao (1952). In fact, all the test procedures, including the last ones mentioned, can be cast in the framework of the general linear hypothesis, as discussed later in Section 5.3, but we are not taking the latter approach at this time, primarily for expository reasons. We now proceed with our discussion in the present subsection by reformulating the problem in a concrete fashion. We first deal with two-sample cases. Later, the results will be generalized.

RECASTING OF THE PROBLEM. Let the model in question be of the form in (5.35). Suppose that two samples at two different time periods or two geographical areas are drawn, then we shall be testing that the two samples are taken from the same population in which the relation (5.35) exists. Let the size of the first sample be n_1 then, according to the model, we have

$$\mathbf{y}_1 = \mathbf{X}_1\boldsymbol{\beta}_1 + \mathbf{u}_1 \tag{5.35a}$$

where $\mathbf{y}_1$ and $\mathbf{u}_1$ are each $n_1 \times 1$, $\mathbf{X}_1$ is $n_1 \times (k + 1)$, and $\boldsymbol{\beta}_1$ is $(k + 1) \times 1$. The LS estimator for $\boldsymbol{\beta}_1$ is then

$$\mathbf{b}_1 = (\mathbf{X}_1'\mathbf{X}_1)^{-1}\mathbf{X}_1'\mathbf{y}_1 \tag{5.37}$$

Notice that our notations now differ from the ones in (5.36) where, for instance, $\mathbf{X}_1$ is $n \times l$. Similar to (5.35a), we have for the second sample of size n_2

$$\mathbf{y}_2 = \mathbf{X}_2\boldsymbol{\beta}_2 + \mathbf{u}_2 \tag{5.35b}$$

where $\boldsymbol{\beta}_2$ is $(k + 1) \times 1$ and is estimated by

$$\mathbf{b}_2 = (\mathbf{X}_2'\mathbf{X}_2)^{-1}\mathbf{X}_2'\mathbf{y}_2 \tag{5.38}$$

We are to use the two samples to test the hypotheses

$$\begin{aligned} &H_0\colon \boldsymbol{\beta}_1 = \boldsymbol{\beta}_2 \ (= \boldsymbol{\beta}^*) \\ &H_a\colon \boldsymbol{\beta}_1 \neq \boldsymbol{\beta}_2 \end{aligned} \tag{5.39}$$

where $\boldsymbol{\beta}^*$ is a vector of preassigned constants.

To understand clearly the nature of the problem, we consider next a simple regression model.

THE TWO-VARIABLE CASE. Let the model be of the form

$$y = \alpha + \beta x + u \tag{5.40}$$

so that for the two samples

$$Y_{i1} = \alpha_1 + \beta_1 X_{i1} + u_{i1} \qquad i = 1, 2, \ldots, n_1 \tag{5.41}$$

and

$$Y_{i2} = \alpha_2 + \beta_2 X_{i2} + u_{i2} \qquad i = 1, 2, \ldots, n_2 \tag{5.42}$$

with, say, $n_1 + n_2 = n$. Now the LS estimates for (5.41) and (5.42) are, respectively,

$$\hat{Y}_{i1} = a_1 + b_1 X_{i1} \tag{5.41e}$$

and

$$\hat{Y}_{i2} = a_2 + b_2 X_{i2} \tag{5.42e}$$

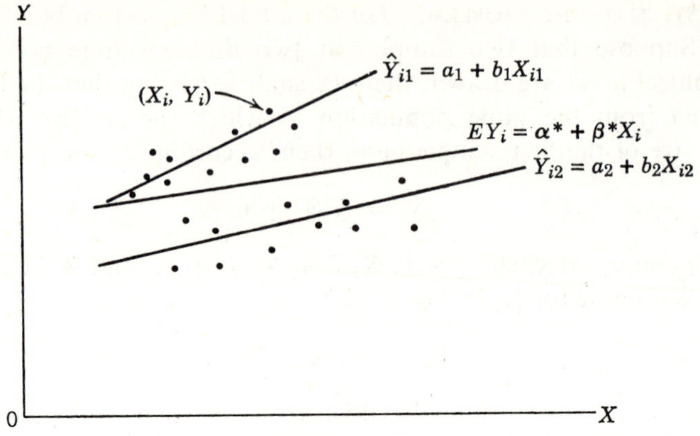

Figure 5.3

which are shown in Figure 5.3 with the hypothesized relation

$$Y_i = \alpha^* + \beta^*X_i + u_i \tag{5.43}$$

Thus, for an arbitrary observation (X_i, Y_i), we have

$$u_i = Y_i - (\alpha^* + \beta^*X_i) \tag{5.43a}$$

Assuming for a moment that (X_i, Y_i) belongs in the first sample, we observe that

$$u_i = [Y_i - (a_1 + b_1X_i)] + [(a_1 + b_1X_i) - (\alpha^* + \beta^*X_i)]$$

Or u_i can be decomposed into two "distances": (1) the "distance" between Y_i and the first sample regression line, and (2) the "distance" between the first sample regression and the hypothesized true regression line. It can be shown easily that for $i = 1, 2, \ldots, n_1$,

$$\sum u_i^2 = \sum [Y_i - (a_1 + b_1X_i)]^2 + \sum [(a_1 + b_1X_i) - (\alpha^* + \beta^*X_i)]^2 \tag{5.44}$$

(prove this as an exercise) and, similarly, that for $i = 1, 2, \ldots, n_2$, coming from the second sample,

$$\sum u_i^2 = \sum [Y_i - (a_2 + b_2X_i)]^2 + \sum [(a_2 + b_2X_i) - (\alpha^* + \beta^*X_i)]^2 \tag{5.45}$$

Notice that the first term in the right-hand side of (5.44) remains unchanged regardless of what we hypothesize to be the values of α^* and β^* in the second term, whose value obviously depends on the null hypothesis. Similarly, this is true for the terms in (5.45). Thus, by taking the sum of the quantities in (5.44) and (5.45), we have

$$\sum_{i=1}^{n} u_i^2 = \sum_{i=1}^{n} [Y_i - (\alpha^* + \beta^*X_i)]^2 \tag{5.46}$$

which is the sum of squares of u_i's in (5.43a) taken over $n_1 + n_2 = n$. Let the sum be Q_1. By the usual assumption of Fixed Model B, Q_1/σ^2 has χ^2 distribution with n degrees of freedom. By combining (5.44) and (5.45) and by defining

$$Q_2 = \sum_{i=1}^{n_1} (Y_i - a_1 - b_1 X_i)^2 + \sum_{i=1}^{n_2} (Y_i - a_2 - b_2 X_i)^2 \tag{5.47}$$

and

$$Q_3 = \sum_{i=1}^{n_1} (a_1 + b_1 X_i - \alpha^* - \beta^* X_i)^2 + \sum_{i=1}^{n_2} (a_2 + b_2 X_i - \alpha^* - \beta^* X_i)^2 \tag{5.48}$$

we see that Q_2/σ^2 has a χ^2 distribution with $(n_1 - 2) + (n_2 - 2) = n - 4$ degrees of freedom and that Q_3/σ^2 has a χ^2 distribution with 4 degrees of feedom. The last distribution follows from the fact that

$$Q_1 = Q_2 + Q_3 \tag{5.49}$$

has n degrees of freedom [see (5.46)], and by Cochran's theorem Q_2/σ^2 and Q_3/σ^2 are independent. Now Q_2 is a measure of variability around individual regression lines, so that if Q_3, a measure of the distances of the individual sample regression lines from the hypothesized line, is large in comparison with Q_2, one would suspect the validity of the hypothesized values α^* and β^*. Seen thus, if the statistic

$$F = \frac{Q_3/4}{Q_2/(n - 4)} \tag{5.50}$$

is large, one is led to reject the null hypothesis.

THE GENERAL CASE. All of the preceding arguments can be restated for multiple regression equations. From (5.35), (5.35a and b), and (5.39), we write, for the general case,

$$Q_1 = (\mathbf{y} - \mathbf{X}\boldsymbol{\beta}^*)'(\mathbf{y} - \mathbf{X}\boldsymbol{\beta}^*) \tag{5.46g}$$

$$Q_2 = (\mathbf{y}_1 - \mathbf{X}_1\mathbf{b}_1)'(\mathbf{y}_1 - \mathbf{X}_1\mathbf{b}_1) + (\mathbf{y}_2 - \mathbf{X}_2\mathbf{b}_2)'(\mathbf{y}_2 - \mathbf{X}_2\mathbf{b}_2) \tag{5.47g}$$

$$Q_3 = (\hat{\mathbf{y}}_1 - \mathbf{X}_1\boldsymbol{\beta}^*)'(\hat{\mathbf{y}}_1 - \mathbf{X}_1\boldsymbol{\beta}^*) + (\hat{\mathbf{y}}_2 - \mathbf{X}_2\boldsymbol{\beta}^*)'(\hat{\mathbf{y}}_2 - \mathbf{X}_2\boldsymbol{\beta}^*) \tag{5.48g}$$

where, or instance, $\hat{\mathbf{y}}_1 = \mathbf{X}_1\mathbf{b}_1$, $\mathbf{y} = \begin{pmatrix}\mathbf{y}_1\\ \mathbf{y}_2\end{pmatrix}$, and $\mathbf{X} = \begin{pmatrix}\mathbf{X}_1\\ \mathbf{X}_2\end{pmatrix}$, with $\mathbf{y}$ being $(n_1 + n_2) \times 1$ and $\mathbf{X}$ being $(n_1 + n_2) \times (k + 1)$. And these quadratic forms have the distributions:

Q_1/σ^2: χ^2 with n d.f.

Q_2/σ^2: χ^2 with $n - 2k - 2$ d.f.

Q_3/σ^2: χ^2 with $2k + 2$ d.f.

It then follows that the test statistic for the null hypothesis is

$$F = \frac{Q_3/(2k + 2)}{Q_2/(n - 2k - 2)} \tag{5.50g}$$

with $F(2k + 2, n - 2k - 2)$ distribution. Contrast (5.50g) with (5.50).

THE USUAL CASE. In practice, however, we frequently have no idea what the true vector $\boldsymbol{\beta}^*$ should be, so that it is often the case that the two samples are pooled for calculating a "common" regression plane. Then we consider this common regression plane as approximating the true population regression plane. Thus, additional $k + 1$ degrees of freedom are lost in Q_1, and instead we have

$$Q_1' = (\mathbf{y} - \mathbf{Xb})'(\mathbf{y} - \mathbf{Xb}) \tag{5.46u}$$

where $\mathbf{b}$ is the LS estimator for $\boldsymbol{\beta}^*$ using the pooled sample. Q_1'/σ^2 then has a χ^2 distribution with $n - k - 1$ degrees of freedom. The test statistic after this change is

$$F = \frac{Q_3'/(k + 1)}{Q_2'/(n - 2k - 2)} \tag{5.50u}$$

where

$$Q_3' = (\hat{\mathbf{y}}_1 - \mathbf{X}_1\mathbf{b})'(\hat{\mathbf{y}}_1 - \mathbf{X}_1\mathbf{b}) + (\hat{\mathbf{y}}_2 - \mathbf{X}_2\mathbf{b})'(\hat{\mathbf{y}}_2 - \mathbf{X}_2\mathbf{b})$$

and

$$Q_2' = Q_1' - Q_3'$$

which is identical with the Q_2 in (5.47g). Since, in general, Q_3 in (5.48g) is greater than or equal to the Q_3' above, the F ratio calculated on the basis of (5.50u) allows for a greater chance of rejecting the null hypothesis than the F from (5.50g). Computationally, one usually finds Q_1' first, then subtract Q_2 from Q_1' to obtain Q_3'.

A general illustration of the test just discussed will be given at the end of this subsection.

A GENERALIZATION. Once the rationale lying behind the derivation of the F ratios in (5.50), (5.50g), and (5.50u) is clear, it is not difficult to generalize the results to the cases of more than two samples. We discuss briefly the three-sample case as an illustration of the possible generalization and leave the understanding of the generality of the test to the reader.

Suppose that these samples of sizes n_1, n_2, and n_3, where $n_1 + n_2 + n_3 = n$, are available and that a test of the constancy of the coefficients of a linear regression equation involving k independent variables (including the constant term) is being sought. Given the regression model, then, the three samples will satisfy

$$\mathbf{y}_1 = \mathbf{X}_1\boldsymbol{\beta}_1 + \mathbf{u}_1 \tag{5.51a}$$

$$\mathbf{y}_2 = \mathbf{X}_2\boldsymbol{\beta}_2 + \mathbf{u}_2 \tag{5.51b}$$

$$\mathbf{y}_3 = \mathbf{X}_3\boldsymbol{\beta}_3 + \mathbf{u}_3 \tag{5.51c}$$

where $\mathbf{y}_1$, $\mathbf{y}_2$, and $\mathbf{y}_3$ are, respectively, $n_1 \times 1$, $n_2 \times 1$, and $n_3 \times 1$, and $\boldsymbol{\beta}_i$, $i = 1, 2, 3$, is $k \times 1$. Consider the usual case of the constancy test which involves the null hypothesis that $\boldsymbol{\beta}_1 = \boldsymbol{\beta}_2 = \boldsymbol{\beta}_3$. We have

$$Q_1' = (\mathbf{y} - \mathbf{X}\mathbf{b})'(\mathbf{y} - \mathbf{X}\mathbf{b}) \tag{5.52}$$

where

$$\mathbf{y} = \begin{pmatrix} \mathbf{y}_1 \\ \mathbf{y}_2 \\ \mathbf{y}_3 \end{pmatrix}$$

$$\mathbf{X} = \begin{pmatrix} \mathbf{X}_1 \\ \mathbf{X}_2 \\ \mathbf{X}_3 \end{pmatrix}$$

$$\mathbf{b} = (\mathbf{X}'\mathbf{X})^{-1}\mathbf{X}'\mathbf{y}$$

By letting

$$\mathbf{b}_1 = (\mathbf{X}_1'\mathbf{X}_1)^{-1}\mathbf{X}_1\mathbf{y}_1$$

$$\mathbf{b}_2 = (\mathbf{X}_2'\mathbf{X}_2)^{-1}\mathbf{X}_2\mathbf{y}_2$$

$$\mathbf{b}_3 = (\mathbf{X}_3'\mathbf{X}_3)^{-1}\mathbf{X}_3\mathbf{y}_3$$

we obtain

$$\begin{aligned} Q_3' = {} & (\hat{\mathbf{y}}_1 - \mathbf{X}_1\mathbf{b})'(\hat{\mathbf{y}}_1 - \mathbf{X}_1\mathbf{b}) + (\hat{\mathbf{y}}_2 - \mathbf{X}_2\mathbf{b})'(\hat{\mathbf{y}}_2 - \mathbf{X}_2\mathbf{b}) \\ & + (\hat{\mathbf{y}}_3 - \mathbf{X}_3\mathbf{b})'(\hat{\mathbf{y}}_3 - \mathbf{X}_3\mathbf{b}) \end{aligned} \tag{5.53}$$

so that

$$\begin{aligned} Q_2' = {} & (\mathbf{y}_1 - \mathbf{X}_1\mathbf{b}_1)'(\mathbf{y}_1 - \mathbf{X}_1\mathbf{b}_1) + (\mathbf{y}_2 - \mathbf{X}_2\mathbf{b}_2)'(\mathbf{y}_2 - \mathbf{X}_2\mathbf{b}_2) \\ & + (\mathbf{y}_3 - \mathbf{X}_3\mathbf{b}_3)'(\mathbf{y}_3 - \mathbf{X}_3\mathbf{b}_3) \end{aligned}$$

where $\hat{\mathbf{y}}_1 = \mathbf{X}_1\mathbf{b}_1$, for instance. From (5.52) and (5.53), we calculate

$$Q_3' = Q_1' - Q_2'$$

and notice that the Q_1'/σ^2 has a χ^2 distribution with $n - k$ degrees of freedom but that Q_2'/σ^2 has a χ^2 distribution with $n - 3k$ degrees of freedom. Thus the appropriate test statistic is the ratio

$$F = \frac{Q_3'/2k}{Q_2'/(n - 3k)} \tag{5.54}$$

which has an F distribution with $(2k, n - 3k)$ degrees of freedom. To recall the meaning of the sums of squares involved here, we observe that Q_2'

measure the variability of the observations about the individual regression planes, but that Q_3' measures the variability of the individual planes around the common regression plane. Furthermore, it is well to be convinced here, as elsewhere, that the numerator of the F statistic is the restricted sum of squares less the unrestricted sum of squares and that the denominator is the unrestricted sum of squares (both divided by appropriate degrees of freedom) when the three samples are combined.

Example 4. We now illustrate the preceding discussion by a cross-section study of the purchase prices of new automobiles over a number of years. The data used are provided through the courtesy of the Board of Governors

Table 5.2 Annual and Pooled Regression Equation Coefficients in the Total Purchase Price of New Cars (TP)

Regressor[a]	1957	1958	1959	Pooled
Constant	242.4*[b]	274.9*	234.5*	269.4*
Y	−1.156	−2.135	0.2633	−5.665
Y^2	149.9*	138.9	46.02	224.2*
SW_1	0.6426*	−0.1335*	0.4752*[b]	0.4504*
SW_2	0.8197*	0.07828	0.3903*	0.4830*
SW_3	0.7264*	−0.1189	0.2962*	0.5006*
W_1	−4.938	−65.11*	20.88	7.58
W_2	−13.42	−57.66*	50.83*	22.34*
W_3	−8.185	−46.12*	24.88	7.505
M	−2.379	69.24*	−15.11	−7.094
N	−1.116	2.769	−4.899*	−2.112
S_e	72.05	75.16	70.09	73.24
$\bar{R}^2$	0.542	0.221	0.437	0.409
n	263	200	254	717
RSS	$1548(10^3)$	$3020(10)^2$	$9259(10)^2$	$2618(10)^3$
ESS	$1308(10)^3$	$1068(10)^3$	$1194(10)^3$	$3787(10)^3$
TSS	$2856(10)^3$	$1370(10)^3$	$2120(10)^3$	$6405(10)^3$

[a] The definition of the regressors is found in Table 5.3. The symbols not defined there are explained as follows.

$\bar{R}^2$ = coefficient of multiple determination, corrected for degrees of freedom.
n = number of observations.
S_e = standard error of estimate.

[b] This symbol indicates that the coefficient is significant at the 5 per cent level, by conventional test.

of the Federal Reserve System, the University of Michigan Survey Research Center, and the Social Systems Research Institute of the University of Wisconsin.

We take the cross-section sample surveys of the United States spending units (call them SU's) for the years 1957, 1958, and 1959 and consider only the SU's that bought new cars in those respective years. Of these SU's, we obtain information on their income, car ownship status, and their family size, the variables for which are defined in Table 5.3. Then the total purchase price (TP) of car (or cars) is regressed on these variables for each annual sample as well as for the entire pooled sample; the results of the regression analysis are shown in Table 5.2. Using these results, we test the null hypothesis

Table 5.3 Definition of Variables that Appear in Table 5.2

Variable	Symbol	Description
Disposable income	Y	Disposable income in thousands of dollars.
	Y^2	Square of the above.
Initial stock	S_2	The estimated market value of all cars the SU owned at the beginning of the period
	W_1	One if the first car SU owned at the beginning of the year was possessed for 3 years or less, 0 otherwise.
	W_2	One if the first car SU owned at the beginning of the year was possessed for 4 up to 7 years, 0 otherwise.
	W_3	One if the first car SU owned at the beginning of the year was possessed for more than 7 years, 0 otherwise.
	SW_1	The estimated market value of all cars owned by SU at the beginning of the year if the SU possessed the first car for 3 years or less, in tens of dollars, 0 otherwise.
	SW_2	The SU's estimated initial value stock, in tens of dollars, if the SU owned the first car for from 4 to 7 years, 0 otherwise.
	SW_3	The SU's estimated initial value stock, in tens of dollars, if the SU owned the first car for more that 7 years, 0 otherwise. (The residual class includes SU's without any initial stock.)
Multiple ownership	M	One if SU owned two or more cars at the beginning of the year, 0 otherwise.
Size of SU	N	The number of people in SU.

that the coefficients of the regression equation

$$TP = \beta_0 + \beta_1 Y + \beta_2 Y^2 + \beta_3 SW_1 + \beta_4 SW_2 + \beta_5 SW_3 + \beta_6 W_1 + \beta_7 W_2 + \beta_8 W_3 + \beta_9 M + \beta_{10} N + u$$

remained stable during the years 1957, 1958, 1959. Here, the *ESS* around the common regression plane Q_1' is $3787(10)^3$ and the ESS's around the individual or annual regression equations sum to $3570(10)^3$, which is Q_2'. The difference $Q_3' = Q_1' - Q_2'$ being a measure of variability of the annual estimated regression planes from the common regression plane, the appropriate test statistic is

$$F = \frac{[3787(10)^3 - 3570(10)^3]/22}{3570(10)^3/(717 - 33)} = 1.89$$

The critical value of F with (22, 684) degrees of freedom at the 5 per cent level is 1.57. Therefore, the null hypothesis is not rejected.

5.2.2. Constancy of a Subset of Regression Coefficients

Consider the general model

$$\mathbf{y} = \mathbf{X}\boldsymbol{\beta} + \mathbf{u}$$

where $\mathbf{y}$ is $n \times 1$, $\mathbf{X}$ is $n \times k$, and $\boldsymbol{\beta}$ is $k \times 1$. Let us rewrite the model for the first and the second samples as

$$\mathbf{y}_1 = (\mathbf{Z}_1 \ \mathbf{W}_1)\begin{pmatrix}\boldsymbol{\gamma}_1 \\ \boldsymbol{\delta}_1\end{pmatrix} + \mathbf{u}_1 \tag{5.55a}$$

and

$$\mathbf{y}_2 = (\mathbf{Z}_2 \ \mathbf{W}_2)\begin{pmatrix}\boldsymbol{\gamma}_2 \\ \boldsymbol{\delta}_2\end{pmatrix} + \mathbf{u}_2 \tag{5.55b}$$

where $\mathbf{y}_1$ and $\mathbf{y}_2$ are, respectively, $n_1 \times 1$ and $n_2 \times 1$, $\mathbf{Z}_1$ and $\mathbf{Z}_2$ are, respectively, $n_1 \times l$ and $n_2 \times l$, $\mathbf{W}_1$ and $\mathbf{W}_2$ are, respectively, $n_1 \times m$ and $n_2 \times m$, $\boldsymbol{\gamma}_1$ and $\boldsymbol{\gamma}_2$ are each $l \times 1$, and $\boldsymbol{\delta}_1$ and $\boldsymbol{\delta}_2$ each $m \times 1$.

By combining (5.55a) and (5.55b), we have

$$\begin{pmatrix}\mathbf{y}_1 \\ \mathbf{y}_2\end{pmatrix} = \begin{pmatrix}\mathbf{Z}_1 & \mathbf{0} & \mathbf{W}_1 & \mathbf{0} \\ \mathbf{0} & \mathbf{Z}_2 & \mathbf{0} & \mathbf{W}_2\end{pmatrix}\begin{pmatrix}\boldsymbol{\gamma}_1 \\ \boldsymbol{\gamma}_2 \\ \boldsymbol{\delta}_1 \\ \boldsymbol{\delta}_2\end{pmatrix} + \begin{pmatrix}\mathbf{u}_1 \\ \mathbf{u}_2\end{pmatrix} \tag{5.56}$$

and the null hypothesis of interest is

$$H_0\colon \boldsymbol{\gamma}_1 = \boldsymbol{\gamma}_2 \ (= \boldsymbol{\beta}, \text{say})$$

Under the null hypothesis, the model is

$$\begin{pmatrix} \mathbf{y}_1 \\ \mathbf{y}_2 \end{pmatrix} = \begin{pmatrix} \mathbf{Z}_1 & \mathbf{W}_1 & \mathbf{0} \\ \mathbf{Z}_2 & \mathbf{0} & \mathbf{W}_2 \end{pmatrix} \begin{pmatrix} \boldsymbol{\beta} \\ \boldsymbol{\delta}_1 \\ \boldsymbol{\delta}_2 \end{pmatrix} + \begin{pmatrix} \mathbf{u}_1 \\ \mathbf{u}_2 \end{pmatrix} \tag{5.56a}$$

The rationale of the test procedure is much the same as the one for the tests discussed in the preceding subsections. The derivation of the test here would be quite tedious, and we dispense with it. We discuss only the test procedure, pointing out that the test is not difficult.

The LS estimate of the coefficient vector in (5.56a) is

$$\begin{pmatrix} \mathbf{b} \\ \tilde{\mathbf{d}}_1 \\ \tilde{\mathbf{d}}_2 \end{pmatrix} = \left[\begin{pmatrix} \mathbf{Z}_1 & \mathbf{W}_1 & \mathbf{0} \\ \mathbf{Z}_0 & \mathbf{0} & \mathbf{W}_2 \end{pmatrix}' \begin{pmatrix} \mathbf{Z}_1 & \mathbf{W}_1 & \mathbf{0} \\ \mathbf{Z}_2 & \mathbf{0} & \mathbf{W}_2 \end{pmatrix} \right]^{-1} \begin{pmatrix} \mathbf{Z}_1 & \mathbf{W}_1 & \mathbf{0} \\ \mathbf{Z}_2 & \mathbf{0} & \mathbf{W}_2 \end{pmatrix}' \begin{pmatrix} \mathbf{y}_1 \\ \mathbf{y}_2 \end{pmatrix} \tag{5.57}$$

If we fit (5.55a) and (5.55b) individually, their LS estimates of the coefficients will be

$$\begin{pmatrix} \mathbf{c}_1 \\ \mathbf{d}_1 \end{pmatrix} = [(\mathbf{Z}_1\ \mathbf{W}_1)'(\mathbf{Z}_1\ \mathbf{W}_1)]^{-1}(\mathbf{Z}_1\ \mathbf{W}_1)'\mathbf{y}_1 \tag{5.57a}$$

and

$$\begin{pmatrix} \mathbf{c}_2 \\ \mathbf{d}_2 \end{pmatrix} = [(\mathbf{Z}_2\ \mathbf{W}_2)'(\mathbf{Z}_2\ \mathbf{W}_2)]^{-1}(\mathbf{Z}_2\ \mathbf{W}_2)'\mathbf{y}_2 \tag{5.57b}$$

where $\mathbf{c}_i$ is the LS estimate of $\boldsymbol{\gamma}_i$.

The sums of squares necessary for computing test statistics can then be obtained by using the results in (5.57), (5.57a), and (5.57b). The sum of squares that measures the distances of individual observations from the common regression plane is

$$Q_1 = \left[\begin{pmatrix} \mathbf{y}_1 \\ \mathbf{y}_2 \end{pmatrix} - \begin{pmatrix} \mathbf{Z}_1 & \mathbf{W}_1 & \mathbf{0} \\ \mathbf{Z}_2 & \mathbf{0} & \mathbf{W}_1 \end{pmatrix} \begin{pmatrix} \mathbf{b} \\ \tilde{\mathbf{d}}_1 \\ \tilde{\mathbf{d}}_2 \end{pmatrix} \right]' \left[\begin{pmatrix} \mathbf{y}_1 \\ \mathbf{y}_2 \end{pmatrix} - \begin{pmatrix} \mathbf{Z}_1 & \mathbf{W}_1 & \mathbf{0} \\ \mathbf{Z}_2 & \mathbf{0} & \mathbf{W}_2 \end{pmatrix} \begin{pmatrix} \mathbf{b} \\ \tilde{\mathbf{d}}_1 \\ \tilde{\mathbf{d}}_2 \end{pmatrix} \right] \tag{5.58a}$$

And it can be shown that Q_1/σ^2 has a χ^2 distribution with $(n - 2m - l)$ degrees of freedom (we assume that $\mathbf{u}_1$ and $\mathbf{u}_2$ have a common variance σ^2). Now Q_1 can be decomposed into two sums of squares Q_2 and Q_3; Q_2 will measure the distances of observations from the individual estimated regression planes, and Q_3 will measure the distances of the individual estimated planes from the common regression plane. Thus

$$Q_2 = \left[\mathbf{y}_1 - (\mathbf{Z}_1 \quad \mathbf{W}_1)\begin{pmatrix}\mathbf{c}_1\\ \mathbf{d}_1\end{pmatrix}\right]'\left[\mathbf{y}_1 - (\mathbf{Z}_1 \quad \mathbf{W}_1)\begin{pmatrix}\mathbf{c}_1\\ \mathbf{d}_1\end{pmatrix}\right]$$
$$+ \left[\mathbf{y}_2 - (\mathbf{Z}_2 \quad \mathbf{W}_2)\begin{pmatrix}\mathbf{c}_2\\ \mathbf{d}_2\end{pmatrix}\right]'\left[\mathbf{y}_2 - (\mathbf{Z}_2 \quad \mathbf{W}_2)\begin{pmatrix}\mathbf{c}_2\\ \mathbf{d}_2\end{pmatrix}\right] \qquad (5.58\text{b})$$

and

$$Q_3 = Q_1 - Q_2$$

It can be shown further that Q_2/σ^2 has a χ^2 distribution with $[(n_1 - m) + (n_2 - m)] = n - 2m$ degrees of freedom. To determine what Q_3 is, we write

$$Q_3 = \left[(\mathbf{Z}_1 \quad \mathbf{W}_1)\begin{pmatrix}\mathbf{c}_1\\ \mathbf{d}_1\end{pmatrix} - (\mathbf{Z}_1 \quad \mathbf{W}_1)\begin{pmatrix}\mathbf{b}\\ \tilde{\mathbf{d}}_1\end{pmatrix}\right]'\left[(\mathbf{Z}_1 \quad \mathbf{W}_1)\begin{pmatrix}\mathbf{c}_1\\ \mathbf{d}_1\end{pmatrix} - (\mathbf{Z}_1 \quad \mathbf{W}_1)\begin{pmatrix}\mathbf{b}\\ \tilde{\mathbf{d}}_1\end{pmatrix}\right]$$
$$+ \left[(\mathbf{Z}_2 \quad \mathbf{W}_2)\begin{pmatrix}\mathbf{c}_2\\ \mathbf{d}_2\end{pmatrix} - (\mathbf{Z}_2 \quad \mathbf{W}_2)\begin{pmatrix}\mathbf{b}\\ \tilde{\mathbf{d}}_2\end{pmatrix}\right]'\left[(\mathbf{Z}_2 \quad \mathbf{W}_2)\begin{pmatrix}\mathbf{c}_2\\ \mathbf{d}_2\end{pmatrix} - (\mathbf{Z}_2 \quad \mathbf{W}_2)\begin{pmatrix}\mathbf{b}\\ \tilde{\mathbf{d}}_2\end{pmatrix}\right],$$

Notice, for example, that $\mathbf{c}_1$ is the estimate of $\boldsymbol{\gamma}_1$ obtained from the first sample regression and that $\tilde{\mathbf{d}}_2$ is the estimate of $\boldsymbol{\delta}_2$ obtained from the "pooled" regression plane. In forming the ratio

$$F = \frac{Q_3/l}{Q_2/(n - 2m - 2l)} \qquad (5.59)$$

we have an F distribution with $(l, n - 2m - 2l)$ degrees of freedom. We again observe that Q_3 is the restricted sum of squares (by hypothesis) less the unrestricted sum of squares and that Q_2 is the unrestricted sum of squares.

Example 5. In a production function study of the Pacific halibut fishing fleet, these data for the following variables representing the 1963 and 1964 annual operations of ten boats were collected: q (catch of halibut in number of pounds), c (catch per skate, representing fish population density), k (inputed capital input in dollars), and L (labor input in number of days). The natural logarithms of the observations on these variables are shown in Table 5.4.

An industry production function of the following specification was formulated.

$$\ln q = \beta_0 + \beta_1 \ln C + \beta_2 \ln K + \beta_3 \ln L + u \qquad (5.59\text{a})$$

Then, by using the data shown in Table 5.4, the regression analysis of (5.59a) was made for each of the two cross sections (1963 and 1964), with the results provided in Table 5.5. Now in the literature β_2 and β_3 in (5.59a) are usually referred to as output elasticities of capital and labor, respectively. Perhaps, we might want to know if the constant term along with the effect of the ln C

Table 5.4 Data on Annual Operations of Ten Halibut Fishing Boats

Boat Number	ln *q*	ln *C*	ln *K*	ln *L*
1963				
1.	8.0159	2.7147	7.3085	6.7742
2.	7.5229	2.7143	6.9713	5.9269
3.	7.8559	3.4046	6.3256	6.2106
4.	8.4554	3.1610	7.3476	6.8024
5.	7.9170	2.4480	7.4678	7.1608
6.	7.4745	2.4599	6.5169	6.1225
7.	8.0501	2.6868	7.4067	6.8669
8.	8.5484	3.0259	7.6996	7.0876
9.	8.4745	2.8800	7.7096	7.0012
10.	7.9899	3.1380	7.0783	6.3026
1964				
1.	7.9977	2.5953	7.4782	6.9632
2.	7.8204	2.9460	6.8406	5.8171
3.	7.8612	3.0716	6.2245	6.1356
4.	8.2995	2.7585	7.3503	6.8480
5.	6.6656	2.4219	6.3374	6.0730
6.	7.6062	2.6404	6.8214	6.4739
7.	8.0338	2.7857	7.2232	6.7262
8.	8.6017	2.6779	7.7859	7.1982
9.	8.3479	2.6963	7.7181	7.0544
10.	7.9831	3.0168	7.1776	6.4739

Table 5.5 The Results of Regression Analysis on Annual Cross Sections for Halibut Fishing Boats

Coefficient of	1963	1964	Pooled[a]	
1	0.8974*[b]	−3.306*[b]	−5.045*	
ln *C*	0.6757*	1.564*	0.9107*	
ln *K*	0.3889*	0.4511*	0.4695*	0.5528*
ln *L*	0.3629*	0.5637*	0.3853*	0.3028
			(1963)	(1964)
R^2	0.938	0.924	0.878	
n	10	10	20	
ESS	0.077996	0.190000	0.465796	

[a] For the explanation about the pooled equation, see the text.

[b] This symbol indicates significance at the 5 percent level, by conventional test. R^2 is the coefficient of multiple determination.

did not change for the two years, although β_2 and β_3 might vary over time. Thus the test discussed in the subsection immediately preceding is relevant.

To perform the test, we must write the model (5.59a) for the pooled sample, keeping in mind that here the constant and the ln C coefficient are the same for both years but that the β_2 and β_3 are allowed to be different for the two years. Then the pooled relation to be estimated becomes

$$\ln q = \beta_0 + \beta_1 \ln C + \beta_2{}' (\ln K)D_1 + \beta_3{}' (\ln L)D_1 + \beta_2{}'' (\ln K)D_2 + \beta_3{}'' (\ln L)D_2 + u \quad (5.59b)$$

where $D_1 = 1$, if the observation falls in 1963, and zero otherwise, and $D_2 = 1$ if the observation falls in 1964, and zero otherwise [compare the joint expression in (5.56a)]. Thus notice in the last column of Table 5.5 that the pooled relation is estimated with the same coefficient estimates for the constant and ln C and that the coefficient estimates for ln K and ln L for 1963 differ from the ones for 1964.

The test statistic can now be obtained by reference to the *ESS* for the individual cross sections and for the combined sample. Q_1 is 0.465796 while $Q_2 = 0.077996 + 0.190000$; it follows that $Q_3 = 0.197800$, the latter's degrees of freedom being $[(20 - 6) - (20 - 8)] = 2$. Thus

$$F = \frac{0.197800/2}{0.267996/(20 - 8)} = 4.43$$

so that the null hypothesis that the constant and the ln C coefficient remained stable during the years studied is rejected. The value of F with (2, 12) degrees of freedom at the 5 percent level is 3.36 (Table A-2, Appendix A).

5.3. General Linear Hypothesis

The t and F tests that we discussed in the preceding sections are only a handful of the specific cases of the test of general linear hypothesis. The Student t test about the individual regression coefficient, the tests of a linear relation among these coefficients (say, the equality of two coefficients in a given equation), the tests of the equality of two sets of regression coefficients and, also, the related tests are all implied by a general procedure in which one places a restriction on the coefficient vector in the form of a linear combination or a number of linear combinations. That is, the null hypotheses under consideration have a representation the form of which is a linear combination or a linear transform of the elements of the coefficient vector.

To observe this, we consider a few examples. For a simple regression model

$$y = \beta_0 + \beta_1 x + u$$

a sample of n observations can be written as

$$\begin{pmatrix} Y_1 \\ \cdot \\ \cdot \\ \cdot \\ Y_n \end{pmatrix} = \begin{pmatrix} 1 & X_1 \\ 1 & X_2 \\ \cdot & \cdot \\ \cdot & \cdot \\ \cdot & \cdot \\ 1 & X_n \end{pmatrix} \begin{pmatrix} \beta_0 \\ \cdot \\ \cdot \\ \cdot \\ \beta_1 \end{pmatrix} + \begin{pmatrix} u_1 \\ \cdot \\ \cdot \\ \cdot \\ u_n \end{pmatrix}$$

If the null hypothesis about β_1 is that

$$\beta_1 = \beta_1{}^*$$

where $\beta_1{}^*$ is a specific preassigned scalar, we can write the hypothesis as

$$\boldsymbol{\delta}'\boldsymbol{\beta} = \beta_1{}^* \tag{5.70}$$

where $\boldsymbol{\delta}' = (0\ 1)$ and $\boldsymbol{\beta}' = (\beta_0\ \beta_1)$. Thus the null hypothesis is in the form of a linear combination of β_0 and β_1, the constants of the combination being the elements of $\boldsymbol{\delta}$.

A situation that illustrates the preceding discussion but that was not discussed previously is the hypothesis about the elasticities of output with respect to different factors. For the function

$$X = AL^{\theta_1}K^{\theta_2}e^u \tag{5.71}$$

appropriate assumptions about u can lead to the least-squares estimation of the relation

$$\ln X = \theta_0 + \theta_1 \ln L + \theta_2 \ln K + u \tag{5.71a}$$

In (5.71a) a hypothesis of interest might be the presence of constant returns to scale

$$\theta_1 + \theta_2 = 1$$

In the form of a linear combination, this hypothesis is

$$\boldsymbol{\delta}'\boldsymbol{\theta} = 1 \tag{5.72}$$

where $\boldsymbol{\delta}' = (0\ 1\ 1)$ and $\boldsymbol{\theta}' = (\theta_0\ \theta_1\ \theta_2)$.

If a further hypothesis is that $\theta_0 = 0$, in addition to the preceding restriction we can write

$$\mathbf{R}\boldsymbol{\theta} = \mathbf{r} \tag{5.73}$$

where

$$\mathbf{R} = \begin{pmatrix} 1 & 0 & 0 \\ 0 & 1 & 1 \end{pmatrix}$$

$$\mathbf{r} = \begin{pmatrix} 0 \\ 1 \end{pmatrix}$$

with $\boldsymbol{\theta}$ as defined before. Here, we place two independent linear restrictions on the coefficients in (5.71a). It is the expressions like (5.72) and (5.73) that will allow a general consideration of the tests in a linear model defined in (4.33).

Notice that if restrictions of the form (5.73) come as known information prior to the estimation, this knowledge should be incorporated in the estimation of the coefficients. Indeed, a gain in efficiency results from such a procedure, as is discussed under the mixed estimation by Theil and Goldberger (1961). The casting of the restrictions in the form (5.73) for purposes of estimation and hypothesis testing are available in standard works such as Graybill (1961), Kempthorne (1952), and Chipman and Rao (1964).

We first proceed to discuss the hypothesis testing that involves a linear combination of coefficients.

5.3.1. A Linear Combination of the Coefficients

Consider the general linear model

$$\mathbf{y} = \mathbf{X}\boldsymbol{\beta} + \mathbf{u} \tag{5.74}$$

for a sample of size n, where $\mathbf{y} = n \times 1$, $\mathbf{X}$ is $n \times (k + 1)$, $\boldsymbol{\beta}$ is $(k + 1) \times 1$, and $E\mathbf{u} = \mathbf{0}$ and $E\mathbf{uu}' = \sigma^2\mathbf{I}$, with $\mathbf{u}$ a normal random vector. Now suppose that a hypothesis is entertained that

$$\boldsymbol{\delta}'\boldsymbol{\beta} = \pi \tag{5.74a}$$

where $\boldsymbol{\delta}$ is a $(k + 1)$ element column vector and π is a scalar. We can show that the BLUE estimator of $\boldsymbol{\delta}'\boldsymbol{\beta}$ is $\boldsymbol{\delta}'\mathbf{b}$ or that

$$\boldsymbol{\delta}'\mathbf{b} = \boldsymbol{\delta}'(\mathbf{X}'\mathbf{X})^{-1}\mathbf{X}'\mathbf{y} \tag{5.74b}$$

Call this statistic p, the linear combination of the least-squares estimator of $\boldsymbol{\beta}$. Then

$$p - \pi = \boldsymbol{\delta}'(\mathbf{b} - \boldsymbol{\beta}) \tag{5.74c}$$

and

$$\begin{aligned} E(p - \pi)^2 &= E\{\boldsymbol{\delta}'(\mathbf{X}'\mathbf{X})^{-1}\mathbf{X}'\mathbf{uu}'\mathbf{X}(\mathbf{X}'\mathbf{X})^{-1}\boldsymbol{\delta}\} \\ &= \sigma^2\boldsymbol{\delta}'(\mathbf{X}'\mathbf{X})^{-1}\boldsymbol{\delta} \\ &= \sigma_p^{\,2} \end{aligned} \tag{5.74d}$$

If σ^2 is known the test of (5.74a) against the alternative hypothesis, $\boldsymbol{\delta}'\boldsymbol{\beta} \neq \pi$ can be performed by using the z-distribution; namely

$$z = \frac{p - \pi}{\sigma_p} \tag{5.74e}$$

is normally distributed with mean 0 and variance 1. If σ^2 is unknown, then the Student-t distribution may be invoked. Thus

$$t_{n-k-1} = \frac{p - \pi}{s_p} \tag{5.74f}$$

where $s_p^{\,2}$ is an unbiased estimate of $\sigma_p^{\,2}$.

Example 6. In Table 5.6, we show a set of hypothetical data for the production relation in an industry, the units of measurement being assumed properly determined. Suppose that we consider the natural logarithm of X, L, K, and M, and call them, respectively, X', L', K', and M'. Then we assume that in the model

$$X' = \beta_0 + \beta_1 L' + \beta_2 K' + \beta_3 M' + u \tag{5.74g}$$

u is normally distributed with $Eu = 0$, $Eu^2 = \sigma^2$. A test for constant returns to scale may be carried out by setting the null hypothesis as

$$H_0: \beta_1 + \beta_2 + \beta_3 = 1$$

so that the linear restriction appropriate for the test is in the form

$$\boldsymbol{\delta}'\boldsymbol{\beta} = (0\ 1\ 1\ 1)\begin{pmatrix}\beta_0\\ \beta_1\\ \beta_2\\ \beta_3\end{pmatrix} = 1$$

Now the estimate of the relation (5.74g) is

$$X' = -5.7102 + 0.6326L' + 0.6285K' + 0.8212M'$$
$$R^2 = 0.601$$
$$s^2 = 0.36991$$
$$\text{Unrestricted } ESS = 12.5771$$

and the estimate of $\sigma_p{}^2$, according to (5.74d), is

$$s^2\boldsymbol{\delta}'(X'X)^{-1}\boldsymbol{\delta}$$

$$= s^2(0\ 1\ 1\ 1)\begin{pmatrix} 34.344 & -4.312 & -0.08992 & -0.5452 \\ -4.312 & 0.6170 & -0.05628 & 0.01840 \\ 0.08992 & -0.05628 & 0.06345 & -0.01883 \\ -0.5452 & 0.01840 & -0.01883 & 0.1039 \end{pmatrix}\begin{pmatrix}0\\1\\1\\1\end{pmatrix}$$

$$= 0.2482$$

Therefore, $s_p = 0.4993$, while $p = 2.0796$, and the hypothesized value of π is 1. On forming the ratio

$$\frac{p - \pi}{s_p} = \frac{2.0796 - 1}{0.4993} = 2.162$$

Table 5.6 Hypothetical Data on a Production Relation

Date	Output (X)	Labor (L)	Capital (K)	Other (M)
1930	727.8	1853.0	1367.0	293.0
1931	1547.2	1686.0	1150.0	178.0
1932	5300.4	1657.0	1154.0	111.0
1933	1137.2	1792.0	889.0	249.0
1934	2886.0	960.0	874.0	164.0
1935	1091.9	1487.0	1174.0	90.0
1936	1359.2	1230.0	1471.0	172.0
1937	576.8	1142.0	959.0	108.0
1938	251.2	1713.0	565.0	49.0
1939	454.7	1499.0	368.0	180.0
1940	419.8	1498.0	252.0	121.0
1941	684.1	1510.0	353.0	115.0
1942	926.6	1539.0	262.0	145.0
1943	750.0	1583.0	279.0	151.0
1944	974.9	1470.0	252.0	200.0
1945	600.8	1566.0	239.0	167.0
1946	1081.9	1436.0	376.0	147.0
1947	494.2	1705.0	252.0	101.0
1948	3377.0	2716.0	1191.0	91.0
1949	2004.9	2439.0	1398.0	111.0
1950	1814.8	2130.0	1323.0	141.0
1951	1557.2	2546 0	795.0	55.0
1952	691.9	2028 0	1691.0	72.0
1953	3296.8	1504.0	1724.0	122.0
1954	6228.7	1597.0	2051.0	221.0
1955	7877.2	2004.0	978.0	392.0
1956	3612.9	2122.0	1881.0	244.0
1957	5172.7	1874.0	2337.0	325.0
1958	6106.0	2425.0	1726.0	315.0
1959	4126 1	2133.0	1426.0	264.0
1960	2489.0	1940.0	475.0	284.0
1961	1557.0	1874.0	1007.0	233.0
1962	1528.7	1870.0	844.0	253.0
1963	2481.3	1237.0	1341.0	317.0
1964	3755.5	1751.0	420.0	275.0
1965	2383.5	1914.0	906.0	287.0
1966	6391.0	1613.0	1181.0	311.0
1967	8310.6	1715.0	1486.0	310.0

and referring to Table A-1 in Appendix A, we notice that the t value for 34 degrees of freedom for $\alpha = 0.05$ is approximately 2.042—a number less than the t value yielded by the computed statistics. Therefore, we conclude that the null hypothesis is rejected.

An alternative way for making the same test is available which illustrates the fact that the square of the t statistic corresponding to, say, n degrees of freedom is equal to the F statistic with $(1, n)$ degrees of freedom. Given the null hypothesis, we first estimate the error sum of squares under the hypothesized constraint, say, $\beta_3 = 1 - \beta_1 - \beta_2$. Thus the relation (5.74g) becomes

$$X' = \beta_0 + \beta_1 L' + \beta_2 K' + (1 - \beta_1 - \beta_2)M' + u$$

which resolves into

$$X' - M' = \beta_0 + \beta_1(L' - M') + \beta_2(K' - M') + u$$

Now define new variables $X' - M' = y$, $L' - M' = x_1$, and $K' - M' = x_2$, so that we have

$$y = \beta_0 + \beta_1 x_1 + \beta_2 x_2 + u$$

On least-squares estimation of this relation, we find that

$$y = 1.9798 - 0.3015x_1 + 0.6473x_2$$

$$R^2 = 0.328$$

$$\text{Restricted } ESS = 14.3229$$

The appropriate F statistic is

$$F = \frac{(14.3229 - 12.5771)/1}{12.5771/34} = 4.70$$

which is identical to $(t = 2.162)^2$ except for the rounding error. The F value with $(1, 34)$ degrees of freedom at $\alpha = 0.05$, from Table A-2 in Appendix A is about 4.20—less than the above F value just calculated. Therefore, the same conclusion as obtained in the t test follows for the F test.

5.3.2. A Number of Independent Linear Combinations—Estimation

At the risk of being repetitious, we reconstruct some of the null hypotheses discussed earlier, so that we may better understand the powerful procedure that will be discussed in the next section. Later in this section, we also consider estimation under restrictions.

Consider the partial joint test in Section 5.1.3. The model (5.26) is

$$\mathbf{y} = (\mathbf{X}_1\ \mathbf{X}_2)\begin{pmatrix}\boldsymbol{\beta}_1 \\ \boldsymbol{\beta}_2\end{pmatrix} + \mathbf{u}$$

where $\boldsymbol{\beta}_1$ is $l \times 1$ and $\boldsymbol{\beta}_2$ is $m \times 1$, and $\boldsymbol{\beta} = \begin{pmatrix} \boldsymbol{\beta}_1 \\ \boldsymbol{\beta}_2 \end{pmatrix}$ is $(l + m) \times 1$. The null hypothesis entertained there is that

$$\boldsymbol{\beta}_2 = \mathbf{0}$$

which in the form of (5.73) is

$$\mathbf{R}\boldsymbol{\beta} = \mathbf{r} \tag{5.75}$$

where

$$\mathbf{R} = \begin{pmatrix} 0 & 0 & \cdots & 0 & 1 & 0 & 0 & \cdots & 0 \\ 0 & 0 & \cdots & 0 & 0 & 1 & 0 & \cdots & 0 \\ \cdot & \cdot & \cdot & \cdot & \cdot & \cdot & \cdot & \cdot & \cdot \\ \cdot & \cdot & \cdot & \cdot & \cdot & \cdot & \cdot & \cdot & \cdot \\ \cdot & \cdot & \cdot & \cdot & \cdot & \cdot & \cdot & \cdot & \cdot \\ 0 & 0 & \cdots & 0 & 0 & 0 & 0 & \cdots & 1 \end{pmatrix}$$

is an $m \times (l + m)$ matrix whose first row, for instance, has 1 as the $(l + 1)$st element; and $\mathbf{r}$ is an m-element zero vector.

Also, consider the test of equality of two subsets of coefficients in two regressions. For the model in (5.56), the null hypothesis is

$$\boldsymbol{\gamma}_1 = \boldsymbol{\gamma}_2$$

or

$$\boldsymbol{\gamma}_1 - \boldsymbol{\gamma}_2 = \mathbf{0}$$

That is, the subvectors $\boldsymbol{\gamma}_1$ and $\boldsymbol{\gamma}_2$ are hypothesized to be identical. Recalling that the coefficient vector in (5.56) is of $(2l + 2m)$ elements, we may write the null hypothesis as follows

$$\mathbf{R}\begin{pmatrix} \boldsymbol{\gamma}_1 \\ \boldsymbol{\gamma}_2 \\ \boldsymbol{\delta}_1 \\ \boldsymbol{\delta}_2 \end{pmatrix} = \mathbf{r} \tag{5.76}$$

where $\mathbf{r}$ is l-element zero vector and $\mathbf{R}$ is $l \times (2l + 2m)$ and is equal to

$$\mathbf{R} = \begin{pmatrix} 1 & 0 & 0 & \cdots & 0 & -1 & 0 & 0 & \cdots & 0 & 0 & \cdots & 0 \\ 0 & 1 & 0 & \cdots & 0 & 0 & -1 & 0 & \cdots & 0 & 0 & \cdots & 0 \\ 0 & 0 & 1 & \cdots & 0 & 0 & 0 & -1 & \cdots & 0 & 0 & \cdots & 0 \\ \cdot & \cdot & \cdot & \cdot & \cdot & \cdot & \cdot & \cdot & \cdot & \cdot & \cdot & \cdot & \cdot \\ \cdot & \cdot & \cdot & \cdot & \cdot & \cdot & \cdot & \cdot & \cdot & \cdot & \cdot & \cdot & \cdot \\ \cdot & \cdot & \cdot & \cdot & \cdot & \cdot & \cdot & \cdot & \cdot & \cdot & \cdot & \cdot & \cdot \\ \underbrace{0 \quad 0 \quad 0 \quad \cdots \quad 1}_{l\text{ columns}} & & & & & \underbrace{0 \quad 0 \quad 0 \quad \cdots \quad -1}_{l\text{ columns}} & & & & & \underbrace{0 \quad \cdots \quad 0}_{2m\text{ columns}} & & \end{pmatrix}$$

If the validity of linear restrictions on the coefficient vector is given and is not in question, then such information should be used in the estimation of the parameters of the model. In general, a model such as (5.74) can be estimated with the least-squares technique as follows. Determine the estimate of $\boldsymbol{\beta}$ so as to minimize

$$S = (\mathbf{y} - \mathbf{X}\boldsymbol{\beta})'(\mathbf{y} - \mathbf{X}\boldsymbol{\beta}) \tag{5.77}$$

subject to the side constraint that $\mathbf{R}\boldsymbol{\beta} = \mathbf{r}$ or $\mathbf{R}\boldsymbol{\beta} - \mathbf{r} = \mathbf{0}$. This can be accomplished by the method of Lagrange's multipliers. Thus the objective function in question is

$$S = (\mathbf{y} - \mathbf{X}\boldsymbol{\beta})'(\mathbf{y} - \mathbf{X}\boldsymbol{\beta}) - 2\boldsymbol{\lambda}'(\mathbf{R}\boldsymbol{\beta} - \mathbf{r}) \tag{5.78}$$

where $\boldsymbol{\lambda}'$ is the vector of the Langrangean multipliers of the number of elements equal to the number of rows of $\mathbf{R}$. It is assumed that the linear restrictions on $\boldsymbol{\beta}$ are independent so that the rank of $\mathbf{R}$ is the number of rows of $\mathbf{R}$. It can be shown* that the minimization of S in (5.78) gives the solution for the coefficient estimator under the constraint

$$\tilde{\boldsymbol{\beta}} = \mathbf{b} + (\mathbf{X}'\mathbf{X})^{-1}\mathbf{R}'[\mathbf{R}(\mathbf{X}'\mathbf{X})^{-1}\mathbf{R}']^{-1}(\mathbf{r} - \mathbf{R}\mathbf{b}) \tag{5.79}$$

where $\mathbf{b}$ is the usual least-squares estimator of $\boldsymbol{\beta}$ *without* the side constraint. Thus the LS estimation of $\boldsymbol{\beta}$ subject to constraint involves (1) calculating the usual LS estimate $\mathbf{b}$ without the constraint, and (2) adding to the result the second term in the right-hand member of (5.79). Furthermore, the covariance matrix of $\tilde{\boldsymbol{\beta}}$ is

$$\mathbf{Var}\ (\tilde{\boldsymbol{\beta}}) = \mathbf{V} - \mathbf{V}\mathbf{R}'(\mathbf{R}\mathbf{V}\mathbf{R}')^{-1}\mathbf{R}\mathbf{V} \tag{5.80}$$

where $\mathbf{V} = \sigma^2(\mathbf{X}'\mathbf{X})^{-1}$.

5.3.3. *A Number of Independent Linear Combinations—Testing*

In the preceding section, our discussion of the LS estimation of the regression coefficient vector $\boldsymbol{\beta}$ proceeded on the assumption that the set of linear restrictions $\mathbf{R}\boldsymbol{\beta} = \mathbf{r}$ was given and known. As earlier discussions make clear, however, it is possible that the restraints are mere hypotheses about the regression coefficients. In the latter case, the set of restrictions form the null hypothesis to be tested against the alternative that $\mathbf{R}\boldsymbol{\beta} \neq \mathbf{r}$. (We consider exact linear restrictions only.) The test of linear restrictions is usually referred to as the test of linear hypothesis.

Consider the usual model (5.74)

$$\mathbf{y} = \mathbf{X}\boldsymbol{\beta} + \mathbf{u}$$

* See, for example, Goldberger (1964, pp. 256–257), Chipman and Rao (1964, pp. 198–207).

and entertain the null hypothesis

$$\mathbf{R}\boldsymbol{\beta} = \mathbf{r} \tag{5.81}$$

against the alternative that

$$\mathbf{R}\boldsymbol{\beta} \neq \mathbf{r} \tag{5.81a}$$

where $\mathbf{R}$ is $p \times (k + 1)$, rank of $\mathbf{R} = p$, (and p is at most equal to $k + 1$). Test procedures under the present framework of notation can be found in Chipman and Rao (1964) and Theil (1970).* Here we follow Theil and briefly discuss the test statistic.

Following the earlier notation, let the unstricted LS estimator of $\boldsymbol{\beta}$ be $\mathbf{b}$, and the LS estimator under the null hypothesis be $\tilde{\boldsymbol{\beta}}$, as shown in (5.79). The relevant sums of squares, then, are

$$Q_1 = (\mathbf{y} - \mathbf{X}\tilde{\boldsymbol{\beta}})'(\mathbf{y} - \mathbf{X}\tilde{\boldsymbol{\beta}}) \tag{5.82a}$$

$$Q_2 = (\mathbf{y} - \mathbf{X}\mathbf{b})'(\mathbf{y} - \mathbf{X}\mathbf{b}) \tag{5.82b}$$

$$\begin{aligned} Q_3 &= (\mathbf{X}\mathbf{b} - \mathbf{X}\tilde{\boldsymbol{\beta}})'(\mathbf{X}\mathbf{b} - \mathbf{X}\tilde{\boldsymbol{\beta}}) \\ &= (\mathbf{b} - \tilde{\boldsymbol{\beta}})'\mathbf{X}'\mathbf{X}(\mathbf{b} - \tilde{\boldsymbol{\beta}}) \end{aligned} \tag{5.82c}$$

Easily, we observe that Q_2/σ^2 has a χ^2 distribution with $(n - k - 1)$ degrees of freedom. To indicate the distribution of Q_3/σ^2, we notice that

$$\begin{aligned} Q_3 &= (\mathbf{b} - \tilde{\boldsymbol{\beta}})'\mathbf{X}'\mathbf{X}(\mathbf{b} - \tilde{\boldsymbol{\beta}}) \\ &= \mathbf{u}'\mathbf{X}(\mathbf{X}'\mathbf{X})^{-1}\mathbf{R}'[\mathbf{R}(\mathbf{X}'\mathbf{X})^{-1}\mathbf{R}']^{-1}\mathbf{R}(\mathbf{X}'\mathbf{X})^{-1}\mathbf{X}'\mathbf{u} \end{aligned} \tag{5.82d}$$

This results from the decomposition of the sum of squares Q_1:

$$\begin{aligned} Q_1 &= (\mathbf{y} - \mathbf{X}\tilde{\boldsymbol{\beta}})'(\mathbf{y} - \mathbf{X}\tilde{\boldsymbol{\beta}}) \\ &= (\mathbf{y} - \mathbf{X}\tilde{\boldsymbol{\beta}})'[\mathbf{I} - \mathbf{X}(\mathbf{X}'\mathbf{X})^{-1}\mathbf{X}'](\mathbf{y} - \mathbf{X}\tilde{\boldsymbol{\beta}}) + (\mathbf{y} - \mathbf{X}\tilde{\boldsymbol{\beta}})'\mathbf{X}(\mathbf{X}'\mathbf{X})^{-1}\mathbf{X}'(\mathbf{y} - \mathbf{X}\tilde{\boldsymbol{\beta}}) \end{aligned}$$

where the last term is equal to $(\mathbf{b} - \tilde{\boldsymbol{\beta}})'\mathbf{X}'\mathbf{X}(\mathbf{b} - \tilde{\boldsymbol{\beta}})$. The last indicated equality can be shown by noting that $\mathbf{X}(\mathbf{X}'\mathbf{X})^{-1}\mathbf{X}'$ is idempotent. Since

$$(\mathbf{b} - \tilde{\boldsymbol{\beta}}) = (\mathbf{X}'\mathbf{X})^{-1}\mathbf{R}'[\mathbf{R}(\mathbf{X}'\mathbf{X})^{-1}\mathbf{R}']^{-1}\mathbf{R}(\mathbf{X}'\mathbf{X})^{-1}\mathbf{X}'\mathbf{u}$$

it follows that (5.82d) holds. Now, the quantity in (5.82d) is a quadratic form in $\mathbf{u}$ and, as is clear, the matrix of the form is

$$\mathbf{X}(\mathbf{X}'\mathbf{X})^{-1}\mathbf{R}'[\mathbf{R}(\mathbf{X}'\mathbf{X})^{-1}\mathbf{R}']^{-1}\mathbf{R}(\mathbf{X}'\mathbf{X})^{-1}\mathbf{X}' \tag{5.84}$$

This matrix is idempotent (show this as an exercise) and has the following rank and trace:

$$\begin{aligned} \operatorname{tr} \mathbf{X}(\mathbf{X}'\mathbf{X})^{-1}\mathbf{R}'[\mathbf{R}(\mathbf{X}'\mathbf{X})^{-1}\mathbf{R}']^{-1}\mathbf{R}(\mathbf{X}'\mathbf{X})^{-1}\mathbf{X}' \\ &= \operatorname{tr} [\mathbf{R}(\mathbf{X}'\mathbf{X})^{-1}\mathbf{R}']^{-1}\mathbf{R}(\mathbf{X}'\mathbf{X})^{-1}\mathbf{X}'\mathbf{X}(\mathbf{X}'\mathbf{X})^{-1}\mathbf{R}' \\ &= p \end{aligned}$$

* When attempting the proof by Chipman and Rao, it is advisable to study first the theory of generalized least squares in Chapter 6.

By recalling that $\mathbf{u}$ is a normal random vector with mean zero and covariance $\sigma^2\mathbf{I}$, we conclude that (5.82c) has $\sigma^2\chi^2$ distribution with p degrees of freedom. Thus the test statistic of the null hypothesis $\mathbf{R}\boldsymbol{\beta} = \mathbf{r}$, provided that it is true, is

$$F = \frac{Q_3/p}{Q_2/(n-k-1)} \tag{5.85}$$

It is well to recall that Q_3 is the restricted sum of squares less the unrestricted sum of squares and that Q_2 is the unrestricted sum of squares.

To illustrate the general nature of the test of general linear hypothesis, we return to some of the F tests previously discussed.

First, consider the test of hypothesis that the regression coefficient vector is a vector of preassigned values, say, $\boldsymbol{\beta} = \boldsymbol{\beta}^*$, as in (5.17) or in (5.18). We observe that the null hypothesis in (5.18) can be alternatively written as

$$\mathbf{R}\boldsymbol{\beta} = \boldsymbol{\beta}^* \tag{5.18a}$$

where $\mathbf{R} = \mathbf{I}_{k+1}$. The substitution of this last equality into the expression for $\tilde{\boldsymbol{\beta}}$ in (5.79) and, then, in turn, into Q_3 in (5.82c) gives rise to the test statistic

$$F = \frac{(\mathbf{b} - \boldsymbol{\beta}^*)'\mathbf{X}'\mathbf{X}(\mathbf{b} - \boldsymbol{\beta}^*)/(k+1)}{Q_2/(n-k-1)} \tag{5.85a}$$

the numerator of which is identical to the Q_3 in (5.21g) except for the degrees of freedom. Since Q_2 is the sum of squares without the restrictions, it is the same as the Q_2 in the general form of the F statistic shown in (5.21).

Consider next the partial joint test. Here we have the model

$$\mathbf{y} = (\mathbf{X}_1\ \mathbf{X}_2)\begin{pmatrix}\boldsymbol{\beta}_1\\ \boldsymbol{\beta}_2\end{pmatrix} + \mathbf{u}$$

where $\boldsymbol{\beta}_1$ is $l \times 1$ and $\boldsymbol{\beta}_2$ is $m \times 1$, sample size is n, and the null hypothesis, according to (5.75), has the restriction matrix

$$\mathbf{R} = (\mathbf{0}\ \mathbf{I}_m) \tag{5.75a}$$

where $\mathbf{0}$ is an $m \times l$ zero matrix and $\mathbf{I}_m$ is an identity matrix of order m. The substitution of this matrix into (5.82d) yields

$$\mathbf{u}'\mathbf{X}(\mathbf{X}'\mathbf{X})^{-1}(\mathbf{0}\ \mathbf{I}_m)'[(\mathbf{0}\ \mathbf{I}_m)(\mathbf{X}'\mathbf{X})^{-1}(\mathbf{0}\ \mathbf{I}_m)']^{-1}(\mathbf{0}\ \mathbf{I}_m)(\mathbf{X}'\mathbf{X})^{-1}\mathbf{X}'\mathbf{u} \tag{5.86}$$

under the null hypothesis that $(\mathbf{0}\ \mathbf{I}_m)\begin{pmatrix}\boldsymbol{\beta}_1\\ \boldsymbol{\beta}_2\end{pmatrix} = \boldsymbol{\beta}_2^*$.

Before proceeding, let $\mathbf{X} = (\mathbf{X}_1\ \mathbf{X}_2)$ according to (5.26). Then, let

$$\mathbf{S} = (\mathbf{X}'\mathbf{X}) = \begin{pmatrix} \mathbf{X}_1'\mathbf{X}_1 & \mathbf{X}_1'\mathbf{X}_2 \\ \mathbf{X}_2'\mathbf{X}_1 & \mathbf{X}_2'\mathbf{X}_2 \end{pmatrix} = \begin{pmatrix} \mathbf{S}_{11} & \mathbf{S}_{12} \\ \mathbf{S}_{21} & \mathbf{S}_{22} \end{pmatrix} \tag{5.87a}$$

$$\mathbf{S}^{-1} = \begin{pmatrix} \mathbf{S}_{11} & \mathbf{S}_{12} \\ \mathbf{S}_{21} & \mathbf{S}_{22} \end{pmatrix}^{-1} = \begin{pmatrix} \mathbf{S}^{11} & \mathbf{S}^{12} \\ \mathbf{S}^{21} & \mathbf{S}^{22} \end{pmatrix} \tag{5.87b}$$

Notice that $\mathbf{S}^{22} = \mathbf{S}_{22} - \mathbf{S}_{21}(\mathbf{S}_{11})^{-1}\mathbf{S}_{12}$, so that $\mathbf{S}^{22} = \mathbf{D}^{-1}$ for $\mathbf{D}$ involved in the development leading to and including the test statistic (5.30). Furthermore, on closer scrutiny, one verifies that

$$\begin{aligned}(\mathbf{0}\ \mathbf{I}_m)(\mathbf{X}'\mathbf{X})^{-1}\mathbf{X}'\mathbf{u} &= \mathbf{S}^{22}\mathbf{X}_2'\mathbf{M}_1\mathbf{u} \\ &= (\mathbf{b}_2 - \boldsymbol{\beta}_2^*)\end{aligned}$$

where $\mathbf{M}_1 = \mathbf{I} - \mathbf{X}_1(\mathbf{X}_1'\mathbf{X}_1)^{-1}\mathbf{X}_1'$. And for (5.86), we can write the equivalent expression

$$(\mathbf{b}_2 - \boldsymbol{\beta}_2^*)'\mathbf{D}(\mathbf{b} - \boldsymbol{\beta}_2^*)$$

which is precisely (5.29), that is Q_3, discussed in Section 5.1.3.

Summarizing, we observe that, in calculating the test statistic under the null hypothesis $\mathbf{R}\boldsymbol{\beta} = \mathbf{r}$, two types of the null hypothesis can be distinguished: (1) the one in which it is hypothesized that $\boldsymbol{\beta}$ is a vector of coefficients with preassigned values, and (2) the one in which $\mathbf{R}$ defines a set of hypothesized relations among the coefficients. In the former, the calculation of the desired test statistic is easier done with the procedures discussed in Section 5.1 than with the procedure obtained in this section. In the latter, the calculation of Q_3 is achieved by first calculating $\tilde{\boldsymbol{\beta}}$ according to (5.79), the estimate subject to the constraints under the null hypothesis, and then substituting the $\tilde{\boldsymbol{\beta}}$ as obtained in (5.82c). In either case, the calculation of Q_2, a measure of variability of the observations around the unconstrained regression hyperplane, is unaltered. Notice further that if the linear restrictions under the null hypothesis are complex in nature, the calculation of Q_3 depends crucially on correct specification of the $\mathbf{R}$ matrix and the $\mathbf{r}$ vector, but then it is an easy matter to substitute these expressions into (5.79) and, in turn, into (5.82c) to find the desired statistic. In the cases where the test involves two or more samples, the specification of $\mathbf{R}$ requires the combining of the samples into one. Generally, the use of computers is advisable here, since regression analysis programs incorporating linear restrictions on the coefficients are usually available.

CHAPTER 6

GENERALIZED LEAST SQUARES

In the preceding chapter, we were primarily concerned with the hypotheses regarding the regression coefficients in the model $\mathbf{y} = \mathbf{X}\boldsymbol{\beta} + \mathbf{u}$. However, questions about the coefficients $\boldsymbol{\beta}$ are not the only ones that can be asked in regard to the entire model. The reader will recall that we made certain assumptions about the observation matrix $\mathbf{X}$ and about the distribution of the disturbances $\mathbf{u}$. The main objectives of this and the next chapters are to probe into the adequacy of these assumptions when the model is applied to the analysis of economic behavioral relations and to the relations arising in other fields. It is frequently the case that the basic relationship under analysis does not conform to the assumptions of the Fixed Models A and B and that, when this occurs, the inferences made with the statistics obtained through the LS methods thus far discussed are not always valid. For convenience in later reference, we shall call the LS procedure discussed up to this point ordinary least squares (OLS) or, alternatively, direct least squares (DLS).

The basic construct of the generalized linear regression model is significant not only as a generalization of the classical linear regression model but also as a framework within which the later discussions of two-stage and three-stage least-squares procedures can be conducted.

6.1. The Generalized Linear Regression Model

The generalized linear regression model due to Aitken (1935) can be given as follows. Instead of $\mathbf{y} = \mathbf{X}\boldsymbol{\beta} + \mathbf{u}$, assumptions about which are given in (4.32) or (4.33), we now work with the model

$$\mathbf{y} = \mathbf{X}\boldsymbol{\beta} + \mathbf{u} \tag{6.1}$$

where we assume that

$$E\mathbf{u} = \mathbf{0} \tag{6.2a}$$

$$E\mathbf{u}\mathbf{u}' = \sigma^2\boldsymbol{\Omega} \tag{6.2b}$$

with $\mathbf{\Omega}$ being a positive definite matrix

$$\mathbf{X} \text{ is a nonstochastic observation matrix, say, of dimensionality } n \times (k+1) \tag{6.2c}$$

and

$$\mathbf{X} \text{ is of full rank} \tag{6.2d}$$

We notice that these assumptions are identical to the ones in (4.32) except for the difference that the covariance matrix of $\mathbf{u}$ now is $\sigma^2\mathbf{\Omega}$, whereas earlier it was $\sigma^2\mathbf{I}$. That is, the earlier assumption of a scalar matrix* for the covariance of $\mathbf{u}$ is now replaced by the assumption that

$$\sigma^2\mathbf{\Omega} = \sigma^2 \begin{pmatrix} w_{11} & w_{12} & \cdots & w_{1n} \\ w_{21} & w_{22} & \cdots & \cdot \\ \cdot & \cdot & \cdot & \cdot \\ \cdot & \cdot & \cdot & \cdot \\ \cdot & \cdot & \cdot & \cdot \\ w_{n1} & w_{n2} & \cdots & w_{n2} \end{pmatrix} \tag{6.3a}$$

which says that $\mathbf{\Omega}$ is a full matrix and that $\sigma^2\mathbf{I}$ is only a special case of $\sigma^2\mathbf{\Omega}$. In Aitken's original work it was assumed that $E\mathbf{uu}' = \mathbf{\Lambda} = \sigma^2\mathbf{\Omega}$ where $\mathbf{\Lambda}$ is known only up to a scalar involving σ^2 which is unknown. That is, the form of $\mathbf{\Omega}$ can be known but not, so to speak, its size. In what follows we treat only the cases where the scale of $\mathbf{\Omega}$ can be fixed. [For a treatment of the generalized least squares at a more general level, see Theil (1970, Chapter 6).]

One such case is heteroskedasticity in the disturbances so that the covariance of $\mathbf{u}$ would be

$$E\mathbf{uu}' = \sigma^2 \begin{pmatrix} k_1 & 0 & \cdots & 0 \\ 0 & k_2 & \cdots & 0 \\ \cdot & \cdot & \cdot & \cdot \\ \cdot & \cdot & \cdot & \cdot \\ \cdot & \cdot & \cdot & \cdot \\ 0 & 0 & \cdots & k_n \end{pmatrix} \tag{6.3b}$$

Another example would be the existence of correlation among the successive u_i's (of which we have more to say later in this chapter) so that we could have

$$E\mathbf{uu}' = \frac{\sigma_\epsilon^2}{1-\rho^2} \begin{pmatrix} 1 & \rho & \rho^2 & \cdots & \rho^{n-1} \\ \rho & 1 & \rho & \cdots & \rho^{n-2} \\ \cdot & \cdot & \cdot & \cdot & \cdot \\ \cdot & \cdot & \cdot & \cdot & \cdot \\ \cdot & \cdot & \cdot & \cdot & \cdot \\ \rho^{n-1} & \rho^{n-2} & \rho^{n-3} & \cdots & 1 \end{pmatrix} \tag{6.3c}$$

* A scalar matrix is a diagonal matrix with the diagonal elements all identical to a scalar or a real number.

Here, σ_ϵ^2 is the variance of the error term in the first-order auto regressive scheme

$$u_i = \rho u_{i-1} + \epsilon_i \tag{6.3d}$$

These cases are sufficient to indicate that we now are dealing with a general case (6.3a), and here the OLS procedures for inference are appropriate only with some modification. We now elaborate on this last point.

6.2. Aitken's Generalized Least Squares (GLS)

To motivate the discussion of generalized least squares, let us observe briefly what will happen when OLS is applied. Suppose that the researcher is not aware that the regression model he is analyzing is (6.1) and that he uses OLS for estimation. He then obtains the estimators for $\boldsymbol{\beta}$ and the covariance of the estimator of $\boldsymbol{\beta}$

$$\hat{\boldsymbol{\beta}} = (\mathbf{X}'\mathbf{X})^{-1}\mathbf{X}'\mathbf{y} \tag{6.5a}$$

and

$$\text{Var}\,(\hat{\boldsymbol{\beta}}) = \sigma^2(\mathbf{X}'\mathbf{X})^{-1} \tag{6.5b}$$

respectively. Now consider these quantities in the light of the assumptions in (6.2a to d). We see that

$$\begin{aligned} \hat{\boldsymbol{\beta}} &= (\mathbf{X}'\mathbf{X})^{-1}\mathbf{X}'(\mathbf{X}\boldsymbol{\beta} + \mathbf{u}) \\ &= \boldsymbol{\beta} + (\mathbf{X}'\mathbf{X})^{-1}\mathbf{X}'\mathbf{u} \end{aligned} \tag{6.5c}$$

and that, because of (6.2a) and (6.2c), $\hat{\boldsymbol{\beta}}$ in (6.5a) is unbiased for $\boldsymbol{\beta}$. However, from (6.5c),

$$\hat{\boldsymbol{\beta}} - \boldsymbol{\beta} = (\mathbf{X}'\mathbf{X})^{-1}\mathbf{X}'\mathbf{u}$$

and

$$\begin{aligned} E(\hat{\boldsymbol{\beta}} - \boldsymbol{\beta})(\hat{\boldsymbol{\beta}} - \boldsymbol{\beta})' &= E[(\mathbf{X}'\mathbf{X})^{-1}\mathbf{X}'\mathbf{u}\mathbf{u}'\mathbf{X}(\mathbf{X}'\mathbf{X})^{-1}] \\ &= \sigma^2(\mathbf{X}'\mathbf{X})^{-1}(\mathbf{X}'\boldsymbol{\Omega}\mathbf{X})(\mathbf{X}'\mathbf{X})^{-1} \end{aligned} \tag{6.5d}$$

This last expression differs from (6.5b) unless $\boldsymbol{\Omega} = \mathbf{I}$, and illustrates why the OLS methods may not be always appropriate in analyzing the generalized linear regression model. A way out of this problem is given by Aitken's generalized least-squares method. Essentially this involves a transformation of the model (6.1) and then an application of OLS.

Suppose that we premultiply the model in (6.1) by an $n \times n$ transformation matrix $\mathbf{T}$ which is nonsingular. We then have

$$\mathbf{T}\mathbf{y} = \mathbf{T}\mathbf{X}\boldsymbol{\beta} + \mathbf{T}\mathbf{u} \tag{6.6}$$

Suppose further that we choose $\mathbf{T}$ to be such that

$$\mathbf{T}'\mathbf{T} = \boldsymbol{\Omega}^{-1} \tag{6.6a}$$

then*

$$E\mathbf{Tuu'T'} = \sigma^2\mathbf{T\Omega T'} \qquad (6.6b)$$
$$= \sigma^2\mathbf{I}$$

Thus, if we let the transformed variables in (6.6) be new variables, say,

$$\mathbf{Ty} = \mathbf{z} \qquad \mathbf{TX} = \mathbf{W} \qquad (6.6c)$$

and

$$\mathbf{Tu} = \mathbf{v}$$

we have the usual classical linear regression model of the variety for which OLS methods are suitable, since $E\mathbf{Tuu'T'} = E\mathbf{vv'} = \sigma^2\mathbf{I}$. That is, the model

$$\mathbf{z} = \mathbf{W}\boldsymbol{\beta} + \mathbf{v} \qquad (6.7)$$

now possesses the properties of the classical linear regression model (4.5).

By further extending the preceding argument, we can derive an estimator of $\boldsymbol{\beta}$ in (6.1) by the OLS. Let the OLS estimator for $\boldsymbol{\beta}$ in (6.7) be $\mathbf{b}^*$, then noticing (6.6a to c), we have

$$\mathbf{b}^* = (\mathbf{W'W})^{-1}\mathbf{W'z}$$
$$= (\mathbf{X'T'TX'})^{-1}\mathbf{X'T'Ty}$$
$$= (\mathbf{X'\Omega^{-1}X'})^{-1}\mathbf{X'\Omega^{-1}y} \qquad (6.8)$$

and

$$E(\mathbf{b}^* - \boldsymbol{\beta})(\mathbf{b}^* - \boldsymbol{\beta})' = \sigma^2(\mathbf{W'W})^{-1}$$
$$= \sigma^2(\mathbf{X'\Omega^{-1}X'})^{-1} \qquad (6.9)$$

We can show that the estimator $\mathbf{b}^*$ is BLUE for $\boldsymbol{\beta}$ with the covariance matrix as given.† This estimation procedure is referred to as Aitken's generalized least squares and, as is clear from the generality of the covariance of $\mathbf{u}$, it is a generalization of the Gauss-Markov theorem on least squares.

Another parameter in the model (6.17) is σ^2, for which an unbiased estimator is available. Let the GLS estimated residuals be denoted by $\tilde{\mathbf{u}}$, then

$$\tilde{\mathbf{u}} = \mathbf{y} - \mathbf{Xb}^*$$
$$= \mathbf{X}\boldsymbol{\beta} + \mathbf{u} - \mathbf{X}(\mathbf{X'\Omega^{-1}X})^{-1}\mathbf{X'\Omega^{-1}y}$$
$$= \mathbf{X}\boldsymbol{\beta} + \mathbf{u} - \mathbf{X}\boldsymbol{\beta} - \mathbf{X}(\mathbf{X'\Omega^{-1}X})^{-1}\mathbf{X'\Omega^{-1}u}$$
$$= [\mathbf{I} - \mathbf{X}(\mathbf{X'\Omega^{-1}X})^{-1}\mathbf{X'\Omega^{-1}}]\mathbf{u}$$
$$= \mathbf{Mu} \qquad (6.10)$$

* A theorem in matrix algebra says that if $\mathbf{\Omega}$ is positive definite, then there exists a matrix $\mathbf{T}$ so that $\mathbf{T\Omega T'} = \mathbf{I}$ and $\mathbf{T'T} = \mathbf{\Omega}^{-1}$. Students unfamiliar with this theorem should read elsewhere to convince themselves of the result.

† See, for example, Goldberger (1964, pp. 232–233).

where we define

$$\mathbf{M} = \mathbf{I} - \mathbf{X}(\mathbf{X}'\boldsymbol{\Omega}^{-1}\mathbf{X})^{-1}\mathbf{X}'\boldsymbol{\Omega}^{-1} \tag{6.11}$$

Noticing from (6.6c) that $\mathbf{v} = \mathbf{Tu}$, we form the sum of squares of the GLS estimated residuals

$$\begin{aligned}\hat{\mathbf{v}}'\hat{\mathbf{v}} &= \tilde{\mathbf{u}}'\boldsymbol{\Omega}^{-1}\tilde{\mathbf{u}} \\ &= \mathbf{u}'\mathbf{M}'\boldsymbol{\Omega}^{-1}\mathbf{M}\mathbf{u} \\ &= \mathbf{u}'\boldsymbol{\Omega}^{-1}\mathbf{M}\mathbf{u}\end{aligned} \tag{6.12}$$

(direct calculation shows that $\mathbf{M}'\boldsymbol{\Omega}^{-1}\mathbf{M} = \boldsymbol{\Omega}^{-1}\mathbf{M}$) and by taking the expectation

$$\begin{aligned}E\tilde{\mathbf{u}}'\boldsymbol{\Omega}^{-1}\tilde{\mathbf{u}} &= E\mathbf{u}'\boldsymbol{\Omega}^{-1}\mathbf{M}\mathbf{u} \\ &= E \operatorname{tr} (\mathbf{u}'\boldsymbol{\Omega}^{-1}\mathbf{M}\mathbf{u}) \\ &= E \operatorname{tr} (\mathbf{M}\mathbf{u}\mathbf{u}'\boldsymbol{\Omega}^{-1}) \\ &= \operatorname{tr} E(\mathbf{M}\mathbf{u}\mathbf{u}'\boldsymbol{\Omega}^{-1}) \\ &= \sigma^2 \operatorname{tr} (\mathbf{M}) \\ &= \sigma^2(n - k - 1)\end{aligned} \tag{6.13}$$

It follows that the unbiased estimator for σ^2 is

$$\tilde{s}^2 = \frac{\tilde{\mathbf{u}}'\boldsymbol{\Omega}^{-1}\tilde{\mathbf{u}}}{n - k - 1} \tag{6.14}$$

so that the unbiased estimator for $E(\mathbf{b}^* - \boldsymbol{\beta})(\mathbf{b}^* - \boldsymbol{\beta})'$ is

$$\widehat{\mathbf{Var}}\,(\mathbf{b}^*) = \tilde{s}^2(\mathbf{X}'\boldsymbol{\Omega}^{-1}\mathbf{X})^{-1} \tag{6.15}$$

6.3. Further Estimation Problems

One fundamental assumption in the discussions of the preceding section is that the matrix $\boldsymbol{\Omega}$ can be known. If the elements of the $\boldsymbol{\Omega}$ matrix cannot be known, under general conditions, they cannot be estimated from the sample of size n, since $\boldsymbol{\Omega}$ is $n \times n$ and there are altogether $n(n + 1)/2$ elements in $\boldsymbol{\Omega}$. However, it is possible to obtain a consistent estimator of $\boldsymbol{\Omega}$, say $\hat{\boldsymbol{\Omega}}$, by applying restrictions to $\boldsymbol{\Omega}$. One way in which such restrictions can arise is to assume existence of an autoregressive scheme for the disturbances such as in (6.3d) and to find from the data a consistent estimate of ρ, the population first-order autocorrelation coefficient. The estimate can then be used to form $\hat{\boldsymbol{\Omega}}$ which, in turn, can be substituted for $\boldsymbol{\Omega}$ in (6.8), (6.9), and (6.14) to yield consistent estimators for $\boldsymbol{\beta}$, **Var** $(\mathbf{b}^*)$, and σ^2.

Returning to the question of applying the OLS method to analyze what, in fact, is a generalized linear regression model, we notice first that either OLS or GLS will yield an unbiased estimator of $\boldsymbol{\beta}$. This is clear from the assumption that $E\mathbf{u} = \mathbf{0}$ so that in (6.8) $E\mathbf{b}^* = \boldsymbol{\beta}$.

However, in estimating the covariance of $\mathbf{b}^*$, the OLS estimator $\mathbf{Var}\,(\mathbf{b}) = s^2(\mathbf{X}'\mathbf{X})^{-1}$ is biased in two ways. First $Es^2 \neq \sigma^2$, and second $(\mathbf{X}'\mathbf{X})^{-1} \neq (\mathbf{X}'\mathbf{X})^{-1}(\mathbf{X}'\boldsymbol{\Omega}\mathbf{X})(\mathbf{X}'\mathbf{X})^{-1}$. This is elaborated on. First, consider

$$\begin{aligned} E\hat{\mathbf{u}}'\hat{\mathbf{u}} &= E\mathbf{u}'\mathbf{M}_1{}'\mathbf{M}_1\mathbf{u} \\ &= \text{tr}\,(E\mathbf{u}'\mathbf{M}_1{}'\mathbf{M}_1\mathbf{u}) \\ &= \text{tr}\,(\mathbf{M}_1 E\mathbf{u}\mathbf{u}') \end{aligned} \tag{6.16}$$

where

$$\mathbf{M}_1 = \mathbf{I} - \mathbf{X}(\mathbf{X}'\mathbf{X})^{-1}\mathbf{X}' \tag{6.16a}$$

but now $E\mathbf{u}\mathbf{u}' = \sigma^2\boldsymbol{\Omega}$ so that

$$\begin{aligned} E\mathbf{u}'\mathbf{u} &= \sigma^2\,\text{tr}\,(\mathbf{M}_1\boldsymbol{\Omega}) \\ &= \sigma^2\,\text{tr}\,[\boldsymbol{\Omega} - \mathbf{X}(\mathbf{X}'\mathbf{X})^{-1}\mathbf{X}'\boldsymbol{\Omega}] \end{aligned} \tag{6.17}$$

Recall that the OLS estimator for σ^2 is

$$s^2 = \frac{\hat{\mathbf{u}}'\hat{\mathbf{u}}}{n - k - 1}$$

so that

$$\begin{aligned} Es^2 &= \frac{1}{n - k - 1}\,E\hat{\mathbf{u}}'\hat{\mathbf{u}} \\ &= \frac{\sigma^2}{n - k - 1}\,[\text{tr}\,\boldsymbol{\Omega} - \text{tr}\,(\mathbf{X}'\mathbf{X})^{-1}\mathbf{X}'\boldsymbol{\Omega}\mathbf{X})] \end{aligned} \tag{6.18}$$

Thus to the extent that the last bracketed expression differs from $(n - k - 1)$, bias will exist in the OLS estimate of σ^2.

Next, consider the last expression in (6.5d) which is the OLS estimator of $\mathbf{Var}\,(\hat{\boldsymbol{\beta}})$ but which allows for the fact that now $E\mathbf{u}\mathbf{u}' = \sigma^2\boldsymbol{\Omega}$. That is

$$\begin{aligned} E(\hat{\boldsymbol{\beta}} - \boldsymbol{\beta})(\hat{\boldsymbol{\beta}} - \boldsymbol{\beta})' &= E(\mathbf{b} - \boldsymbol{\beta})(\mathbf{b} - \boldsymbol{\beta})' \\ &= \sigma^2(\mathbf{X}'\mathbf{X})^{-1}(\mathbf{X}'\boldsymbol{\Omega}\mathbf{X})(\mathbf{X}'\mathbf{X})^{-1} \end{aligned} \tag{6.19}$$

Thus, unless $(\mathbf{X}'\boldsymbol{\Omega}\mathbf{X})(\mathbf{X}'\mathbf{X})^{-1} = \mathbf{I}$, the OLS estimator $s^2(\mathbf{X}'\mathbf{X})^{-1}$ will be biased by the amount $[(\mathbf{X}'\boldsymbol{\Omega}\mathbf{X})(\mathbf{X}'\mathbf{X})^{-1} - \mathbf{I}]$. Notice further that in both (6.18) and (6.19) the direction of bias cannot be told until one knows the nature of the matrices $(\mathbf{X}'\mathbf{X})^{-1}$ and $\boldsymbol{\Omega}$.

Thus far in this chapter we have been concerned with some general problems involved in the estimation of a generalized linear regression model. Our objective in the next few sections is to examine the practical considerations of the general results and some of the specific cases of hypothesis testing about the disturbances.

6.4. Analysis of Residuals

Since usually the $\mathbf{\Omega}$ matrix is not known, an effort must be made to estimate $\mathbf{\Omega}$ or, where estimates cannot be obtained, to find the approximate pattern of the distribution of the disturbances. These efforts necessarily rely on the OLS estimated residuals, and the procedures often are not analytical, although they may be very informative.

In practice, when the researcher suspects that a departure from the assumption of spherical disturbances, namely, $E\mathbf{uu}' = \sigma^2\mathbf{I}$, exists in the model he works with, he, nonetheless, may use the OLS method to estimate his model. Then take the OLS estimated residuals and plot them against a number of variables, such as time, the independent variables, the estimates of the values of the dependent variable, and sometimes the variables not previously included in the model. Such an analysis can point to (1) the existence of autocorrelation in the disturbances, (2) the presence of heteroskedasticity in the disturbances, (3) the use of wrong functional form, and (4) the possible exclusion of significant variables. We exhibit a few hypothetical situations that illustrate the last statement.

A plot of OLS residuals against time, if one is dealing with time-series data, may be as shown in Figure 6.1. Notice the cyclic pattern of the residuals as time (t) increases. That is, the successive residuals tend to show a pattern

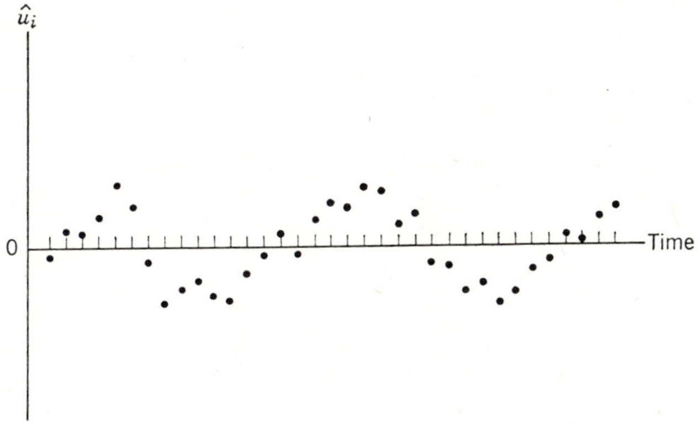

Figure 6.1

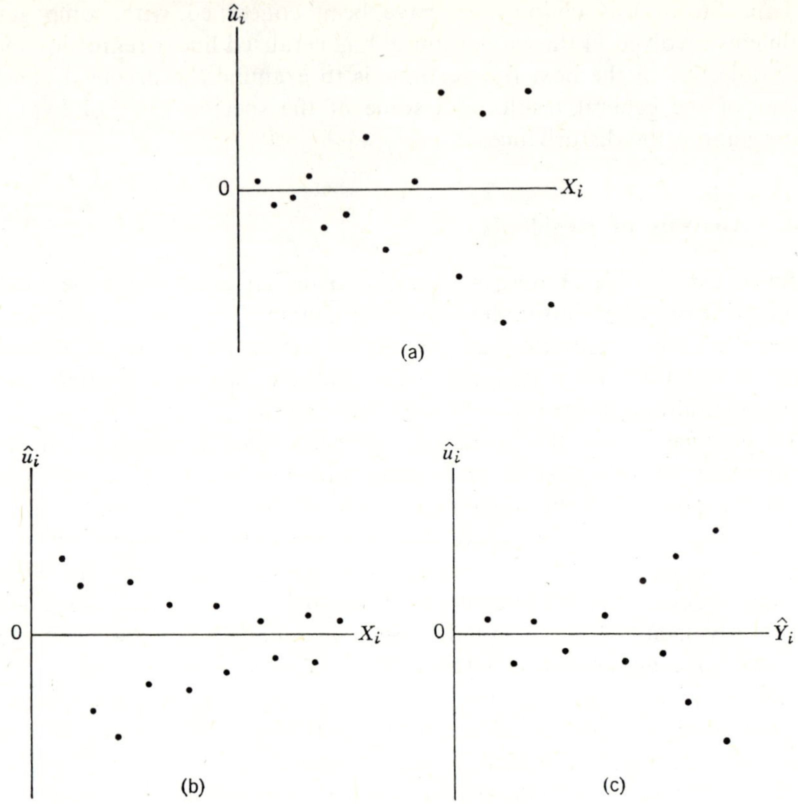

Figure 6.2*a*, *b*, and *c*

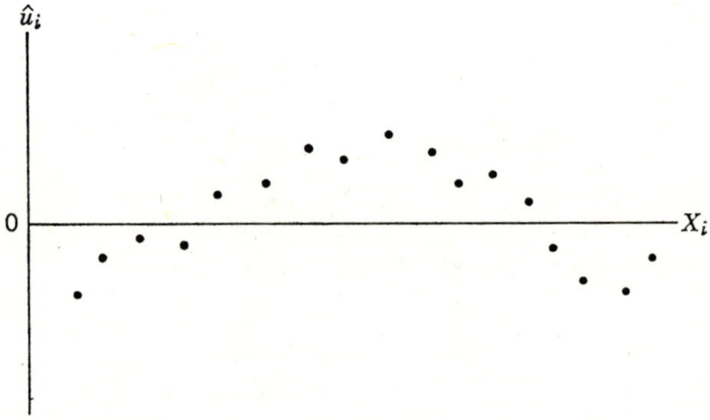

Figure 6.3

where the size of a residual this period depends on the size of the one preceding. It is also possible that a similar pattern is found for the residuals when they are plotted against one independent variable. Graphically this is a situation where the t in Figure 6.1 is replaced by the ith independent variable, X_i.

The case of heteroskedasticity was considered briefly in Section 3.3.2. The typical patterns of heterogeneous disturbances are shown in Figure 6.2*a* and *b* where the residuals are plotted against a given independent variable. Heterogeneous variance may also exist as the estimated value of the dependent variable increases (Figure 6.2*c*).

Sometimes a researcher may proceed with the OLS analysis without plotting the data beforehand and, on plotting the residuals, finds something like that shown in Figure 6.3. Apparently, as X_i increases in value, the dependent variable first increases and then decreases, suggesting that a quadratic form, say,

$$y = \beta_0 + \beta X + \beta_2 X^2 + u$$

may be appropriate.

Notice that the sort of residual analysis just discussed are all specific to the framework chosen and are only suggestive of the direction in which the researcher might move to achieve correct inference.

For a more definitive approach to residual analysis, see the next section Anscombe (1961) and, for further exposition, Draper and Smith (1966).

We now examine a more systematic treatment of the cases that represent a departure from spherical disturbances. In particular, we consider the testing and the estimation of first-order autocorrelation in the disturbances.

6.5. Autocorrelation in the Disturbances

In the usual linear regression model

$$\mathbf{y} = \mathbf{X}\boldsymbol{\beta} + \mathbf{u} \tag{6.22}$$

the successive disturbances may be correlated for various reasons, thereby violating the assumption of spherical disturbances which is very important in the application of the OLS method. Autocorrelation of first and higher orders in the disturbances can arise because of (1) the faulty functional form assumed for the model, (2) the omission of variables in the analysis of a relation either as a result of the lack of data or of mere ignorance, (3) the tendency for observation errors in economic variables to persist, (4) the estimation of missing data by either averaging or extrapolating, and (5) the lagged effect of temporary shocks distributed over a number of time periods.

6.5.1. The OLS Estimators

Given that first-order autocorrelation in the disturbances is a special case of nonspherical disturbances, it can be shown that, for the model (6.1) with the assumptions (6.2a) and (6.3c), the OLS estimator of the coefficient vector $\boldsymbol{\beta}$ is unbiased.

It is instructive to determine what the covariance matrix of $\mathbf{u}$ is when the latter is a vector of disturbances that have first-order autocorrelation. From the assumption of the existence of the first-order autoregressive scheme in the disturbances, say, in $\mathbf{y} = \mathbf{X}\boldsymbol{\beta} + \mathbf{u}$,

$$u_i = \rho u_{i-1} + \epsilon_i \tag{6.23}$$

where we assume that $E\epsilon_i = 0$, $E\epsilon_i^2 = \sigma_\varepsilon^2$ for all i (that is, this process is assumed to go on always, over time or space), and that $Eu_iu_j = 0$ for $i \neq j$, through successive substitution, we can write

$$\begin{aligned} u_i &= \rho u_{i-1} + \epsilon_i \\ &= \rho(\rho u_{i-2} + \epsilon_{i-1}) + \epsilon_i \\ &= \cdots \\ &= \epsilon_i + \rho\epsilon_{i-1} + \rho^2\epsilon_{i-2} + \cdots \\ &= \sum_{\alpha=0}^{\infty} \rho^\alpha \epsilon_{t-\alpha} \end{aligned} \tag{6.24}$$

It then follows that, by virtue of the assumption $E\epsilon_i = 0$,

$$Eu_i = 0 \tag{6.24a}$$

$$Eu_i^2 = E\epsilon_i^2 + \rho^2 E\epsilon_{i-1}^2 + \rho^4 E\epsilon_{i-2}^2 + \cdots \tag{6.24b}$$

because of the lack of autocorrelation assumed for the ϵ_i's. Therefore, we have

$$\begin{aligned} Eu_i^2 = \sigma_u^2 &= (1 + \rho^2 + \rho^4 + \cdots)\sigma_\epsilon^2 \\ &= \frac{\sigma_\epsilon^2}{1 - \rho^2} \qquad \text{for all } i \end{aligned} \tag{6.25}$$

Furthermore

$$\begin{aligned} Eu_iu_{i-1} &= E(\epsilon_i + \rho\epsilon_{i-1} + \rho^2\epsilon_{i-2} + \cdots)(\epsilon_{i-1} + \rho\epsilon_{i-2} + \rho^2\epsilon_{i-3} + \cdots) \\ &= E[\epsilon_i + \rho(\epsilon_{i-1} + \rho\epsilon_{i-2} + \cdots)](\epsilon_{i-1} + \rho\epsilon_{i-2} + \cdots) \\ &= \rho E(\epsilon_{i-1} + \rho\epsilon_{i-2} + \cdots)^2 \\ &= \rho\sigma_u^{\,2} \end{aligned}$$

In a similar fashion, we deduce that

$$Eu_iu_{i-2} = \rho^2\sigma_u^{\,2}$$

and that, in general, for any $\alpha \neq 0$,

$$Eu_iu_{i-\alpha} = \rho^\alpha\sigma_u^{\,2} \tag{6.26}$$

Thus we observe that in the first-order autogressive case

$$\mathbf{Euu}' = \frac{\sigma_\epsilon^{\,2}}{1-\rho^2}\begin{bmatrix} 1 & \rho & \rho^2 & \cdots & \rho^{n-1} \\ \rho & 1 & \rho & \cdots & \rho^{n-2} \\ \rho^2 & \rho & 1 & \cdots & \rho^{n-3} \\ \cdot & \cdot & \cdot & \cdot & \cdot \\ \cdot & \cdot & \cdot & \cdot & \cdot \\ \cdot & \cdot & \cdot & \cdot & \cdot \\ \rho^{n-1} & \rho^{n-2} & \rho^{n-3} & \cdots & 1 \end{bmatrix} \tag{6.27}$$

as given earlier in (6.3c).

As discussed in Section 6.3, the covariance matrix of the usual OLS estimator of $\boldsymbol{\beta}$ when, in fact, $E\mathbf{uu}' = \sigma^2\mathbf{\Omega}$ is $\sigma^2(\mathbf{X}'\mathbf{X})^{-1}(\mathbf{X}'\mathbf{\Omega}\mathbf{X})(\mathbf{X}'\mathbf{X})^{-1}$ Suppose now that a researcher applies the OLS method to a linear regression model that has first-order autocorrelated disturbances so that he, in effect, uses the covariance of the coefficient vector estimator $\sigma^2(\mathbf{X}'\mathbf{X})^{-1}$. From (6.19) we notice that, unless $\mathbf{\Omega} = \mathbf{I}$, we have

$$(\mathbf{X}'\mathbf{X})^{-1} \neq (\mathbf{X}'\mathbf{X})^{-1}(\mathbf{X}'\mathbf{\Omega}\mathbf{X})(\mathbf{X}'\mathbf{X})^{-1} \tag{6.28}$$

Since many economic time-series are positively serially correlated, it is of interest to illustrate with a small sample the point that the usual OLS formula for the sampling variance of a regression coefficient tends to underestimate the true variance.

Consider the model

$$y_i = \beta x_i + u_i \qquad i = 1, 2, 3, 4$$

with

$$u_i = \rho u_{i-1} + \epsilon_i$$

where $E\epsilon_i = 0$, $E\epsilon_i^{\,2} = \sigma_\epsilon^{\,2}$. Then the variance of the usual OLS estimator of β is

$$\mathbf{Var}\,(\mathbf{b}) = \sigma^2(\mathbf{X}'\mathbf{X})^{-1}(\mathbf{X}'\boldsymbol{\Omega}\mathbf{X})(\mathbf{X}'\mathbf{X})^{-1}$$

$$= \sigma^2\left(\sum_{i=1}^{4} x_i^{\,2}\right)^{-1}(x_1\ x_2\ x_3\ x_4)\begin{bmatrix} 1 & \rho & \rho^2 & \rho^3 \\ \rho & 1 & \rho & \rho^2 \\ \rho^2 & \rho & 1 & \rho \\ \rho^3 & \rho^2 & \rho & 1 \end{bmatrix}\begin{bmatrix} x_1 \\ x_2 \\ x_3 \\ x_4 \end{bmatrix}\left(\sum_{i=1}^{4} x_i^{\,2}\right)^{-1}$$

$$= \frac{\sigma^2}{\left(\sum_{i=1}^{4} x_i^{\,2}\right)^2}(x_1\ x_2\ x_3\ x_4)\begin{bmatrix} x_1 + \rho x_2 + \rho^2 x_3 + \rho^3 x_4 \\ \rho x_1 + x_2 + \rho x_3 + \rho^2 x_4 \\ \rho^2 x_1 + \rho x_2 + x_3 + \rho x_4 \\ \rho^3 x_1 + \rho^2 x_2 + \rho x_3 + x_4 \end{bmatrix}$$

$$= \frac{\sigma^2}{\left(\sum_{i=1}^{4} x_i^{\,2}\right)^2}\left(\sum_{i=1}^{4} x_i^{\,2} + 2\rho\sum_{i=1}^{3} x_i x_{i+1} + 2\rho\sum_{i=1}^{2} x_i x_{i+2} + 2\rho^2 x_1 x_4\right)$$

$$= \frac{\sigma^2}{\left(\sum_{i=1}^{4} x_i^{\,2}\right)}\left[1 + \left(\sum_{i=1}^{4} x_i^{\,2}\right)^{-1}\left(2\rho\sum_{i=1}^{3} x_i x_{i+1} + 2\rho^2\sum_{i=1}^{2} x_i x_{i+2} + 2\rho^3 x_1 x_4\right)\right] \tag{6.29}$$

The first factor of the last expression is the variance of the OLS estimator of β in the absence of first-order autocorrelation in u_i's, and the expression in the square brackets measures the extent of the bias in the OLS estimator because of the autocorrelation in u_i's. An examination of the second term in the brackets indicates that, if $\rho > 0$ and if the independent variable x is positively serially correlated, the term as a whole is positive. Furthermore, the value of this same term in the brackets depends critically on both ρ and whatever lagged correlation that may exist in x_i's. If x_i's are pure noise with mean zero, the entire expression reduces to $\sigma^2/\sum x_i^{\,2}$.

The underestimation of **Var (b)** by the usual OLS can be fairly serious and, consequently, the inferences drawn on the basis of the OLS estimator would be rather poor. This suggests that, when the disturbances are autocorrelated, the usual OLS statistics, such as t and F, may give rise to invalid tests of hypotheses.

A problem also arises in using the OLS estimate of an equation for prediction when, in fact, the GLS model underlies the behavior. That is, the prediction error based on the true OLS estimator (where allowance is given for the autocorrelation) may be quite large. But the problem is not as serious as it seems, since any knowledge about the size of ρ can be exploited in prediction. Actually, if a good estimate of ρ can be obtained, this estimate can be inserted into expressions such as (6.28) and (6.29) for correction. For example, refer to Section 6.5.4 where we apply the GLS model explicitly to the autoregressive model and where we also perform the autoregressive transformation of the type (6.6) except for the first observation.

6.5.2. *Testing for First Order Autocorrelation*

In the preceding subsection, we discussed some of the inferential problems that may arise in connection with the use of the OLS method when, in fact, there is a first-order autocorrelation in the disturbances. In this subsection, we consider how we might statistically test for the existence of a first-order autocorrelation in the disturbances.

First, we consider the well-known von Neumann ratio. If $u_1, u_2, \ldots, u_n$ are successive random drawings from a normal population, the mean-square successive difference to the variance ratio

$$D = \frac{\sum_{i=2}^{n} (u_i - u_{i-1})^2}{\sum_{i=1}^{n} u_i^2} \frac{n}{n-1} \tag{6.35}$$

will approach a normal distribution, as n increases, with the following mean and variance

$$ED = \frac{2n}{n-1} \tag{6.35a}$$

$$\text{Var}\,(D) = \frac{4n^2(n-2)}{(n-1)^3(n+1)} \tag{6.35b}$$

The ratio D is known as the von Neumann ratio, often denoted by δ^2/s^2, and the significant points of D have been computed for small sample sizes. For the first discussions of the ratio, see J. von Neumann (1941, 1942) and, for derivation of small sample distributions, see B. I. Hart (1942).

In principle, this ratio is not useful for the kind of problems that we encounter. What we usually obtain in regression analysis is the set of the OLS estimated residuals, $\hat{u}_i$, since the random observations u_i are not known. And the use of D is appropriate only when u_i are available. A suitable test statistic when the OLS estimated residuals are used for computation is the Durbin-Watson statistic, d:

$$d = \frac{\sum_{i=2}^{u} (\hat{u}_i - \hat{u}_{i-1})^2}{\sum_{i=1}^{n} \hat{u}_i^2} \tag{6.36}$$

Primarily because of the fact that the exact distribution of d depends on the correlation structure of the regressors in any given problem, Durbin and

Watson derived significant points for the upper bound d_U and the lower bound d_L of d for certain sample sizes and numbers of independent variables.* For instance, under the null hypothesis that the first-order autocorrelation coefficient $\rho = 0$, the significant points for d_L and d_U at the 5 percent level with selected n and k (the number of independent variables) are shown in Table 6.1.†

Table 6.1 Significant Points of d_L and d_U (at the 5 Percent Level)

	$k = 2$		$k = 5$	
n	d_L	d_U	d_L	d_U
20	1.10	1.54	0.79	1.99
30	1.28	1.57	1.07	1.83
50	1.46	1.63	1.34	1.77
100	1.63	1.72	1.57	1.78

Source. See Appendix A.

To illustrate the test let us first contrast the usual sample correlation between the successive residuals $\hat{\rho}$ with the d statistic.

$$\hat{\rho} = \frac{\sum_{i=2}^{n} \hat{u}_i \hat{u}_{i-1}}{\sqrt{\sum_{i=1}^{n} \hat{u}_i^2} \sqrt{\sum_{i=2}^{n} \hat{u}_{i-1}^2}} \tag{6.37}$$

Now approximately, $\sum_{i=1}^{n} \hat{u}_i^2 = \sum_{i=2}^{n} \hat{u}_i^2 = \sum_{i=2}^{n} \hat{u}_{i-1}^2$, so that

$$d = \frac{2\sum_{i=2}^{n} \hat{u}_i^2 - 2\sum_{i=2}^{n} \hat{u}_i \hat{u}_{i-1}}{\sum_{i=1}^{n} \hat{u}_i^2} = 2 - 2\hat{\rho} = 2(1 - \hat{\rho}) \tag{6.36a}$$

It follows then that, when $\hat{\rho} = 0$, $d = 2$; as $\hat{\rho}$ increases from 0 to 1, d falls from 2 to 0; and as $\hat{\rho}$ decreases from 0 to -1, d rises from 2 to 4. These approximate results are useful in appreciating the test procedures that involve the d statistics since, for example, if positive autocorrelation exists ($\hat{\rho} > 0$), the calculated d may be close to 0. Thus, suppose that a researcher obtains

* See J. Durbin and G. S. Watson (1950, 1951).

† Table 4, Durbin and Watson (1950). Also under H_0: $\rho = 0$, the expectation of d is equal to 2.

a d statistic 1.02 on running a regression with two independent variables and a sample of size 50. Suppose further that he wishes to test the null hypothesis $\rho = 0$ against the alternative hypothesis that $\rho > 0$ at the 5 per cent level. From Table A-6, Appendix A, which contains the significant points for the given level of significance, he finds that for $n = 50$ and $k = 2$ the $d_L = 1.46$ is greater than the calculated $d = 1.02$. Therefore, he concludes that the null hypothesis is rejected. If the null hypothesis is to be tested against the alternative that $\rho < 0$ at the 5 per cent level, then the appropriate critical points are $d_U{}^* = 4 - d_U = 4 - 1.63 = 2.37$ and $d_L{}^* = 4 - d_L = 4 - 1.46 = 2.54$. In diagram form, for the one-tail tests at the 5 per cent level we have the critical values of d as shown in Figure 6.4. If the null hypothesis is tested against the alternative that $\rho \neq 0$, then a two-tail test is in order. Then, at the 5 per cent level, one would need to use the table of significant points at the 2.5 per cent level and to combine the one-tail tests each at the latter level of significance.

One shortcoming of the tests using the Durbin-Watson statistic d is that, if a test statistic falls between the upper and the lower bounds, the test is inconclusive. The cases of inconclusive tests can be frequently observed if the sample size is small and the number of independent variables large, relatively speaking. For instance, a d statistic of 1.02 in the interval $0.79 < d < 1.99$ for $n = 20$ and $k = 5$ at the 5 per cent level would mean an inconclusive test. In an effort to reduce the frequency of falling into "regions of ignorance," Theil and Nagar (1961) propose a test that is based on an approximate distribution of the Durbin-Watson d and applicable only in the situations where regressors follow smooth trends (the first and the second differences of the regressors are small in absolute value in comparison with

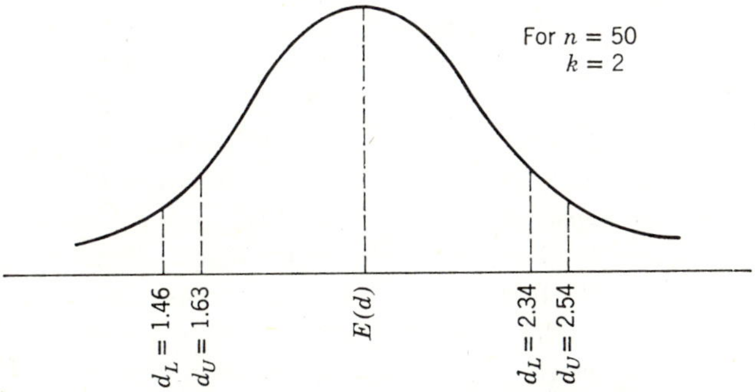

Figure 6.4

Table 6.2 Significance Points of the Von Neumann Ratio (at the 5 Per cent Level)

n	$k' = 3$	$k' = 6$
20	1.53	1.96
30	1.57	1.82
50	1.63	1.78
100	1.72	1.78

Source. See Table A-8, Appendix A.

the range of the corresponding variables). In Table 6.2, we reproduce a part of the table computed by Theil and Nagar for contrast with Table 6.1. For n larger than 100, we can use a normal approximation given in Theil and Nagar (1961, p. 803). Note that the number of coefficients k' in Table A-8 includes the constant term so that $k' = k + 1$, where k is the number of independent variables, and also is the k used in Tables A-6 and A-7

6.5.3. The *BLUS* Residuals

The major difficulty in the derivation of the test procedure regarding autocorrelation is that the OLS estimated residuals have a covariance matrix that is not scalar. Namely

$$E\hat{\mathbf{u}}'\hat{\mathbf{u}} = \sigma^2\mathbf{M}$$

where $\mathbf{M} = \mathbf{I} - \mathbf{X}(\mathbf{X}'\mathbf{X})^{-1}\mathbf{X}'$, as shown in earlier sections. This means that the $\hat{u}_i$ are not necessarily uncorrelated nor do they have homogeneous variance. This then prevents us from applying the classical tests, since these tests are based on the assumed distribution of u_i not on that of $\hat{u}_i$.

This difficulty can be avoided if we are willing to make some extra calculations and to work with $n - k - 1$ residuals instead of n residuals. The procedure, due to Theil (1968), is to obtain a residual vector, say $\mathbf{u}^*$, with the following properties: (1) $\mathbf{u}^*$ is a linear combination (L) of $\mathbf{y}$, (2) $\mathbf{u}^*$ is unbiased (U) in the sense that $E(u_i{}^* - \hat{u}_i) = 0$ for i chosen by the BLUS procedure and where $\hat{u}_i$ is an OLS estimated residual, (3) $\mathbf{u}^*$ has a scalar (S) covariance matrix of the form $\sigma^2\mathbf{I}$,† and (4) given the class defined to have the L, U, and S properties above, $\mathbf{u}^*$ is best (B) in a particular quadratic sense. For the actual steps required for calculation of the *BLUS* residuals the reader is referred to Theil (1965, 1968).

† Notice that $\mathbf{I}$ is of order $n - k - 1$, if the number of independent variables is $k + 1$.

With the assumption of the normality of **u** and, hence, of the normality of **u***, the elements of the latter can be used to calculate the von Neumann ratio to test for autocorrelation. From (6.35) the appropriate statistics is

$$D_{BLUS} = \frac{\sum_{i=2}^{n-k-1} (u_i^* - u_{i-1}^*)^2}{\sum_{i=1}^{n-k-1} u_i^*} \frac{n-k-1}{n-k-2} \tag{6.38}$$

and the significant points can be found in Table A-5, Appendix A.

6.5.4. *Estimating Autocorrelation Coefficient*

If the various tests establish that there is autocorrelation in the disturbances, what can be done about it? Some people suggest that one should look for a model for which the tests will suggest that there is no autocorrelation, since the existence of the latter necessarily imply that something is wrong with the specification of the model that gave rise to it.

In the absence of any better solution, the question turns to how one might estimate the first-order autocorrelation coefficient if such an autocorrelation is known or is assumed to exist. One very frequently used procedure is to estimate the ρ based on the OLS estimates of the disturbances, call them the OLS residuals. Thus, although it is assumed that

$$u_i = \rho u_{i-1} + \epsilon_i \qquad i = 1, 2, \ldots, n$$

the ρ may be estimated from

$$\hat{u}_i = \rho \hat{u}_{i-1} + \epsilon_i \qquad i = 2, 3, \ldots, n$$

by the usual OLS method. The estimator of ρ, say $\hat{\rho}$, so obtained is consistent for ρ as long as $\hat{u}_i$'s are consistent for u_i's. And by using ρ, we can construct a consistent estimator of $\mathbf{\Omega}$, say $\hat{\mathbf{\Omega}}$. This $\hat{\mathbf{\Omega}}$ can then be used to find the OLS coefficient vector estimate

$$\mathbf{b}^* = (\mathbf{X}'\hat{\mathbf{\Omega}}^{-1}\mathbf{X}')^{-1}\mathbf{X}'\hat{\mathbf{\Omega}}^{-1}\mathbf{y} \tag{6.39}$$

and related statistic. In this connection, it is useful to note the inverse of the covariance matrix $\mathbf{\Omega}$, which appears in (6.2b), (6.6b), and elsewhere, when the first-order autocorrelation exists. From (6.27), we let $\sigma^2 = \sigma_\epsilon^2/(1 - \rho^2)$. Then, in general, the first-order autocorrelated disturbances have a covariance matrix

$$E\mathbf{uu}' = \sigma^2\mathbf{\Omega} = \sigma^2 \begin{bmatrix} 1 & \rho & \cdots & \rho^{n-1} \\ \rho & 1 & \cdots & \rho^{n-2} \\ \rho^2 & \rho & \cdots & \rho^{n-3} \\ \cdot & \cdot & \cdot & \cdot \\ \cdot & \cdot & \cdot & \cdot \\ \cdot & \cdot & \cdot & \cdot \\ \rho^{n-1} & \rho^{n-2} & \cdots & 1 \end{bmatrix} \tag{6.40}$$

so that

$$\mathbf{\Omega}^{-1} = \begin{bmatrix} 1 & -\rho & 0 & \cdots & 0 & 0 \\ -\rho & 1+\rho^2 & -\rho & \cdots & 0 & 0 \\ 0 & -\rho & 1+\rho^2 & \cdots & 0 & 0 \\ \cdot & \cdot & \cdot & \cdot & \cdot & \cdot \\ \cdot & \cdot & \cdot & \cdot & \cdot & \cdot \\ \cdot & \cdot & \cdot & \cdot & \cdot & \cdot \\ 0 & 0 & 0 & \cdots & -\rho & 1 \end{bmatrix} \tag{6.41}$$

Now, if we choose a transformation matrix **T** in the sense discussed in Section 6.1, we notice that the $(n-1) \times n$ matrix

$$\mathbf{T} = \begin{bmatrix} -\rho & 1 & 0 & \cdots & 0 & 0 \\ 0 & -\rho & 1 & \cdots & 0 & 0 \\ \cdot & \cdot & \cdot & \cdot & \cdot & \cdot \\ \cdot & \cdot & \cdot & \cdot & \cdot & \cdot \\ \cdot & \cdot & \cdot & \cdot & \cdot & \cdot \\ 0 & 0 & 0 & \cdots & -\rho & 1 \end{bmatrix} \tag{6.42a}$$

yields the $n \times n$ matrix

$$\mathbf{T}'\mathbf{T} = \begin{bmatrix} \rho^2 & -\rho & 0 & \cdots & 0 \\ -\rho & 1+\rho^2 & -\rho & \cdots & 0 \\ \cdot & \cdot & \cdot & \cdot & \cdot \\ \cdot & \cdot & \cdot & \cdot & \cdot \\ \cdot & \cdot & \cdot & \cdot & \cdot \\ 0 & 0 & 0 & \cdots & 1 \end{bmatrix} \tag{6.42b}$$

which is identical with $\mathbf{\Omega}^{-1}$ in (6.41) except for the element of the first row and the first column. Thus, for practical purposes, we can apply transformation **T** to the observed variables according to (6.6). This particular transformation yields, for instance,

$$\mathbf{Ty} = \begin{bmatrix} Y_2 - \rho Y_1 \\ Y_3 - \rho Y_2 \\ \cdot \\ \cdot \\ \cdot \\ Y_n - \rho Y_{n-1} \end{bmatrix} \tag{6.43a}$$

and, similarly, for any independent variable x_i,

$$\mathbf{Tx}_i = \begin{bmatrix} X_{i2} - \rho X_{i1} \\ X_{i3} - \rho X_{i2} \\ \cdot \\ \cdot \\ \cdot \\ X_{in} - \rho X_{i,n-1} \end{bmatrix} \tag{6.43b}$$

The OLS method then can be applied to the transformed variables, and the OLS estimator of $\boldsymbol{\beta}$ (namely, the GLS estimator) will be BLUE as discussed before. Of course, very often, the true ρ is not known, and its estimate $\hat{\rho}$ may be used in (6.43a) and (6.43b) in place of ρ. Recall that in this case the estimator of $\boldsymbol{\beta}$, $\mathbf{b}^*$, by using $\hat{\rho}$ is no longer BLUE.

Referring to (6.43a) and (6.43b), we notice that if $\rho = 1$ the matrix $\mathbf{T}$ performs first-differencing of the variables. We often see empirical works where the first-differenced variables are used in regression analysis. In these cases, the analysts may know or may be assuming that ρ is close to unity; in principle, it is wrong to take first differences as appropriate transformation of the mere finding that the disturbances are positively autocorrelated.

Another method for estimating the autocorrelation coefficient ρ is a search procedure suggested by Hildreth and Lu (1960). For the model

$$\mathbf{y} = \mathbf{X}\boldsymbol{\beta} + \mathbf{u} \tag{6.44a}$$

with autocorrelation in the $\mathbf{u}$

$$\mathbf{u} = \rho\mathbf{u}^0 + \boldsymbol{\epsilon}$$

where $\mathbf{u}^0$ is the vector $\mathbf{u}$ lagged one time period, we can run the ordinary regression of

$$(\mathbf{y} - \rho\mathbf{y}^0) = (\mathbf{X} - \rho\mathbf{X}^0)\boldsymbol{\beta} + (\mathbf{u} - \rho\mathbf{u}^0) \tag{6.44b}$$

for any value of ρ. Here $\mathbf{y}^0$ and $\mathbf{X}^0$ are, respectively, the $\mathbf{y}$ and $\mathbf{X}$ lagged one time period: for instance, $\mathbf{y}' = (Y_2\ Y_3 \cdots Y_{n+1})$, $(\mathbf{y}^0)' = (Y_1\ Y_2 \cdots Y_n)$ and, similarly, $\mathbf{u}' = (u_2\ u_3 \cdots u_{n+1})$. Then allow ρ in (6.44b) to vary in the interval

$(-1, 1)$, calculate the usual OLS residual sum of squares for each ρ chosen, select the ρ that makes the residual sum of squares minimum, *and* designate the corresponding estimates of $\boldsymbol{\beta}$ as the final estimates. Hildreth and Lu show that these estimators for ρ and $\boldsymbol{\beta}$ are consistent if $\mathbf{u}$ is a normal vector.

For other methods of estimating ρ and for a related treatment of autocorrelated disturbances, in general, consult Johnston (1963, Chapter 7), Christ (1966, Chapter 9), and Zellner (1970, Chapter 4).

6.6. Heteroskedasticity

This long word means that the variance of the disturbances are not an identical constant. That is, in the GLS model,

$$E\mathbf{uu}' = \sigma^2\boldsymbol{\Omega} = \sigma^2\begin{bmatrix} k_1 & 0 & 0 & \cdots & 0 \\ 0 & k_2 & 0 & \cdots & 0 \\ 0 & 0 & k_3 & \cdots & 0 \\ \cdot & \cdot & \cdot & \cdot & \cdot \\ \cdot & \cdot & \cdot & \cdot & \cdot \\ \cdot & \cdot & \cdot & \cdot & \cdot \\ 0 & 0 & 0 & \cdots & k_n \end{bmatrix} \tag{6.45}$$

where k''s can vary with i in any way at all, or where there is nonhomogeneity of the variances. Recall the illustrative cases of Figures 6.2*a* to *c*.

Heteroskedasticity arises frequently in the analysis of cross-section data. When we use a linear sampling method to obtain observed data, it is quite possible that the variances of the subgroups (with respect to a variable) in a population may be different, as in the case of sampling linearly across high-income and low-income sections of a city. It is well-known that the savings behavior of the rich and of the poor can differ in the form of the large and small variances of the savings variable: among the high-income people the spending pattern may vary greatly, but among the low-income people most income would be spent for daily necessities.

Clearly, the correct procedure for the estimation of the parameters of a regression model having the disturbances with the covariance (6.45) would be the GLS method. However, the k_i's are almost always unknown so that, whenever an approximate pattern of unequal variances can be learned from the data, assumptions as to such a pattern can be used in applying the GLS methods. For instance, if it can be assumed that in the model

$$y_i = \alpha + \beta x_i + u_i \qquad i = 1, 2, \ldots, n$$

$\text{Var}(u_i) = \sigma^2 x_i^2$ (that is, the variance of u_i is directly proportional to the square of x_i) or all i, then we have

$$\boldsymbol{\Omega} = \begin{bmatrix} x_1^2 & 0 & \cdots & 0 \\ 0 & x_2^2 & \cdots & \cdot \\ \cdot & \cdot & \cdot & \cdot \\ \cdot & \cdot & \cdot & \cdot \\ \cdot & \cdot & \cdot & \cdot \\ 0 & 0 & \cdots & x_n^2 \end{bmatrix}$$

$$\boldsymbol{\Omega}^{-1} = \begin{bmatrix} \frac{1}{x_1^2} & 0 & \cdots & 0 \\ 0 & \frac{1}{x_2^2} & \cdots & 0 \\ \cdot & \cdot & \cdot & \cdot \\ \cdot & \cdot & \cdot & \cdot \\ \cdot & \cdot & \cdot & \cdot \\ 0 & 0 & \cdots & \frac{1}{x_n^2} \end{bmatrix}$$

and it follows that the GLS estimate of α and β is

$$\begin{pmatrix} \hat{\alpha} \\ \hat{\beta} \end{pmatrix} = (\mathbf{X}'\boldsymbol{\Omega}^{-1}\mathbf{X})^{-1}\mathbf{X}'\boldsymbol{\Omega}^{-1}\mathbf{y}$$

This is equivalent to transforming the data according to

$$\frac{y_i}{x_i} = \alpha\left(\frac{1}{x_i}\right) + \beta + \frac{u_i}{x_i}$$

Then Var $(u_i/x_i) = \sigma^2$ for all i.

If Eu_i^2 increases or decreases monotonically, it is possible to use the *BLUS* residuals to make a test for heteroskedasticity. That is, the ratio

$$\frac{\text{mean square of the first } \frac{1}{2}(n - k - 1) \textit{ BLUS} \text{ residuals}}{\text{mean square of the last } \frac{1}{2}(n - k - 1) \textit{ BLUS} \text{ residuals}}$$

has the F distribution with $[\frac{1}{2}(n - k - 1), \frac{1}{2}(n - k - 1)]$ degrees of freedom.

The test for homogeneity of variances is available in standard statistical works. Typically, however, one computes the *OLS* residuals $\hat{u}_i$, classifies these residuals into groups, computes the sample variance of each group, and applies a test for homogeneity. The main difficulty here lies in the fact that there is no analytical procedure to determine the optimal grouping of the $\hat{u}_i$'s once they are estimated.

CHAPTER 7

PROBLEMS AND VARIANTS OF THE STANDARD MODEL

Given the standard model, or the classical linear regression model,

$$\mathbf{y} = \mathbf{X}\boldsymbol{\beta} + \mathbf{u} \tag{7.1}$$

where the assumptions about the model are as given in (4.32), we have discussed in Chapter 6 the various problems of estimation and hypothesis testing when the assumption that $E\mathbf{uu}' = \sigma^2\mathbf{I}$ was replaced by $E\mathbf{uu}' = \sigma^2\boldsymbol{\Omega}$. Although the general nature of $\boldsymbol{\Omega}$ does not limit its applicability to first-order autocorrelation and heteroskedasticity, our discussion centered on them as they are frequently observed phenomenon in economics. While the conceptual and analytical usefulness of the pure Aitken's estimation should not be minimized, we must observe that the estimation of what is usually unknown $\boldsymbol{\Omega}$ is a difficult task, especially, if two or more of the problems such as autocorrelation and heteroskedasticity exist simultaneously. In fact, analytically we are still unable to establish, for instance, the properties of an estimator of $\boldsymbol{\Omega}$ when, indeed, some form of autocorrelation and heteroskedasticity exist in $\mathbf{u}$ at the same time. This difficulty will remain with us for sometime to come, but for now we must continue to study individual problems that arise from the violation of the standard assumptions vis-à-vis economic research and analysis. These problems are indeed numerous and cannot be covered or even properly referred to within the scope of this book. In this chapter, our objectives are to discuss some of the typical problems that develop in economic research and to provide a framework of reference that will help us to understand better the material in the last three chapters of this book.

In particular, we examine closely the nature of the assumption about the observation matrix $\mathbf{X}$ and touch on the problems of multicollinearity and errors in the independent variables. This discussion is followed by a treatment of discrete variables that often appear in regression equations. Because of the frequency of their use in empirical work, models that are not linear in the parameters are discussed. Furthermore, on the properties of the **X** matrix, we generalize them to include situations where we have independent variables that are stochastic and that are correlated with the disturbances with varying degrees of intensity.

7.1. Multicollinearity

Recall that one of the assumptions for the model in (7.1) was that $\rho(\mathbf{X})$ is of full rank, meaning that the rank is $k + 1$, or the number of columns of $\mathbf{X}$ is $k + 1$. If $\mathbf{X}$ is not of full rank, then $|\mathbf{X}'\mathbf{X}| = 0$, so that the OLS estimate $\mathbf{b} = (\mathbf{X}'\mathbf{X})^{-1}\mathbf{X}'\mathbf{y}$ cannot be obtained. But the requirement that $|\mathbf{X}'\mathbf{X}| \neq 0$ is not really a stringent one, since, in effect, it says that no column of $\mathbf{X}$ can be written as an exact linear combination of the other columns. Thus it does not include the situations where the columns of $\mathbf{X}$ are such as to give rise to intercorrelation in the data representing the $(k + 1)$ columns (variables) and where, because of the lack of perfect collinearity, the estimate $(\mathbf{X}'\mathbf{X})^{-1}\mathbf{X}'\mathbf{y}$ can be found. If the columns of $\mathbf{X}$ are so collinear as to be close to the case of "exact linear combinations" but not quite the case, it can be appreciated that the $|\mathbf{X}'\mathbf{X}|$ can be close to zero and the $\mathbf{Var}\ (\mathbf{b}) = \sigma^2(\mathbf{X}'\mathbf{X})^{-1}$ can "blow up," since $(\mathbf{X}'\mathbf{X})^{-1}$ is defined as $|\mathbf{X}'\mathbf{X}|^{-1}$ (adj $\mathbf{X}'\mathbf{X})'$. This is a serious case of multicollinearity. Multicollinearity among the columns can exist in varying degrees; one extreme situation is where the columns of $\mathbf{X}$ are pairwise orthogonal (that is, $\mathbf{x}_i'\mathbf{x}_j = 0$ for all i and j, $i \neq j$), so that there is a complete lack of multicollinearity; at the other extreme is the case of complete collinearity, that is, $|\mathbf{X}'\mathbf{X}| = 0$. In between the two extremes lies the most commonly observed situation—some degree of intercorrelatedness in the independent variables as appear in the data in the columns of $\mathbf{X}$. Since multicollinearity is a data problem, rounding error can also cause it. See the discussion of ill-conditioned matrix in Subsection 4.3.3.

In general, the existence of multicollinearity in the $\mathbf{X}$ matrix results in (1) the inaccurate estimation of the regression coefficients because of the large sample variances of the coefficient estimators, (2) the uncertain specification of the model with respect to the inclusion of variables, and (3) the resulting difficulty in the interpretation of the estimated coefficients. The first point has been amply clarified in the preceding discussion. To elaborate on the second and third points, we first discuss orthogonal regression and the technique of auxiliary regression.

7.1.1. Orthogonal Regression

If the columns of $\mathbf{X}$ are pairwise orthogonal, it will be the case that

$$\mathbf{X}_1'\mathbf{X}_2 = \mathbf{0} \tag{7.2}$$

where $\mathbf{X}_1$ and $\mathbf{X}_2$ result from an arbitrary partition of $\mathbf{X}$ such that

$$\mathbf{X} = (\mathbf{X}_1 \; \mathbf{X}_2)$$

with the dimensionality of $\mathbf{X}_1$ $n \times m$, that of $\mathbf{X}_2$ $n \times l$, and $m + l = k + 1$. Thus the OLS estimator of $\boldsymbol{\beta}$ in (7.1) is

$$\begin{aligned} \mathbf{b} &= [(\mathbf{X}_1 \; \mathbf{X}_2)'(\mathbf{X}_1 \; \mathbf{X}_2)]^{-1}(\mathbf{X}_1 \; \mathbf{X}_2)'\mathbf{y} \\ &= \begin{bmatrix} \mathbf{X}_1'\mathbf{X}_1 & \mathbf{0} \\ \mathbf{0} & \mathbf{X}_2'\mathbf{X}_2 \end{bmatrix}^{-1} \begin{pmatrix} \mathbf{X}_1'\mathbf{y} \\ \mathbf{X}_2'\mathbf{y} \end{pmatrix} \\ &= \begin{bmatrix} (\mathbf{X}_1'\mathbf{X}_1)^{-1}\mathbf{X}_1'\mathbf{y} \\ (\mathbf{X}_2'\mathbf{X}_2)^{-1}\mathbf{X}_2'\mathbf{y} \end{bmatrix} \end{aligned} \tag{7.3}$$

From the above, we observe that the regression of y on the first m independent variables yields m coefficient estimates that do not change on the addition of the l independent variables in a second regression, provided that each column of $\mathbf{X}_1$ is orthogonal to every column of $\mathbf{X}_2$.*

Specializing the preceding discussion to a 2-independent-variable case

$$y = \beta_1 X_1 + \beta_2 X_2 + u$$

we notice that, if the columns of the observations for X_1 and X_2 are orthogonal to each other, the OLS estimate of β_1 through the regression

$$y = \beta_1 X_1 + u$$

is the same as the OLS estimate of β_1 obtained in

$$y = \beta_1 X_1 + \beta_2 X_2 + u$$

Any departure from the pairwise orthogonality means that the existence of multicollinearity and the severity of multicollinearity increases as $|\mathbf{X}'\mathbf{X}|$ moves closer to zero. Keep in mind that the principles of least squares are not invalidated by the existence of multicollinearity; it is not least squares that is at fault. The fact is that the data will simply not allow any method to distinguish between the effects of collinear variables on the dependent variable.

We now investigate in what way multicollinearity affects estimation and specification. To this end, let us consider some aspects of misspecification.

7.1.2. *Effects of Multicollinearity on Specification*

Let us suppose that for the standard model we have properly specified a relation in which $k + 1$ independent variables appear. Suppose further that there is a high collinearity among these independent variables, so that

* It is not required that the columns within $\mathbf{X}_1$ and $\mathbf{X}_2$ be pairwise orthogonal.

the estimates obtained show statistical insignificance. Then one is led to doubt the validity of the original specification since "theory" seems to mesh poorly with "data." Here one solution may be to drop a number of independent variables that are highly collinear with the others and to reestimate the relation with the remaining variables. To see the effect of this reestimation in some generality we employ the device used by Theil (1957), Griliches (1957), and others.

Let the correctly specified model be (7.1)

$$\mathbf{y} = \mathbf{X}\boldsymbol{\beta} + \mathbf{u} \tag{7.1}$$

But, in reality, suppose that we use $\bar{\mathbf{X}}$ in place of $\mathbf{X}$ for estimation. Here we assume, say, $\bar{\mathbf{X}} \neq \mathbf{X}$, that is, $\bar{\mathbf{X}}$ and $\bar{\mathbf{X}}$ are different matrices and, perhaps, $\rho(\bar{\mathbf{X}}) \leq \rho(\mathbf{X})$. Hence, now we have

$$\mathbf{y} = \bar{\mathbf{X}}\bar{\boldsymbol{\beta}} + \bar{\mathbf{u}} \tag{7.1a}$$

and the OLS estimator for $\bar{\boldsymbol{\beta}}$ is

$$\bar{\mathbf{b}} = (\bar{\mathbf{X}}'\bar{\mathbf{X}})^{-1}\bar{\mathbf{X}}'\bar{y} \tag{7.4}$$

The substitution of (7.1) in (7.4) yields

$$\bar{\mathbf{b}} = (\bar{\mathbf{X}}'\bar{\mathbf{X}})^{-1}\bar{\mathbf{X}}'(\mathbf{X}\boldsymbol{\beta} + \mathbf{u}) \tag{7.4a}$$

so that

$$\begin{aligned} E\bar{\mathbf{b}} &= (\bar{\mathbf{X}}'\bar{\mathbf{X}})^{-1}\bar{\mathbf{X}}'\mathbf{X}\boldsymbol{\beta} \\ &= \mathbf{P}\boldsymbol{\beta} \end{aligned} \tag{7.5}$$

where $\mathbf{P} = (\bar{\mathbf{X}}'\bar{\mathbf{X}})^{-1}\bar{\mathbf{X}}'\mathbf{X}$. Consider this last matrix. It is, in effect, the matrix of regression coefficients of the columns of $\mathbf{X}$ on all of the variables included in $\bar{\mathbf{X}}$. Namely, $\mathbf{X} = (\mathbf{x}_1\mathbf{x}_2 \cdots \mathbf{x}_{k+1})$, and the first column of $\mathbf{P}$ is the coefficient estimates of $\mathbf{x}_1$ regressed on $\bar{X}$, or $(\bar{\mathbf{X}}'\bar{\mathbf{X}})^{-1}\bar{\mathbf{X}}'\mathbf{x}_1$. Writing the latter out generally for any i, we have the model

$$\begin{aligned} x_i &= p_{1i}\bar{x}_1 + p_{2i}\bar{x}_2 + \cdots + p_{mi}\bar{x}_m + \epsilon_i \\ i &= 1, 2, \ldots, k+1 \end{aligned} \tag{7.6}$$

where it is assumed that $\bar{X}$ is $n \times m$, $m < k + 1$, and $\bar{x}_i$ are included independent variables. This relation associating the "true" variable x_i with the included variables is referred to as an auxiliary regression equation. Thus if a prior knowledge of $\mathbf{P}$ is available, via (7.5), one can ascertain how the true regression coefficients are related to the "substitute" estimates $\bar{\mathbf{b}}$.

To gain a better insight into the present discussion, we specialize the $\bar{\mathbf{X}}$ to the one where $\bar{\mathbf{X}}$ is equal to $\mathbf{X}$ except the $(k + 1)$st column in $\mathbf{X}$ is absent in $\bar{\mathbf{X}}$, so that the latter consists of the first k columns of $\mathbf{X}$. Let this "substitute" observation matrix be denoted by $\mathbf{X}_1$. Then the matrix of coefficients of regression of $\mathbf{X}$ on $\mathbf{X}_1$ is

$$\begin{aligned}\mathbf{P} &= (\mathbf{X}_1'\mathbf{X}_1)^{-1}\mathbf{X}_1'\mathbf{X} \\ &= [(\mathbf{X}_1'\mathbf{X}_1)^{-1}\mathbf{X}_1'\mathbf{X}_1(\mathbf{X}_1'\mathbf{X}_1)^{-1}\mathbf{X}_1'\mathbf{x}_{k+1}] \\ &= [\mathbf{I}_k\ \mathbf{p}] \end{aligned} \tag{7.7}$$

where $\mathbf{p}$ is the coefficient vector of the regression of x_{k+1} on $x_1, x_2, \ldots, x_k$, or auxiliary regression of x_{k+1} on $x_1, x_2, \ldots, x_k$. Therefore

$$\mathbf{P}\bar{\mathbf{b}} = \begin{pmatrix} 1 & 0 & \cdots & 0 & p_1 \\ 0 & 1 & \cdots & 0 & p_2 \\ \cdot & \cdot & \cdot & \cdot & \cdot \\ \cdot & \cdot & \cdot & \cdot & \cdot \\ \cdot & \cdot & \cdot & \cdot & \cdot \\ 0 & 0 & \cdots & 1 & p_k \end{pmatrix} \begin{pmatrix} \beta_1 \\ \beta_2 \\ \cdot \\ \cdot \\ \cdot \\ \beta_{k+1} \end{pmatrix} \tag{7.8}$$

Remembering that $\bar{\mathbf{b}} = \mathbf{P}\boldsymbol{\beta}$, the correspondence between the estimated coefficients with $\mathbf{X}_1$ in the equation and the true coefficients with $\mathbf{X}$ in the equation is

$$\begin{aligned} E\bar{b}_1 &= \beta_1 + p_1\beta_{k+1} \\ E\bar{b}_2 &= \beta_2 + p_2\beta_{k+1} \\ &\ \ \vdots \\ E\bar{b}_k &= \beta_k + p_k\beta_{k+1} \end{aligned} \tag{7.8a}$$

Here, at least, the direction of the bias of each $\bar{b}_i$, $i = 1, 2, \ldots, k$, can be determined if we know the direction of the effect of x_{k+1} on the dependent variable and the sign of p_i, namely, the direction of the effect of x_i on x_{k+1}. For example, in a production function where only labor (L) and capital (K) inputs appear it is possible that the correct specification is to include managerial input (M). Thus the estimated production function only with L and K would have coefficients that are biased. The direction of bias would be determined as follows: the effect of M on output (β_{k+1}) is positive, although the effect of M on both labor and capital is most likely positive $p_1 > 0$ and $p_2 > 0$. Therefore, both the labor and capital coefficients would be biased upward; that is, they would overestimate the effects of K and L on output if M has positive influences on output and is also positively correlated with K and L.

A general observation that emerges from the above discussion is that as long as the $\mathbf{P}$ matrix in (7.5) is not an identity matrix of order $k + 1$, the estimates obtained through the use of the "substitute" matrix $\bar{\mathbf{X}}$ will be biased.

But, of course, **P** can never be an identity matrix unless the true variables in the correct specification are included in the regression. Thus any departure from the use of variables other than the ones correctly specified would bring about biases in the OLS estimator. So, if high collinearity exists among the data for the correctly included variables, the chances are that the researcher will either drop some variables or will impose linear restrictions (see Section 5.3.2) on the parameters of the true model for reestimation. Estimates obtained in this way are biased, since dropping variables (zero restrictions) or imposing linear restrictions are departures from what is presumed to be a correct original model.

In economic research we often do not know if we have correctly included the relevant variables in a relation, not to mention the functional form. Here, the specification in the sense of exclusion or inclusion of variables is an open pandora's box if the included and the excluded variables suffer from multicollinearity in their data severally or jointly. For any set of included variables, high multicollinearity plays havoc with the accurate determination of the coefficient estimates; hence, the validity of the specific set of included variables can be questioned. Any change in the list of included and excluded variables raises the question of possible bias in the reestimated coefficients. The unbiasedness property of OLS procedure is undermined by poor specification. Multicollinearity in the data cuts across the estimation, the specification, and the interpretation of empirical results. Often, in economic research we are searching for a better and better specification of a relation, but at each alteration of specification we must contend with the problems of estimation and interpretation when we use data that are multicollinear.

7.1.3. Detection and Solution of Multicollinearity

Given the sort of difficulties that can arise when multicollinearity is present in the data for the independent variables, what then are the ways in which we can detect the existence of multicollinearity and, if found, how might we solve the problem? The proper answer here is that the detection is easy but that the solution is difficult.

First about detection. It is well known that most economic variables are correlated with one another, or that they move together. And, since economic data are not generated by experimentally controlled conditions (in which case pairwise orthogonality can be achieved), multicollinearity is a prevalent phenomenon in the observations for independent variables. Thus the problem is not so much in detecting the existence of collinearity but instead that it is rather to find out the severity of multicollinearity. Usually, as part of the standard computations for regression analysis, we calculate (or the machine does) the simple correlations for the pairs of the independent variables and forms the correlation matrix, say,

$$\mathbf{R} = (r_{ij}) \tag{7.9}$$

where r_{ij} = sample correlation between x_i and x_j, say, $i = 1, 2, \ldots, k$. If r_{ij} are large, say, 0.80 or greater, we see that pairwise collinearity is serious. But how high should r_{ij} be before it is intolerable is a question that can be answered only by the investigator according to individual problems. Klein suggests a rule of thumb that multicollinearity is "tolerable" if

$$r_{ij} < R \tag{7.10}$$

where R is the square root of the coefficient of multiple determination, R^2. [See Klein (1962), p. 101.]

But the rule (7.10) is useful only for pairwise considerations. The determinant of $(\mathbf{X}'\mathbf{X})$ can be zero if the several columns of $\mathbf{X}$ can be written in a linear combination where any two columns may or may not be highly collinear. Easily, we can observe that, if pairwise orthogonality exists for all the columns of $\mathbf{X}$, then the sample correlation matrix $\mathbf{R} = \mathbf{I}$, or

$$|\mathbf{R}| = 1 \tag{7.11}$$

Any departure from this orthogonality would mean that

$$0 \le |\mathbf{R}| < 1 \tag{7.12}$$

The determinant is nonnegative because $\mathbf{X}'\mathbf{X}$ is positive semidefinite, and $\mathbf{R}$ is derived by the computation

$$(r_{ij}) = (\mathbf{x}_i'\mathbf{x}_j) \tag{7.12a}$$

where $\mathbf{x}_i$ is the observations on the ith independent variable in the form of derivations from the sample mean divided by the sample standard derivation. The closer $|\mathbf{R}|$ is to 0, the greater the severity of multicollinearity and, the closer $|\mathbf{R}|$ is to 1, the less the degree of multicollinearity. The question here is: What is the "critical" value of $|\mathbf{R}|$ beyond which one cannot accept multicollinearity as tolerable? There is no definite answer to this question generally. Farrar and Glauber (1964) suggest some statistical measure of significant departure from orthogonality among the columns of $\mathbf{X}$, relying on approximate chi-square distribution, but this requires a distributional assumption about the observations on the independent variables. Thus, at present for the fixed observation matrix $\mathbf{X}$ suffering from multicollinearity, there is no analytical measure of its severity except the one suggested in (7.12).

If multicollinearity is intolerable in the sense of high simple correlations among any two independent variables or in the sense of not being able to obtain well-determined estimates of the regression coefficients, what then are the solutions? One frequently used procedure is to drop the variable or variables with which the other independent variables are highly collinear—the zero restriction. The problems of zero restriction have been discussed in Section 7.1.2, but they are not limited to the ones mentioned. When one drops a variable on the basis of the statistical insignificance of that variable, for instance, through the usual t test, and then reestimates the equation, the estimates obtained will suffer from pretesting bias* as well as from the bias of sequential estimators.

Similar to the procedure of zero restriction is to bring in information extraneous to the sample. For instance, if one is faced with a highly collinear set of time-series data, he might bring in some previously obtained results or some cross-section estimates for the estimation, using the time-series data. In particular, let us say that the following relation is being studied

$$E_t = \alpha + \beta Y_t + \gamma p_t + u_t \tag{7.13}$$

where E_t = consumption expenditure on a certain commodity, Y_t = consumer income, p_t = the price of the commodity, and t = annual time subscript. Now, if Y_t and p_t are highly correlated so as to make OLS estimation of the parameters impossible, one might use a cross-section estimate of β, say b, and estimate (7.13) by

$$(E_t - bY_t) = \alpha + \gamma p_t + u_t \tag{7.13a}$$

Of course, the question here is on the validity of the estimate b for substitution in (7.13). How many annual cross sections are used, in relation to the time period over which t runs, for estimating b? Since the variability of E and Y is greater over cross section than over time, is the cross-section estimate of b subject to the same interpretation as the β in (7.13)? These and the related questions must be answered satisfactorily before we can make a proper use of what is called extraneous information. See, for instance, Meyer and Kuh (1957).

Another example of the use of prior or of extraneous information is to impose linear restriction on the parameters. For instance, if in the production function

$$\ln X = \alpha + \beta \ln K + \gamma \ln L + u \tag{7.14}$$

where $\ln K$ and $\ln L$ are highly collinear *and* we have reason to believe that $\beta + \gamma = 1$, then we might incorporate this knowledge in the estimation. Here, in effect, we would be dropping either $\ln K$ or $\ln L$ from the estimation,

* Pretesting bias arises in an estimator when the extimator no longer has the probability distribution implied by the original model. For instance, after a regression equation is estimated by OLS, one may drop a variable, say, because it has a wrong sign, and the regression is rerun. Then the "zero" coefficient for the dropped variable in the second equation is biased because of pretesting.

say, in the form,

$$\begin{aligned} \ln X &= \alpha + \beta\,(\ln K - \ln L) + \ln L + u \\ (\ln X - \ln L) &= \alpha + \beta\,(\ln K - \ln L) + u \end{aligned} \tag{7.14a}$$

or

$$\ln\left(\frac{X}{L}\right) = \alpha + \beta \ln\left(\frac{K}{L}\right) + u \tag{7.14b}$$

Then, on securing the estimate of β, we can infer the value of γ. To the extent that the information used, that is, $\beta + \gamma = 1$, is correct, the OLS estimates of β and γ will be unbiased and fully efficient (recall Section 5.3.2).

7.1.4. The Second Moments Criterion

The imposition of linear restrictions, such as the zero restriction and other restrictions we just discussed from Section 5.3.2, can be treated in a general way. Recall that for the standard model

$$\mathbf{y} = \mathbf{X}\boldsymbol{\beta} + \mathbf{u} \tag{7.1}$$

the OLS estimator of $\boldsymbol{\beta}$, $\tilde{\boldsymbol{\beta}}$, subject to the restriction

$$\mathbf{R}\boldsymbol{\beta} = \mathbf{r} \tag{7.15}$$

was derived as

$$\tilde{\boldsymbol{\beta}} = \mathbf{b} + (\mathbf{X}'\mathbf{X})^{-1}\mathbf{R}'[\mathbf{R}(\mathbf{X}'\mathbf{X})^{-1}\mathbf{R}']^{-1}(\mathbf{r} - \mathbf{R}\mathbf{b}) \tag{7.16}$$

where $\mathbf{b} = (\mathbf{X}'\mathbf{X})^{-1}\mathbf{X}'\mathbf{y}$ and

$$\mathbf{Var}\,(\tilde{\boldsymbol{\beta}}) = \sigma^2(\mathbf{X}'\mathbf{X})^{-1} - \sigma^2(\mathbf{X}'\mathbf{X})^{-1}\mathbf{R}'[\mathbf{R}(\mathbf{X}'\mathbf{X})^{-1}\mathbf{R}']^{-1}\mathbf{R}(\mathbf{X}'\mathbf{X})^{-1} \tag{7.17}$$

Substitution of $\mathbf{b} = \boldsymbol{\beta} + (\mathbf{X}'\mathbf{X})^{-1}\mathbf{X}'\mathbf{u}$ into (7.16) yields

$$\begin{aligned} \tilde{\boldsymbol{\beta}} - \boldsymbol{\beta} = (\mathbf{X}'\mathbf{X})^{-1}\mathbf{X}'\mathbf{u} &+ (\mathbf{X}'\mathbf{X})^{-1}\mathbf{R}'[\mathbf{R}(\mathbf{X}'\mathbf{X})^{-1}\mathbf{R}']^{-1}(\mathbf{r} - \mathbf{R}\boldsymbol{\beta}) \\ &- (\mathbf{X}'\mathbf{X})^{-1}\mathbf{R}'[\mathbf{R}(\mathbf{X}'\mathbf{X})^{-1}\mathbf{R}']^{-1}\mathbf{R}(\mathbf{X}'\mathbf{X})^{-1}\mathbf{X}'\mathbf{u} \end{aligned} \tag{7.18}$$

so that, if $\mathbf{R}\boldsymbol{\beta} = \mathbf{r}$ is true,

$$E(\tilde{\boldsymbol{\beta}} - \boldsymbol{\beta}) = \mathbf{0} \tag{7.19}$$

But, if $\mathbf{R}\boldsymbol{\beta} = \mathbf{r}$ is false, (7.19) does not hold true and $\tilde{\boldsymbol{\beta}}$ will be biased. However, as (7.17) implies, the OLS estimator under linear restrictions is, in general, more efficient than the OLS estimator without the linear restriction, irrespective of the validity of the restrictions. These facts give rise to the idea that, if the restrictions are approximately true, these restrictions may be used to obtain an estimator of $\boldsymbol{\beta}$ that is "better" than the unrestricted estimator in the mean-square-error sense (see Chapter 3 for the definition). That is, in the event that multicollinearity is severe among the independent variables in a correctly specified model, we may be willing to trade some bias (because of not exactly true restrictions) for a smaller variance of the estimator with restrictions if these restrictions are not too far divergent from the correct specification of the model. This is illustrated below. Before we proceed, notice that, if $\mathbf{R\beta} \neq \mathbf{r}$, we have

$$\begin{aligned}\tilde{\boldsymbol{\beta}} - \boldsymbol{\beta} &= (\mathbf{X'X})^{-1}\mathbf{X'u} + (\mathbf{X'X})^{-1}\mathbf{R'}[\mathbf{R}(\mathbf{X'X})^{-1}\mathbf{R'}]^{-1}[\mathbf{r} - \mathbf{R\beta} - \mathbf{R}(\mathbf{X'X})^{-1}\mathbf{X'u}] \\ &= \{\mathbf{I} - (\mathbf{X'X})^{-1}\mathbf{R'}[\mathbf{R}(\mathbf{X'X})^{-1}\mathbf{R'}]^{-1}\mathbf{R}\}(\mathbf{X'X})^{-1}\mathbf{X'u} \\ &\quad + (\mathbf{X'X})^{-1}\mathbf{R'}[\mathbf{R}(\mathbf{X'X})^{-1}\mathbf{R'}]^{-1}(\mathbf{r} - \mathbf{R\beta}) \qquad (7.18)\end{aligned}$$

so that $E(\tilde{\boldsymbol{\beta}} - \boldsymbol{\beta})$ does not vanish. Furthermore, the mean-square error (MSE) or the second moments of the estimator $\boldsymbol{\beta}$ is

$$\begin{aligned}E(\tilde{\boldsymbol{\beta}} - \boldsymbol{\beta})(\tilde{\boldsymbol{\beta}} - \boldsymbol{\beta}) &= \sigma^2\{\mathbf{I} - (\mathbf{X'X})^{-1}\mathbf{R'}[\mathbf{R}(\mathbf{X'X})^{-1}\mathbf{R'}]^{-1}\mathbf{R}\}(\mathbf{X'X})^{-1} \\ &+ (\mathbf{X'X})^{-1}\mathbf{R'}[\mathbf{R}(\mathbf{X'X})^{-1}\mathbf{R'}]^{-1}(\mathbf{r} - \mathbf{R\beta})(\mathbf{r} - \mathbf{R\beta})'[\mathbf{R}(\mathbf{X'X})^{-1}\mathbf{R'}]^{-1}\mathbf{R}(\mathbf{X'X})^{-1} \qquad (7.19)\end{aligned}$$

Now consider the model

$$y = \beta_1 x_1 + \beta_2 x_2 + u \qquad (7.20)$$

where the observation vectors on x_1 and x_2 are, respectively, $\mathbf{x}_1$ and $\mathbf{x}_2$ and are in the form of deviations from their respective sample means divided by their sample standard derivations. Thus for

$$\mathbf{X} = (\mathbf{x}_1\ \mathbf{x}_2)$$

we have

$$(\mathbf{X'X}) = \begin{bmatrix} 1 & \rho \\ \rho & 1 \end{bmatrix}$$

and the covariance matrix of the OLS estimators of β_1 and β_2 is

$$\mathbf{Cov}\,(b_1, b_2) = \frac{\sigma^2}{n(1 - \rho^2)}\begin{bmatrix} 1 & -\rho \\ -\rho & 1 \end{bmatrix} \qquad (7.21)$$

Here, ρ indicates the degree of collinearity between $\mathbf{x}_1$ and $\mathbf{x}_2$. Suppose now that we place the restriction that $\beta_2 = 0$, or $\mathbf{R\beta} = \mathbf{r}$ where $\mathbf{R} = (0\ 1)$ and $\mathbf{r} = 0$, which may turn out to be false. Then it can be shown (Verify this as an exercise) that

$$\text{MSE}\,(\tilde{\beta}_1, \tilde{\beta}_2) = \frac{\sigma^2}{n(1 - \rho^2)}\begin{bmatrix} 1 - \rho^2 & 0 \\ 1 - \rho^2 & 0 \end{bmatrix} + \beta_2{}^2\begin{bmatrix} \rho^2 & -\rho \\ -\rho & 1 \end{bmatrix} \qquad (7.22)$$

Now contrast the variance of b_1 and the mean-square error of $\tilde{\beta}_1$; namely,

$$\text{Var}\,(b_1) = \frac{\sigma^2}{n(1 - \rho^2)} \qquad (7.23\text{a})$$

$$\text{MSE}\,(\tilde{\beta}_1) = \frac{\sigma^2}{n} + \beta_2{}^2\rho^2 \qquad (7.23\text{b})$$

and for some values of ρ the quantity in (7.23b) may be smaller than the one in (7.23a). For further comments on the MSE criterion, see Toro-Vizcarrondo and Wallace (1968).

7.2. Errors in the Variables

In further consideration of the nature of the observation matrix $\mathbf{X}$, we notice that related to the multicollinearity problem of nonexperimental data is the question of error of measurement in economic variables. Measurement errors can be random or systematic and can come about in various ways. At the data collection level, field surveys provide the most basic source of information. But human nature being what it is, there are tendencies for overreporting, underreporting, nonreporting, or for the refusal to respond by the units being surveyed, depending on the questions being asked. Once responses are obtained, there are the problems of errors arising from the processing of the information, such as transferring, copying, computing, and projecting, etc. Besides the technical problems that affect the accuracy of the data we use, there are also conceptual problems that are related to the definition of a variable and its measurement. For instance, in consideration of the consumer income-expenditure relation, it has been suggested that the concept of income appropriate to the explanation of consumption expenditure should be the permanent or expected income but that the measurement of the latter concept is not available and that any attempt at approximating the measurement by the use of available data invariably introduces errors of measurement.

In this section, we treat error in the variables in the regression situation in a limited number of cases in order to elucidate the nature of the problem. For extensive discussions of the subject see, for example, Johnston (1963, Chapter 6), Malinvaud (1966, Chapter 10), and Madansky (1959).

7.2.1. The Classical Case I

Let us assume that the true model is

$$y^t = \alpha + \beta x^t + u \tag{7.24}$$

but that y^t and x^t cannot be observed. Instead, what is observed and used for estimation is the model

$$y = \alpha + \beta x + v \tag{7.25}$$

where v is an arbitrary error term. Between (7.24) and (7.25) the relations among the variables are as follows

$$y = y^t + \epsilon \tag{7.25a}$$

$$x = x^t + \eta \tag{7.25b}$$

where y and x are the observable counterparts of y^t and x^t, the true variables, and where, furthermore, the errors have the properties

$$E\epsilon = E\eta = Eu = 0$$
$$Ey^t\epsilon = Ex^t\eta = Ex^t\epsilon = Ex^tu = 0$$
$$E\epsilon^2 = \sigma_\epsilon^2 \qquad E\eta^2 = \sigma_\eta^2 \qquad Eu^2 = \sigma^2$$
$$Eu\eta = 0$$

That is, among other things, the measurement errors ϵ and η are independent of the true variables x^t and y^t, respectively. Now, if OLS is applied to (7.52) for getting the estimates of α and β, what are the properties of the OLS estimators?

From (7.24), (7.25a), and (7.25b), we have

$$y - \epsilon = \alpha + \beta(x - \eta) + u$$

or

$$y = \alpha + \beta x + (u - \beta\eta + \epsilon) \tag{7.26}$$

If we are to assume that $(u - \beta\eta + \epsilon)$ fulfills the standard assumptions of the fixed linear regression model, the OLS estimator of β is BLUE. But if $(u - \beta\eta + \epsilon)$ is not independent of x, then the OLS estimator will be biased. That the latter is, indeed, the case is seen in the following. The deviation of the OLS estimator b from the true parameter, say, β, is

$$b - \beta = \frac{\sum (x^t + \eta)(u - \beta\eta + \epsilon)}{\sum (x^t + \eta)^2} \tag{7.27}$$

In the numerator x^t is assumed to be independent of u, η, and ϵ, so that

$$E\{\sum \eta(u - \beta\eta + \epsilon)\} = \beta^2\sigma_\eta^2 \neq 0$$

For a moment, if we treat the denominator in (7.27) as a constant, we observe that the expectation of $(b - \beta)$ is a nonzero quantity. This is partial proof of the bias. (For a general approach, see Section 7.5.2.)

The bias of the OLS estimator when the independent variable is subject to measurement error does not vanish when the sample size is increased. It can be shown that [see Johnston (1960, pp. 149–150) for example]

$$\text{plim } b = \beta \bigg/ \left(1 + \frac{\sigma_\eta^2}{\sigma_{x^t}^2}\right) \tag{7.28}$$

for $n \rightarrow \infty$. It follows then that the OLS estimator b is inconsistent and asymptotically downward biased, the degree of the bias depending on σ_η^2 and $\sigma_{x^t}^2$.

Some measurement errors are not necessarily of the random type discussed above. For the case where the errors are systematic, see Malinvaud (1966, p. 328).

7.2.2. *Modified Classical Case I*

Assume that y^t now is measured without error, or that y's error can be consolidated into the error of the equation. Thus, in place of (7.24), we have

$$y = \alpha + \beta x^t + u \tag{7.24}$$

and, assuming (7.25b),

$$y = \alpha + \beta x + (u - \beta\eta) \tag{7.30}$$

But, as long as $x = x^t + \eta$, $E(b - \beta)$ will not be zero and, furthermore, the asymptotic bias still will be the factor

$$1\bigg/\left(1 + \frac{\sigma_\eta^2}{\sigma_{x^t}^2}\right)$$

Notice that the bias is downward toward zero whether β is positive or negative. Also in the two models discussed the basic problem is not in the measurement error of y but in the dependence between the independent variable and the disturbance.

7.2.3. *Permanent Income Hypothesis*

One interesting application of the errors-in-the-variables models is the permanent income hypothesis regarding the consumption function. In empirical estimation of the consumption function customary practice for some time was to regress consumption expenditure (c) on disposable personal income (y) and typically gave result for a model

$$c = \alpha + \beta y + u \tag{7.31}$$

satisfying the conditions $\alpha, \beta > 0$. However, over a long period of time, it has also been observed that there is stability in the savings-income ratio, meaning that the finding, for example, in (7.31) is not necessarily consistent with the stability of the ratio. Friedman (1957) explains the stability of the savings-income ratio from the concept of the expected or the permanent income and hypothesizes the relation

$$c^e = ky^e \tag{7.32}$$

where $c^e = c + u_c$ and $y^e = y + u_y$, c^e and y^e are, respectively, the expected consumption and the expected income, and u_c and u_y are, respectively, the transitory consumption and the transitory income. The transitory components are severally uncorrelated with expected income and consumption and also are uncorrelated with each other. Since the expected variables are not observable, the estimation of k in (7.32) must rely on observed or measured consumption and income. Thus the OLS estimate of k is

$$\hat{k} = \frac{\sum (c - \bar{c})(y - \bar{y})}{\sum (y - \bar{y})^2}$$

From (7.32)

$$\text{var}\,(y) = \text{var}\,(y^e) + \text{var}\,(u_y)$$
$$\text{var}\,(c) = \text{var}\,(c^e) + \text{var}\,(u_c)$$

and, since $E\{\sum (c - \bar{c})(y - \bar{y})\} = E(\sum cy)$, we have

$$\begin{aligned} \text{plim}\,\hat{k} &= \frac{\text{plim}\,n^{-1}(\sum cy)}{\text{plim}\,n^{-1}\sum (y - \bar{y})^2} \\ &= \frac{\text{cov}\,(c^e, y^e)}{\text{var}\,(y)} \\ &= \frac{k \cdot \text{var}\,(y^e)}{\text{var}\,(y^e) + \text{var}\,(u_y)} \\ &= k \cdot P_y \end{aligned} \tag{7.33}$$

where $P_y = 1$ if $\text{var}\,(u_y) = 0$. This means that the estimate of k based on measured c and y are downward biased, with the degree of bias depending on the size of the variance of the transitory income. If the latter's variance is large, relative to the variance of the expected income, the underestimation of the marginal propensity to consume (k) would be large.

The further implications of the preceding result is the reason for the lower empirical estimate of the marginal propensity to consume (MPC) for entrepreneurs than for workers, and the lower empirical MPC for farmers than for urban population, etc. If cross-section data are used for the estimation of k, a better result can be obtained by grouping the "like" individuals in such a way as to make the group mean of the transitory component of y small or zero and to use the group mean of y as a measure of expected income for each of the individuals in that group. It is also possible simply to regress group mean values of c on group mean values of y; this grouping may be achieved by occupation, age, education, and the like. See Watts (1960) for some examples of this discussion.

7.2.4. *Modified Classical Case II*

An extension of the Modified Classical Case I is stated in this question: If there are more than one independent variable and these variables are subject to measurement errors, would the conclusion that the OLS estimate is downward biased hold true? The answer is not necessarily, and we rely heavily on Theil (1958, pp. 326–329) to indicate briefly why this is the answer.

Suppose that the true model under consideration is

$$\mathbf{y} = \mathbf{X}\boldsymbol{\beta} + \mathbf{u} \tag{7.34}$$

where $\mathbf{X}$ is the true observation matrix but not directly observable and, instead, $\bar{\mathbf{X}}$ is its observable counterpart. Hence

$$\bar{\mathbf{X}} = \mathbf{X} + \mathbf{W} \tag{7.35}$$

where $\mathbf{W}$ is a matrix of measurement errors for $\mathbf{X}$. Suppose now that we use $\bar{\mathbf{X}}$ in place of $\mathbf{X}$ for OLS estimation of (7.34), so that we are dealing with the model

$$\mathbf{y} = \bar{\mathbf{X}}\bar{\boldsymbol{\beta}} + \bar{\mathbf{u}} \tag{7.36}$$

On applying OLS to (7.36), we find the estimator of $\bar{\boldsymbol{\beta}}$ as

$$\bar{\mathbf{b}} = (\bar{\mathbf{X}}'\bar{\mathbf{X}})^{-1}\bar{\mathbf{X}}'\mathbf{y}$$

and

$$E\bar{\mathbf{b}} = \mathbf{P}\boldsymbol{\beta}$$

as in Section 7.1.2, Expression 7.5, where $\mathbf{P}$ now can be shown to be*

$$\begin{aligned}\mathbf{P} &= (\mathbf{X}'\mathbf{X} + \mathbf{W}'\mathbf{X} + \mathbf{W}'\mathbf{W} + \mathbf{X}'\mathbf{W})^{-1}(\mathbf{X}'\mathbf{X} + \mathbf{W}'\mathbf{X})\\ &= [\mathbf{I} + (\mathbf{X}'\mathbf{X} + \mathbf{W}'\mathbf{X})^{-1}(\mathbf{W}'\mathbf{W} + \mathbf{X}'\mathbf{W})]^{-1}\\ &\approx [\mathbf{I} + (\mathbf{X}'\mathbf{X})^{-1}\mathbf{W}'\mathbf{W}]^{-1}\end{aligned} \tag{7.37}$$

Thus

$$\mathbf{P} \approx [\mathbf{I} - (\mathbf{X}'\mathbf{X})^{-1}\mathbf{W}'\mathbf{W}] \tag{7.38}$$

If $\mathbf{W}'\mathbf{W}$ is diagonal $(\mathbf{X}'\mathbf{X})^{-1}\mathbf{W}'\mathbf{W}$ will be diagonal and its elements will be each between 0 and 1, that is the ith element of $E\bar{\mathbf{b}}$ is

$$Eb_i = \beta_i(1 - \sum w_i^2/\sum X_i^2)$$

where, for instance, $\sum w_i^2$ denotes the ith diagonal element of $\mathbf{W}'\mathbf{W}$, and where it will be safely assumed that $\sum w_i^2$ is less than $\sum X_i^2$. If $\mathbf{X}'\mathbf{X}$ and $\mathbf{W}'\mathbf{W}$ are not diagonal, which is a real possibility in empirical situations, then $(\mathbf{X}'\mathbf{X})^{-1}\mathbf{W}'\mathbf{W}$ will not be diagonal, and the nature of the intercorrelations among the columns of $\mathbf{X}$ as well as the true variances of x_i and of the errors will all play a role in determining the matrix $\mathbf{P}$. And there is no assurance that the bias will be uniformly downward as in the case of simple regression with errors in the variables.

* Woodbury's theorem states that, for square matrix $\mathbf{A}$,

$$(\mathbf{A} + \mathbf{UBV})^{-1} = \mathbf{A}^{-1} - \mathbf{A}^{-1}\mathbf{UB}(\mathbf{B} + \mathbf{RV}^{-1}\mathbf{UB})^{-1}\mathbf{BVA}^{-1}$$

7.3. Qualitative or Discrete Variables

Thus far we have assumed that the variables that enter the regression models are real and continuous. In empirical applications, however, we encounter variables that are qualitative or discrete in the sense that these variables are basically classificatory in nature, and their effects on the dependent variables may or may not be linear. For instance, in household expenditure surveys one obtains from each household information about the amount of spending in a particular category, about income, asset holdings, and the household head's age, race, and sex. The last three variables are referred to as demographic variables that are often useful as explanatory variables in cross-section studies. Another example is the use of seasonals in the analysis of quarterly or other time-series data. See Expression 5.23a and the data in Table 5.1. Now one may postulate a nonlinear effect of age on the amount of spending and, similarly, for the effect of the sex and race variables. Here, short of using quadratic or other polynomial functions (consider the problem of multicollinearity), one can use the so-called dummy variable technique.

7.3.1. Dummy Variable Technique

In general, this technique can be applied to any variable whose variation is capable of falling into mutually exclusive classes. Thus, in the broadest sense, any real variable can be converted into a dummy variable or dummy system. Consider the age variable above for illustration. Here for possible values of age ranging from, say, 16 to 80 we may assign the observed values of age to the following classes

$$\begin{aligned} A_1&: 16 \leq A < 40 \\ A_2&: 41 \leq A < 59 \\ A_3&: 60 \leq A < 80 \end{aligned} \tag{7.40}$$

where A is age and A_i is the ith age class. Thus the three age classes are mutually exclusive in that an observation on A can fall into only one of the classes. Another illustration is the race variable (R),

$$\begin{aligned} R_1&: \text{white} \\ R_2&: \text{nonwhite} \end{aligned} \tag{7.41}$$

The first step in the application of the dummy variable technique is to assign a classificatory index that has a zero-one scheme. For instance, we may designate the dummy terms as follows:

$$\begin{aligned} A_1 &= 1 \quad \text{if} \quad 16 \leq A < 40, \quad A_1 = 0 \text{ otherwise} \\ A_2 &= 1 \quad \text{if} \quad 41 \leq A < 60, \quad A_2 = 0 \text{ otherwise} \\ A_3 &= 1 \quad \text{if} \quad 61 \leq A \leq 80, \quad A_3 = 0 \text{ otherwise} \end{aligned} \tag{7.40a}$$

Sometimes an analyst might meet the situation by scaling as follows:

$$\begin{aligned} A_1 &= 1 \quad \text{if} \quad 16 \leq A < 40 \\ A_2 &= 2 \quad \text{if} \quad 41 \leq A < 60 \\ A_3 &= 3 \quad \text{if} \quad 61 \leq A \leq 80 \end{aligned} \tag{7.40b}$$

However, remember that scaling is not equivalent to dummy variable technique; in the example given immediately above a linear effect of age is assumed. As for the race variable, we may have*

$$\begin{aligned} R_1 &= 1 \quad \text{if white,} \quad R_1 = 0 \text{ otherwise} \\ R_2 &= 1 \quad \text{if nonwhite,} \quad R_2 = 0 \text{ otherwise} \end{aligned} \tag{7.41a}$$

For pedagogical reasons we shall restrict the discussion to zero-one dummy systems, such as (7.40a) and (7.41a). It is useful to observe that by a dummy system we mean a variable that is "decomposed" into more than one dummy term and that a dummy variable is interchangeable with a dummy system. Thus a dummy term is a regressor and a dummy variable can appear as a number of regressors.

Having completed the zero-one assignment, we are ready to apply the usual least-squares analysis on, say,

$$y = \beta_0 + \beta_1 A_1 + \beta_2 A_2 + \beta_3 A_3 + u \tag{7.42}$$

But (7.42) cannot be estimated as it stands because the observation matrix will have linearly dependent columns; here, that is, the first column of $\mathbf{X}$ will be equal to the sum of columns 2, 3, and 4. In other words, an observation on the variable for which the dummy system is created must fall into one of the classes, since they are collectively exhaustive. The usual procedure, then, is to drop a dummy term or the constant term to avoid the singularity of $\mathbf{X}'\mathbf{X}$. But what of the interpretation of the estimated coefficients?

To discuss this last question, let us assume that the constant term was dropped from (7.42); consequently, we have

$$y = \beta_1 A_1 + \beta_2 A_2 + \beta_3 A_3 + u \tag{7.43}$$

* This classification allows an inference that there may be a residual class that is neither white nor nonwhite. For now, we assume that white and nonwhite are two exhaustive classes.

Here for, say, a sample of 8 the **X** matrix may be

$$\mathbf{X} = \begin{bmatrix} 1 & 0 & 0 \\ 1 & 0 & 0 \\ 1 & 0 & 0 \\ 0 & 1 & 0 \\ 0 & 1 & 0 \\ 0 & 0 & 1 \\ 0 & 0 & 1 \\ 0 & 0 & 1 \end{bmatrix}$$

That is, there are three observations falling into the A_1 class, two into the A_2 class, and three into the A_3 class. In general, for a sample of size n, we can say that n_1, n_2, and n_3 observations will fall into the A_1, A_2, and A_3 classes, respectively, with $n = n_1 + n_2 + n_3$. Thus it is easy to verify that, in general,

$$(\mathbf{X}'\mathbf{X}) = \begin{pmatrix} n_1 & 0 & 0 \\ 0 & n_2 & 0 \\ 0 & 0 & n_3 \end{pmatrix}$$

so that

$$\mathbf{b} = \begin{pmatrix} \frac{1}{n_1} & 0 & 0 \\ 0 & \frac{1}{n_2} & 0 \\ 0 & 0 & \frac{1}{n_3} \end{pmatrix} \begin{pmatrix} \sum_{n_1} y_{1j} \\ \sum_{n_2} y_{2j} \\ \sum_{n_3} y_{3j} \end{pmatrix} = \begin{pmatrix} \bar{y}_1 \\ \bar{y}_2 \\ \bar{y}_3 \end{pmatrix} \tag{7.44}$$

where $\bar{y}_i$ is the average of the values of the dependent variables who belong to the A_i class. Or, graphically, we may have the situation shown in Figure 7.1, and the OLS estimate b_i is the average effects of being in the A_i class on the dependent variable. To state it in another way

$$E(y \mid A_i = 1) = \beta_i \qquad i = 1, 2, 3 \tag{7.44a}$$

which are estimated by b_i, $i = 1, 2, 3$.

But we must ask what would be the result if we dropped the A_3 term instead of the constant term? Thus

$$y = \beta_0 + \beta_1 A_1 + \beta_2 A_2 + u \tag{7.45}$$

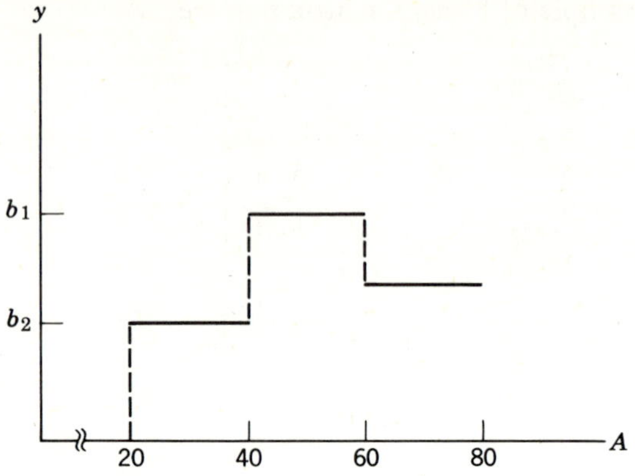

Figure 7.1

must be estimated. We observe that in (7.45)

$$\begin{aligned} E(y \mid A_1 = 1) &= \beta_0 + \beta_1 \\ E(y \mid A_2 = 1) &= \beta_0 + \beta_2 \\ E(y \mid A_3 = 1) &= \beta_0 \end{aligned} \tag{7.54a}$$

so that the OLS estimates b_1 and b_2 are now interpreted as the effect on y of being in the A_1 and A_2 classes relative to the effect of being in the A_3 class. This relative-effect interpretation of the coefficients of the dummy terms in a variable is quite common and is particularly appropriate when there are more than one dummy system in the explanatory variables.

This last remark deserves an elaboration. Suppose that both the age and race variables are to enter as explanatory variables, say,

$$y = \beta_0 + \beta_1 A_1 + \beta_2 A_2 + \beta_3 A_3 + \gamma_1 R_1 + \gamma_2 R_2 + u \tag{7.46}$$

We observe that (7.46) cannot be estimated since, again, $\mathbf{X}'\mathbf{X}$ will be singular. Furthermore, the dropping of the constant term will not allow the estimation because, then, in the observation matrix $\mathbf{X}$, the first column (A_1) is always equal to the sum of the fourth and the fifth columns less the second and the third columns. The rule of thumb is that, whenever there are two or more dummy systems, drop one dummy term from each system (preserving the constant, say) for OLS estimation. The interpretation of the coefficient estimates is similar to the one discussed earlier. That is, if the following relation is estimated

$$y = \beta_0 + \beta_1 A_1 + \beta_2 A_2 + \gamma_1 R_1 + u \tag{7.46}$$

then

$$
\begin{aligned}
&E(y \mid A_1 = 1, R_1 = 1) = \beta_0 + \beta_1 + \gamma_1 \\
&E(y \mid A_2 = 1, R_1 = 1) = \beta_0 + \beta_2 + \gamma_1 \\
&E(y \mid A_3 = 1, R_1 = 1) = \beta_0 + \gamma_1 \\
&E(y \mid A_1 = 1, R_1 = 0 \quad \text{or} \quad R_2 = 1) = \beta_0 + \beta_1 \\
&E(y \mid A_2 = 1, R_2 = 1) = \beta_0 + \beta_2 \\
&E(y \mid A_3 = 1, R_2 = 1) = \beta_0
\end{aligned}
$$

Thus the effect of y of being in A_1 and R_1 is $(\beta_0 + \beta_1 + \gamma_1)$, although that on y of being in A_1 and R_2 is $(\beta_0 + \beta_1)$, with the difference being γ_1, a reasonable result.

7.3.2. *Additivity Versus Interaction*

When we specify a linear regression model, we make the assumption, among others, that the regressor variables have linearly additive effects on the dependent variable. There are situations where the assumption of the interaction effect of any two or more variables is appropriate—for example, in studying the returns from education. These returns, for instance, in the form of income, may depend on the number of years of education and race but, in addition to the effects of these two variables being additive, it is possible that the effect of education may be different for difference in race. That is, the effect of education may depend on whether one is white or nonwhite. The rationale here is that education effects income, perhaps, because of training. Race causes a difference in income, perhaps, because one race is disadvantaged generally. But same amount of training for one race may produce more income than for another—that is, the compounding of education and race effects. Another example would be the expenditure pattern of two age groups. It is possible that the marginal propensity to consume of the age group of 20 to 25 is greater than the one of the age group of 40 to 50 in addition to the usual income and age effects on spending.

Returning to the case of the income difference in a cross-section study, we may have a model

$$
\begin{aligned}
y = {} & \alpha_0 + \alpha_1 E_1 + \alpha_2 E_2 + \alpha_3 E_3 + \alpha_4 R_1 + \alpha_5 R_2 \\
& + \beta_1 E_1 R_1 + \beta_2 E_2 R_1 + \beta_3 E_3 R_1 \\
& + \gamma_1 E_1 R_2 + \gamma_2 E_2 R_2 + \gamma_3 E_3 R_2 + u
\end{aligned} \tag{7.47}
$$

so that, for example,

$$E(y \mid E_1 = 1, R_1 = 1) = \alpha_0 + \alpha_1 + \alpha_4 + \beta_1 \tag{7.49a}$$

and

$$E(y \mid E_1 = 1, R_2 = 1) = \alpha_0 + \alpha_1 + \alpha_5 + \gamma_1 \tag{7.49b}$$

Here, α_1 measures the effect on income of being in the E_1 class and, similarly, α_4 and α_5 are, respectively, the effects of being in the race classes R_1 and R_2; they are all additive effects of education and race. Now notice that the interaction terms allow for the effect of E_1 on income to be different for different race classes; namely, these effects are measured by β_1 and γ_1.

The model as it stands in (7.47) cannot be estimated for the same reasons as before and the dropping of terms becomes necessary. One form for estimation is

$$\begin{aligned} y = \alpha_0 + \alpha_2 E_2 + \alpha_3 E_3 + \alpha_5 R_2 \\ + \gamma_2 E_2 R_2 + \gamma_3 E_3 R_2 + u' \end{aligned} \tag{7.48a}$$

Notice that E_1 and R_1 have been dropped or that the constraint that $\alpha_1 = \alpha_4 = 0$ has been imposed. Then the interpretation of the estimates of the coefficients in (7.48a) must take this constraint into account. For details on this interpretation, see Goldberger (1964, Chapter 5, Section 2).

Interaction terms can be formulated for any two dummy systems or between a dummy system and a continuous variable. An example of the latter would be the interaction effects of income and education on the consumer durables purphase. Specifically, one may have the model:

$$\begin{aligned} \text{expenditure} = \alpha_0 + \alpha_1 E_1 + \alpha_2 E_2 + \alpha_3 E_3 + \alpha_4 y \\ + \beta_1 E_1 y + \beta_2 E_2 y + \beta_3 E_3 y + u \end{aligned}$$

For a sample of 12, a typical observation matrix on the regressors would be

$$\mathbf{X} = \begin{array}{c} \\ \end{array}\begin{bmatrix} & (E_1) & (E_2) & (E_3) & (y) & (E_1 y) & (E_2 y) & (E_3 y) \\ 1 & 1 & 0 & 0 & 500 & 500 & 0 & 0 \\ 1 & 0 & 0 & 1 & 380 & 0 & 0 & 380 \\ 1 & 0 & 1 & 0 & 420 & 0 & 420 & 0 \\ 1 & 0 & 1 & 0 & 570 & 0 & 570 & 0 \\ 1 & 0 & 0 & 1 & 290 & 0 & 0 & 290 \\ 1 & 1 & 0 & 0 & 720 & 720 & 0 & 0 \\ 1 & 1 & 0 & 0 & 800 & 800 & 0 & 0 \\ 1 & 0 & 1 & 0 & 900 & 0 & 900 & 0 \\ 1 & 0 & 1 & 0 & 250 & 0 & 250 & 0 \\ 1 & 0 & 0 & 1 & 860 & 0 & 0 & 860 \\ 1 & 1 & 0 & 0 & 725 & 725 & 0 & 0 \\ 1 & 0 & 0 & 1 & 605 & 0 & 0 & 605 \end{bmatrix}$$

Here the moment matrix $(\mathbf{X}'\mathbf{X})$ is singular; column 1 of $\mathbf{X}$ is a linear combination of columns 2, 3, and 4; column 5 is a linear combination of columns

6, 7, and 8. Thus one should drop one of the E_i columns and the corresponding $(E_i y)$ column to form $(\mathbf{X}'\mathbf{X})$ for estimation.

Although the dummy variable techniques discussed in this section are flexible schemes, they are expensive in degrees of freedom. Furthermore, the restriction on the effect of an independent variable to remain constant over a range of the variable may be a disadvantage. Added to this is the empirically messy problem of determining what may be the optimal boundaries of the dummy classes in a dummy system.

7.3.3. Dichotomous Dependent Variables

In the regression analysis of cross-section data models are frequently observed where the dependent variable is a qualitative variable having only two classes or alternatives. For instance, the only response to a survey question is yes or no; a consumer choice is to buy or not to buy, and so forth. This type of qualitative variable is generally referred to as a dichotomous dependent variable and usually is assigned the value of 0 or 1 for the purpose of estimation, although other methods of the assignment of value are possible.

Consider a cross section of consumers whose purchase behavior with respect to a major durable we want to study. Taking automobiles as an example, we might want to describe, for a sample of spending units, how the purchase of an automobile in one time period may be related to income (Y) and age (A). Let a spending unit that bought an automobile be given a value of 1 for the purchase variable (B) and let $B = 0$ for a spending unit not buying. Then we have

$$B = \beta_0 + \beta_1 Y + \beta_2 A + u \tag{7.52}$$

assuming that $Eu = 0$. By taking the expectation

$$E(B) = \beta_0 + \beta_1 Y + \beta_2 A$$

and noticing that the values of $E(B)$ can range from 0 to 1, we determine that $E(B_i)$ for the ith spending unit is the probability of that spending unit purchasing an automobile during the time period of study. Or

$$\Pr(B_i = 1) = \beta_0 + \beta_1 Y_i + \beta_2 A_i$$

while

$$\Pr(B_i = 0) = 1 - (\beta_0 + \beta_1 Y_1 + \beta_2 A_i)$$

This interpretation allows reference to a function of the type (7.52) as a probability function. When the relationship is linear, we call it a linear probability function.

In general, we can have the probability model

$$y_i = f(x_{1i}, x_{2i}, \ldots, x_{ki}; \boldsymbol{\theta}) + u_i \tag{7.53}$$

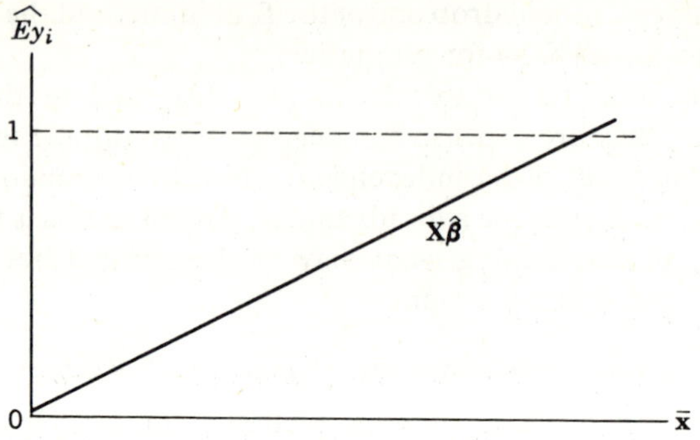

Figure 7.2

for the ith observation with $\boldsymbol{\theta}$ the parameter vector of certain size, and $Eu_i = 0$ for all i. Specifically, we can have the linear probability model

$$\mathbf{y} = \mathbf{X}\boldsymbol{\beta} + \mathbf{u} \tag{7.54}$$

with the assumptions that $E\mathbf{u} = \mathbf{0}$ *and* that $y_i = 0$ or 1, so that

$$Ey_i = \Pr(y_i = 1) = P_i \tag{7.55}$$

where y_i is the ith element of $\mathbf{y}$. Some statistical problems arise when OLS is applied to (7.54).

When the least-squares fit is made on (7.54), in effect, we have a linear estimate as in Figure 7.2. That is, if we can "collapse" the independent variables into, say, $\bar{\mathbf{x}}$, then the line designated in Ey_i is an estimate of the linear probability function. Since there is no provision in the estimation procedure to prevent this estimate from going out of the interval [0, 1], there will be some values of $\bar{\mathbf{x}}$ that will give estimates $\widehat{Ey_i} > 1$ or $\widehat{Ey_i} < 0$. Customarily, when such "outside" estimates are obtained, an arbitrary rule is applied whereby $\widehat{Ey_i}$ larger than 1 is set equal to 1 and $\widehat{Ey_i}$ less than 0 is set equal to 0.

Another problem is the fact that the disturbances are heteroskedastic. This is shown as follows. The true value of u_i is either P_i or $(1 - P_i)$ with the respective probabilities $P_i = \Pr(y_i = 1)$ and $(1 - P_i) = \Pr(y_i = 0)$, so that although

$$Eu_i = (1 - P_i)P_i - P_i(1 - P_i) = 0 \tag{7.55a}$$

we have

$$\begin{aligned} Eu_i^2 &= (1 - P_i)^2 P_i + P_i^2(1 - P_i) \\ &= P_i(1 - P_i) \end{aligned} \tag{7.55b}$$

which depends on individual observations. Zellner and Lee (1965) suggest the Aitken's procedure to meet this problem, noticing that

$$E\mathbf{uu}' = \begin{bmatrix} P_1(1-P_1) & 0 & \cdots & 0 \\ 0 & P_2(1-P_2) & \cdots & 0 \\ 0 & 0 & \cdot & 0 \\ \cdot & \cdot & \cdot & \cdot \\ \cdot & \cdot & \cdot & \cdot \\ \cdot & \cdot & \cdot & \cdot \\ 0 & 0 & \cdots & P_n(1-P_n) \end{bmatrix} \tag{7.56}$$

A two-step procedure can be applied here. First, run the regression of y on the X's, calculate the LS estimated $\hat{y}_i$, and equate $\hat{y}_i = \hat{P}_i$. Second, substitute $\hat{P}_i$ in place of P_i in (7.56) and obtain the Zellner estimate* of the $\boldsymbol{\beta}$ vector. In this connection, it is necessary to apply the arbitrary assignment of the P_i value in the event that $\hat{y}_i$ is outside of [0, 1]. It is not known what is to be gained from the application of Aitken's procedure, or the two-step procedure, when the cases of estimates of y_i lying outside [0, 1] are numerous.

Another choice of the form of f in (7.35) is an S-shaped curve, assumed in the probit analysis model, that allows for the estimate of Ey_i to be contained in the [0, 1] interval, as shown in Figure 7.3. For the estimation procedures and the interpretation involved in probit analysis, see Finney (1947) for the simple regression model and Tobin (1955) and Rosett (1959) for the multiple regression model. For other choices of the form of f in (7.53), such as logit model and Gompit model, see Zellner and Lee (1965).

For further discussion of the applications of linear probability, see Goldberger (1964, Chapter 5, Section 5) and Huang (1963).

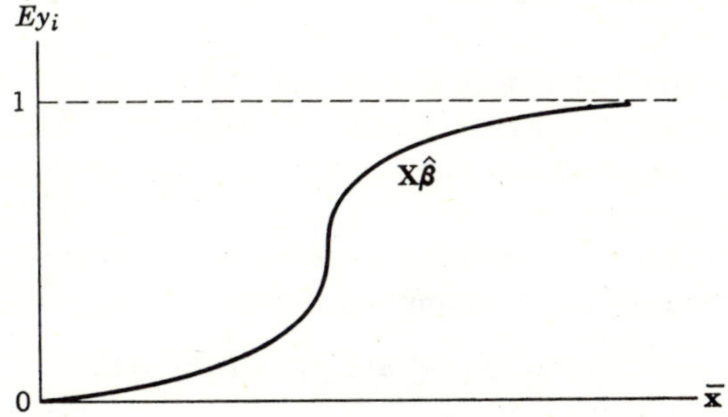

Figure 7.3

* See Aitken estimator, page 199.

7.4. Introduction to Nonlinear Regression

A natural extension of the consideration of the alternative functional form in (7.53) is to allow nonlinearity in the variables or in the parameters or both to enter f. Generally, if y is continuous and the behavioral relation is

$$y = f(x_1, x_2, \ldots, x_k; \theta_1, \theta_2, \ldots, \theta_p) + u \tag{7.57}$$

the form of f is not limited to the one that is linear in the parameters, although this is assumed in the standard regression model. That economic relationships abound in nonlinearity can be easily observed in Engel curve fitting, production function study, and in many other areas of work. The purpose of this section is to introduce some of the elements of nonlinear estimation. First, however, let us clarify some of the terminology.

7.4.1. *Some Preliminaries*

The model that is linear in the parameters and the variables would be

$$y = \beta_1 x_1 + \beta_2 x_2 + \cdots + \beta_k x_k + u \tag{7.57a}$$

the standard model that we have thus far considered in the preceding chapters. That is, in terms of (7.57) $\theta_i = \beta_i$ for $i = 1, 2, \ldots, k$, and $k = p$. If, on the other hand, we have a logarithmic transformation of a nonlinear function,

$$y = x_1^{\alpha_1} x_2^{\alpha_2} \cdots x_k^{\alpha_k} e^v$$

so as to obtain

$$\ln y = \alpha_1 \ln x_1 + \alpha_2 \ln x_2 + \cdots + \alpha_k \ln x_k + v \tag{7.57b}$$

we then have a model linear in the parameters and nonlinear in the variables. Another variant model may be

$$y = \delta_1 x_1 + \delta_2 \sin x_2 + \delta_3 \frac{x_3}{x_4} \cdots + \delta_p \ln x_k + w \tag{7.57c}$$

where p is arbitrary but reflects the number of terms that are functions of any number of x_i's, $i \leq p$. Models such as (7.57b) and (7.57c) are intrinsically linear, and the OLS procedure can be applied if the disturbances satisfy the usual standard assumptions.

For some models, the nonlinear estimation procedure is desirable. For instance, if we specify the consumption function to be given by

$$c_t = ky_t^e + \alpha(L_{t-1} - L_t^*) + u_t \tag{7.57d}$$

$$L_t^* = \eta y_t^e \tag{7.57e}$$

where y_t^e is the expected income for period t, L_t^* is the desired level of liquid assets for period t, and L_{t-1} is the opening stock of liquid assets. Although

y_t^e is not observable, we may assume that it is related to measured income (y_t) in the following way:

$$y^e{}_t = (1 - \lambda)(y_t + \lambda y_{t-1} + \lambda^2 y_{t-2} + \cdots) \tag{7.57f}$$

That is, the expected income is generated by applying a set of geometrically falling weights to the current and to the past measured incomes. Now, by combining (7.57d), (7.57e), and (7.57f), we arrive at

$$c_t = (k - \alpha\eta)(1 - \lambda)(y_t + \lambda y_{t-1} + \lambda^2 y_{t-2} + \cdots) + \alpha L_{t-1} + u_t \tag{7.57g}$$

This relation is not estimable because of the long series of lagged measured incomes. To derive an estimable relation, we use Koyck's device (what mathematicians often use in working with repeating decimals) by lagging (7.57g) one time period, by multiplying through the lagged relation by λ, and by subtracting it from (7.57g), and find the resulting expression:

$$c_t = \lambda c_{t-1} + (k - \alpha\eta)(1 - \lambda)y_t + \alpha(L_{t-1} - \lambda L_{t-2}) + (u_t - \lambda u_{t-1}) \tag{7.58}$$

Now, even if the disturbance terms $(u_t - \lambda u_{t-1})$ satisfy the standard assumptions, the OLS method of estimating this relation is not satisfactory because, for instance, λ will be overidentified (for a discussion of identification, see Chapter 9) in that there will be three estimated values for it. This is one strong reason why a nonlinear procedure, in which λ will be uniquely estimated, is recommended.

In general, models that are nonlinear in the parameters require nonlinear procedures for estimation, unless a method for the transformation of variables is available to reduce the model to one that is linear in the parameters, or unless there are other specifying assumptions of the model that make OLS applicable.

7.4.2. An Iterative Estimation Procedure

Various computation procedures are available to find the parameter estimates of nonlinear functions, for example, see Draper and Smith (1966, Chapter 10). Some procedures rely on an iterative process and yield only approximate results. Many procedures utilizing computer programs involve first linearizing the nonlinear function and then use the least-squares method on the linearized relation. For the detailed discussion of these procedures, see Meeter (1964), Marquardt (1963), Booth and Peterson (1960). The material below is merely intended to give an idea of what linearization and iterative process entail.

Given the model in (7.57), we can write the model in terms of the observations as follows

$$\mathbf{y} = f(\mathbf{X}; \boldsymbol{\theta}) + \mathbf{u} \tag{7.59}$$

where $\mathbf{y}$ is say $n \times 1$, $\mathbf{X}$ is $n \times k$, and $\boldsymbol{\theta}$ is $p \times 1$. We denote the sum of squares by S, for example, $\mathbf{u}'\mathbf{u} = S(\boldsymbol{\theta})$, $\boldsymbol{\theta}$ being the parameter vector associated with the disturbance vector $\mathbf{u}$. To linearize (7.59), we must first pick a certain parameter vector, say $\boldsymbol{\theta}_0$, around which we write the Taylor series expansion of (7.59). Thus

$$f(\mathbf{X}; \boldsymbol{\theta}) = f(\mathbf{X}; \boldsymbol{\theta}_0) + \left.\frac{\partial f}{\partial \boldsymbol{\theta}}\right|_{\theta=\theta_0} (\boldsymbol{\theta} - \boldsymbol{\theta}_0) + \frac{1}{2!}\left.\frac{\partial}{\partial \boldsymbol{\theta}}\frac{\partial f}{\partial \boldsymbol{\theta}}\right|_{\theta=\theta_0} (\boldsymbol{\theta} - \boldsymbol{\theta}_0)^2 + \cdots \tag{7.60a}$$

Dropping the second- and higher-order terms and letting $\boldsymbol{\eta}_0 = f(\mathbf{X}; \boldsymbol{\theta}_0)$, we have from (7.60a)

$$\mathbf{y} = \boldsymbol{\eta}_0 + \left.\frac{\partial f}{\partial \boldsymbol{\theta}}\right|_{\theta=\theta_0} (\boldsymbol{\theta} - \boldsymbol{\theta}_0) + \mathbf{u}_0 \tag{7.60b}$$

Notice that $\left.\dfrac{\partial f}{\partial \boldsymbol{\theta}}\right|_{\theta=\theta_0}$ means that the partials of f with respect to $\boldsymbol{\theta}$ are evaluated at $\boldsymbol{\theta}_0$ for each observation of x_i. To fix ideas about these quantities, we observe that

$$\left.\frac{\partial f}{\partial \boldsymbol{\theta}}\right|_{\theta=\theta_0} = \left.\begin{pmatrix} \dfrac{\partial f}{\partial \theta_1} \\ \dfrac{\partial f}{\partial \theta_2} \\ \cdot \\ \cdot \\ \cdot \\ \dfrac{\partial f}{\partial \theta_p} \end{pmatrix}\right|_{\theta=\theta_0}$$

$$= \begin{bmatrix} \left(\dfrac{\partial f(1)}{\partial \theta_1}\right)_0 & \left(\dfrac{\partial f(1)}{\partial \theta_2}\right)_0 & \cdots & \left(\dfrac{\partial f(1)}{\partial \theta_p}\right)_0 \\ \cdot & \cdot & \cdot & \cdot \\ \cdot & \cdot & \cdot & \cdot \\ \cdot & \cdot & \cdot & \cdot \\ \left(\dfrac{\partial f(n)}{\partial \theta_1}\right)_0 & \left(\dfrac{\partial f(n)}{\partial \theta_2}\right)_0 & \cdots & \left(\dfrac{\partial f(n)}{\partial \theta_p}\right)_0 \end{bmatrix} \tag{7.60c}$$

where $\partial f(i)/\partial \theta_j$ is the partial of f with respect to θ_j evaluated with the ith observation on the independent variables. The zero subscript outside the large parentheses indicates the initial (zeroth) iteration using $\boldsymbol{\theta}_0$. The matrix in (7.60c) is, of course, $n \times p$.

Now let

$$\mathbf{r}_0 = \mathbf{y} - \boldsymbol{\eta}_0 \tag{7.61a}$$

and

$$\mathbf{D}_0 = \left.\frac{\partial f}{\partial \boldsymbol{\theta}}\right|_{\theta=\theta_0} \tag{7.61b}$$

By combining (7.60b), (7.61a), and (7.61b), we arrive at

$$\mathbf{r}_0 = \mathbf{D}_0(\boldsymbol{\theta} - \boldsymbol{\theta}_0) + \mathbf{u}_0 \tag{7.62}$$

where $\mathbf{r}_0$ is the vector of the dependent variable, $\mathbf{D}_0$ is the observation matrix, and $(\boldsymbol{\theta} - \boldsymbol{\theta}_0)$ is the coefficient vector in the usual linear regression model. Thus, by choosing a starting value of $\boldsymbol{\theta}$, here $\boldsymbol{\theta}_0$, and linearizing the model, we come up with a usual linear regression model. The OLS estimate of $(\boldsymbol{\theta} - \boldsymbol{\theta}_0)$, given $\boldsymbol{\theta}_0$, then determines the new coefficient vector, say $\boldsymbol{\theta}_1$, since

$$\widehat{(\boldsymbol{\theta} - \boldsymbol{\theta}_0)} = (\mathbf{D}_0'\mathbf{D}_0)^{-1}\mathbf{D}_0'\mathbf{r}_0 \tag{7.62a}$$

and letting $\widehat{(\boldsymbol{\theta} - \boldsymbol{\theta}_0)} = \boldsymbol{\delta}_0$, we get $\boldsymbol{\theta}_1 = \boldsymbol{\delta}_0 + \boldsymbol{\theta}_0$.

With $\boldsymbol{\theta}_1$ determined from the first round of estimation, we are ready to start on the second. We form

$$\boldsymbol{\eta}_1 = f(\mathbf{X}; \boldsymbol{\theta}_1)$$

$$\mathbf{r}_1 = \mathbf{y} - \boldsymbol{\eta}_1$$

$$\mathbf{D}_1 = \left.\frac{\partial f}{\partial \boldsymbol{\theta}}\right|_{\theta=\theta_1}$$

and have the linear relation

$$\mathbf{r}_1 = \mathbf{D}_1(\boldsymbol{\theta} - \boldsymbol{\theta}_1) + \mathbf{u}_1$$

to which OLS is applied. This way the iteration is carried out until some predetermined criteria are satisfied. For example, the iteration may be stopped (1) if $S(\boldsymbol{\theta}_i)$ attains a certain value, or (2) if the percentage change in $S(\boldsymbol{\theta}_i)$ defined by $[S(\boldsymbol{\theta}_i) - S(\boldsymbol{\theta}_{i-1})]/S(\boldsymbol{\theta}_{i-1})$ reaches a small value, or (3) the percentage change in the vector $\boldsymbol{\theta}_i$ from $\boldsymbol{\theta}_{i-1}$ is practically zero, or (4) a combination of (2) and (3), and so forth.

The iterative procedures using OLS in each step of calculation in principle converge and yield consistent estimates. The procedure yields maximum likelihood estimators if the disturbances are normally distributed. For the asymptotic variance-covariance matrix of nonlinear estimaters of coefficients and other analytic results, see Hartley and Booker (1965).

7.5. Stochastic Explanatory Variables

Through Chapter 6 and much of Chapter 7, the model we have worked with has been the standard model in which the observation matrix **X** is fixed. This is a rather stringent assumption in view of the frequently observed phenomenon that economic variables tend to be mutually determining or interdependent, so that the variables that enter an equation as explanatory variables may themselves be affected by the dependent variables as well as by the other explanatory variables. This possibility aside, there exists a variety of situations in behavioral and other sciences where some or all of the explanatory variables in a linear regression model are random variables, stochastic and uncontrollable from sample to sample. This is a generalization of the classical or standard linear regression model.

That $x_0, x_1, x_2, \ldots, x_k$ are random means, for one thing, that the rows of **X** must now be given probability consideration. For another, the possibility now exists that the disturbance term may be correlated with or may not be independent of one or more of the x_i's. Our task in this section is to consider the applicability of the OLS estimation method to models with varying degrees of dependency between the disturbance and the explanatory variables, as it relates to the optimal properties established for the OLS estimator. To anticipate the development, we shall show that most of the nice properties of the OLS estimator carry over to the stochastic-explanatory variable model if the disturbance is independent of the explanatory variables. We shall learn that the OLS estimator is "bad" when there is a full dependence of the disturbance and the explanatory variables. In the sequel, we shall introduce the instrumental variable technique.

7.5.1. The Special Case

Where **X** is random and the disturbance term is fully independent of **X**, we have the special case of the stochastic explanatory variable model. Thus for the model

$$\mathbf{y} = \mathbf{X}\boldsymbol{\beta} + \mathbf{u} \tag{7.63}$$

the assumption of full independence means that in

$$\mathbf{X} = \begin{pmatrix} \mathbf{X}_1 \\ \mathbf{X}_2 \\ \cdot \\ \cdot \\ \cdot \\ \mathbf{X}_n \end{pmatrix} \quad \text{and} \quad \mathbf{u} = \begin{pmatrix} u_1 \\ u_2 \\ \cdot \\ \cdot \\ \cdot \\ u_n \end{pmatrix} \tag{7.63a}$$

$E\mathbf{X}_i u_j = 0$ for all i and j. This assumption together with the ones in (4.32) other than the fixity of $\mathbf{X}$ means that

$$E(\mathbf{u} \mid \mathbf{X}) = E\mathbf{u} = 0 \tag{7.64a}$$

$$\begin{aligned} E(\mathbf{y} \mid \mathbf{X}) &= E(\mathbf{X}\boldsymbol{\beta} + \mathbf{u} \mid \mathbf{X}) \\ &= \mathbf{X}\boldsymbol{\beta} + E(\mathbf{u} \mid \mathbf{X}) \\ &= \mathbf{X}\boldsymbol{\beta} \end{aligned} \tag{7.64b}$$

$$E(\mathbf{u}\mathbf{u}' \mid \mathbf{X}) = E\mathbf{u}\mathbf{u}' = \sigma^2\mathbf{I} \tag{7.64c}$$

It follows then that the disturbance vector **u** has the distributional properties conditional on the value of **X**. What would happen to the properties of the OLS estimators of the parameters $\boldsymbol{\beta}$, σ^2, and $\mathbf{V}(\mathbf{b})$, in this case? The answer is that the estimators are unbiased and consistent.

To see unbiasedness, we first consider the OLS estimate of $\boldsymbol{\beta}$, **b**. As previously

$$\begin{aligned} \mathbf{b} &= (\mathbf{X}'\mathbf{X})^{-1}\mathbf{X}'\mathbf{y} \\ &= \boldsymbol{\beta} + (\mathbf{X}'\mathbf{X})^{-1}\mathbf{X}'\mathbf{u} \end{aligned}$$

so that

$$\begin{aligned} E\mathbf{b} &= \boldsymbol{\beta} + E[(\mathbf{X}'\mathbf{X})^{-1}\mathbf{X}'\mathbf{u}] \\ &= \boldsymbol{\beta} + E(\mathbf{X}'\mathbf{X})^{-1}\mathbf{X}'E\mathbf{u} \\ &= \boldsymbol{\beta} \end{aligned} \tag{7.65}$$

Now the estimator for σ^2 is

$$s^2 = (\hat{\mathbf{u}}'\hat{\mathbf{u}})/(n - k - 1) \tag{7.66}$$

and

$$E\hat{\mathbf{u}}'\hat{\mathbf{u}} = E(\mathbf{u}'\mathbf{M}\mathbf{u} \mid \mathbf{X}) = \sigma^2(n - k - 1) \tag{7.67}$$

Therefore, by combining (7.66) and (7.67), we have the expectation

$$Es^2 = E\{E(s^2 \mid \mathbf{X})\} = \sigma^2 \tag{7.68}$$

As for the sample covariance of **b**, we notice that

$$\begin{aligned} E\mathbf{S}(\mathbf{b}) &= E\{E(\mathbf{S}(\mathbf{b}) \mid \mathbf{X})\} \\ &= E\{E[s^2(\mathbf{X}'\mathbf{X})^{-1} \mid \mathbf{X}]\} \\ &= \sigma^2 E[(\mathbf{X}'\mathbf{X})^{-1} \mid \mathbf{X}]\} \\ &= \sigma^2(\mathbf{X}'\mathbf{X})^{-1} \end{aligned} \tag{7.69}$$

We conclude that the OLS estimators **b**, s^2, and $\mathbf{S}(\mathbf{b})$ are unbiased for $\boldsymbol{\beta}$, σ^2, and $\mathbf{V}(\mathbf{b})$, respectively.

For consistency, we first must make the assumptions about how the $\mathbf{X}_i$'s and the u_i's in (7.63a) are generated, since we shall be considering the properties of the OLS estimators as the sample size becomes infinite. Thus, for the model (7.63), we assume that

u is a sample generated by a probability law so that $Eu_i = 0$, $Eu_i^2 = \sigma^2$, and $Eu_i u_j = 0$ for $i \neq j$, for all i and j (7.63b)

X is a sample generated by a multivariate probability law with the second-moment matrix for $\mathbf{X}_i$ equal to $\mathbf{\Sigma}$ for all i, $|\mathbf{\Sigma}| \neq 0$ (7.63c)

The laws generating **X** and **u** are independent (7.63d)

These assumptions allow the specifications:

$$E\mathbf{u} = \mathbf{0} \tag{7.70a}$$

$$E\mathbf{uu}' = \sigma^2\mathbf{I} \tag{7.70b}$$

$$\text{plim}\, n^{-1}(\mathbf{u}'\mathbf{u}) = \sigma^2 \tag{7.70c}$$

$$\text{plim}\, n^{-1}(\mathbf{X}'\mathbf{X}) = \mathbf{\Sigma} \tag{7.70d}$$

$$\text{plim}\, n^{-1}(\mathbf{X}'\mathbf{u}) = 0 \tag{7.70e}$$

Now we proceed to take the probability limits of the OLS estimators. First

$$\begin{aligned} \text{plim}\, \mathbf{b} &= \text{plim}\, (\mathbf{X}'\mathbf{X})^{-1}\mathbf{X}'\mathbf{y} \\ &= \text{plim}\, [\boldsymbol{\beta} + (\mathbf{X}'\mathbf{X})^{-1}\mathbf{X}'\mathbf{u}] \\ &= \boldsymbol{\beta} + \text{plim}\, (\mathbf{X}'\mathbf{X})^{-1}\mathbf{X}'\mathbf{u} \\ &= \boldsymbol{\beta} + \text{plim}\, (n^{-1}\mathbf{X}'\mathbf{X})^{-1}\, \text{plim}\, (n^{-1}\mathbf{X}'\mathbf{u}) \\ &= \boldsymbol{\beta} + \mathbf{\Sigma}^{-1} \cdot \mathbf{0} \\ &= \boldsymbol{\beta} \end{aligned} \tag{7.71}$$

Second, noticing that

$$\hat{\mathbf{u}}'\hat{\mathbf{u}} = \mathbf{u}'\mathbf{u} - \mathbf{u}'\mathbf{X}(\mathbf{X}'\mathbf{X})^{-1}\mathbf{X}'\mathbf{u}$$

we have

$$\begin{aligned} \text{plim}\, (n^{-1}\mathbf{u}'\mathbf{u}) &= \text{plim}\, (n^{-1}\mathbf{u}'\mathbf{u}) - \text{plim}\, (n^{-1}\mathbf{u}'\mathbf{X}(\mathbf{X}'\mathbf{X})^{-1}\mathbf{X}'\mathbf{u}) \\ &= \sigma^2 - \text{plim}\, (n^{-1}\mathbf{u}'\mathbf{X})\, \text{plim}\, (n^{-1}\mathbf{X}'\mathbf{X})^{-1}\, \text{plim}\, (n^{-1}\mathbf{X}'\mathbf{u}) \\ &= \sigma^2 - \mathbf{0} \cdot \mathbf{\Sigma} \cdot \mathbf{0} \\ &= \sigma^2 \end{aligned}$$

Therefore, the OLS estimator of σ^2 has the probability limit

$$\begin{aligned} \text{plim}\, s^2 &= \text{plim}\, (n - k - 1)^{-1}\hat{\mathbf{u}}'\hat{\mathbf{u}} \\ &= \text{plim}\, [(n - k - 1)^{-1}n]\, \text{plim}\, (n^{-1}\hat{\mathbf{u}}'\hat{\mathbf{u}}) \\ &= 1 \cdot \sigma^2 \\ &= \sigma^2 \end{aligned} \tag{7.72}$$

Consider now the estimator $\mathbf{S}(\mathbf{b})$ for the asymptotic covariance of $\mathbf{b}$, $\bar{\mathbf{V}}(\mathbf{b})$, which from (7.69) is $\bar{\mathbf{V}}(\mathbf{b}) = n^{-1}\sigma^2\boldsymbol{\Sigma}^{-1}$. We have

$$\begin{aligned}\text{plim } \mathbf{S}(\mathbf{b}) &= \text{plim } s^2(\mathbf{X}'\mathbf{X})^{-1} \\ &= \text{plim } [n^{-1}s^2(n^{-1}\mathbf{X}'\mathbf{X})^{-1}] \\ &= n^{-1}\sigma^2\boldsymbol{\Sigma}^{-1} \end{aligned} \tag{7.73}$$

and we observe that $\mathbf{S}(\mathbf{b})$ is consistent for $\bar{\mathbf{V}}(\mathbf{b})$.

7.5.2. The General Case

When there is dependence the probability laws generating the samples $\mathbf{X}$ and $\mathbf{u}$, the assumptions (7.64b to c) and (7.70e) are no longer appropriate. As a result, in the general case the expectation of the OLS estimator for $\boldsymbol{\beta}$ is

$$E\mathbf{b} = \boldsymbol{\beta} + E[(\mathbf{X}'\mathbf{X})^{-1}\mathbf{X}'\mathbf{u}] \tag{7.74}$$

where the last term does not vanish even though $E\mathbf{u} = \mathbf{0}$. This is because $\mathbf{X}$ and $\mathbf{u}$ are not independent. Thus, $\mathbf{b}$ is biased for $\boldsymbol{\beta}$ as long as $E[(\mathbf{X}'\mathbf{X})^{-1}\mathbf{X}'\mathbf{u}]$ is nonzero. Also

$$\begin{aligned}\text{plim } \mathbf{b} &= \boldsymbol{\beta} + \text{plim } (n^{-1}\mathbf{X}'\mathbf{X})^{-1}(n^{-1}\mathbf{X}'\mathbf{u}) \\ &= \boldsymbol{\beta} + \text{plim } (n^{-1}\mathbf{X}'\mathbf{X})^{-1} \text{ plim } (n^{-1}\mathbf{X}'\mathbf{u}) \end{aligned} \tag{7.75}$$

where the last term does not approach zero because plim $(n^{-1}\mathbf{X}'\mathbf{u}) \neq \mathbf{0}$. Therefore, $\mathbf{b}$ is both biased and inconsistent for $\boldsymbol{\beta}$. Similarly, the OLS estimators s^2 and $\mathbf{S}(\mathbf{b})$ are biased for σ^2 and $\mathbf{V}(\mathbf{b})$, respectively; and s^2 and $\mathbf{S}(\mathbf{b})$ are inconsistent for σ^2 and $\mathbf{V}(\mathbf{b})$. We have the situation that Goldberger aptly describes as follows. "If a regressor is correlated with the disturbance, the least-squares estimation—which attempts to give as much credit to regressors and as little to disturbance as possible—will give a misleading estimate of the influence of variations in the regressor on variations in the regressand."

The case of full dependence between $\mathbf{X}$ and $\mathbf{u}$ arises in economics frequently, mostly through the interdependence of the variables that appear in an equation. This falls in the realm of simultaneous equation systems and is discussed in Chapters 9 and 10.

7.5.3. Instrumental Variable Technique

However, there is an approach called instrumental variable technique that circumvents the difficulties of the OLS estimators in the general case. This technique relies on choosing a set of instrumental variables, say, $z_0, z_1, z_2, \ldots, z_k$, which are uncorrelated with *u and* are correlated with the x_i's. That is, for

$$\mathbf{Z} = (\mathbf{z}_0\ \mathbf{z}_1\ \mathbf{z}_2 \cdots \mathbf{z}_k)$$

an $n \times (k + 1)$ observation matrix on the z_i's, we require that

$$\text{plim } n^{-1}\mathbf{Z}'\mathbf{u} = \mathbf{0} \tag{7.76}$$

and that

$$\text{plim } n^{-1}\mathbf{Z}'\mathbf{X} = \mathbf{Var}\,(\mathbf{z}, \mathbf{x}) \tag{7.77}$$

is finite and nonsingular. Then the instrumental variable estimator of $\boldsymbol{\beta}$, say,

$$\mathbf{b}^* = (\mathbf{Z}'\mathbf{X})^{-1}\mathbf{Z}'\mathbf{y} \tag{7.78}$$

can be shown to be consistent for $\boldsymbol{\beta}$. This is indicated as follows. Noting that $\mathbf{y} = \mathbf{X}\boldsymbol{\beta} + \mathbf{u}$, we have

$$\begin{aligned} \mathbf{b}^* &= (\mathbf{Z}'\mathbf{X})^{-1}\mathbf{Z}'(\mathbf{X}\boldsymbol{\beta} + \mathbf{u}) \\ &= \boldsymbol{\beta} + (\mathbf{Z}'\mathbf{X})^{-1}\mathbf{Z}'\mathbf{u} \end{aligned} \tag{7.79}$$

so that

$$\begin{aligned} \text{plim } \mathbf{b}^* &= \boldsymbol{\beta} + \text{plim } (n^{-1}\mathbf{Z}'\mathbf{X})^{-1}(n^{-1}\mathbf{Z}'\mathbf{u}) \\ &= \boldsymbol{\beta} + \text{plim } (n^{-1}\mathbf{Z}'\mathbf{X})^{-1} \text{ plim } (n^{-1}\mathbf{Z}'\mathbf{u}) \\ &= \boldsymbol{\beta} \end{aligned} \tag{7.80}$$

by (7.76) and (7.77). Notice that, although $\mathbf{b}^*$ is consistent for $\boldsymbol{\beta}$, it is not necessarily unbiased; see (7.79) where $E\mathbf{Z}'\mathbf{u}$ is not necessarily $\mathbf{0}$ for small samples.

Since

$$(\mathbf{b}^* - \boldsymbol{\beta})(\mathbf{b}^* - \boldsymbol{\beta})' = (\mathbf{Z}'\mathbf{X})^{-1}\mathbf{Z}'\mathbf{u}\mathbf{u}'\mathbf{Z}(\mathbf{X}'\mathbf{Z})^{-1}$$

the asymptotic covariance matrix of $\mathbf{b}^*$ is

$$\mathbf{Var}\,(\mathbf{b}^*) = n^{-1}\sigma^2[\mathbf{Var}\,(\mathbf{z}, \mathbf{x})]^{-1}[\mathbf{Var}\,(\mathbf{z})][\mathbf{Var}\,(\mathbf{z}, \mathbf{x})']^{-1}$$

which can be consistently estimated by

$$\frac{(\mathbf{y} - \mathbf{X}\mathbf{b}^*)'(\mathbf{y} - \mathbf{X}\mathbf{b}^*)}{n - k - 1}(\mathbf{Z}'\mathbf{X})^{-1}(\mathbf{Z}'\mathbf{Z})[(\mathbf{Z}'\mathbf{X})^{-1}]'$$

See, for example, Goldberger (1964, pp. 284–286) for this derivation.

7.6. Lagged Variables as Regressors

Another problem that frequently complicates life for empirical workers is the use of lagged variables as regressors. It is the purpose of this section to make a brief survey of two types of lagged-variable models: distributed lag models and lagged dependent variable models.

7.6.1. Distributed Lag Models

In the analysis of cause-and-effect relations, say, between two variables, it is sometimes proper to assume that one variable affects the other with a time-lag. For instance, the receipt of income in the time period $t - 1$, x_{t-1}, determines the consumption expenditure on services in period t, y_t; or one may assume a model

$$y_t = \alpha + \beta x_{t-1} + u_t \tag{7.85}$$

where all the assumptions of the standard regression model are met. Under certain conditions, it also may be proper to entertain the assumption that the lagged effect of income on the services expenditure is distributed over a number of time periods, say,

$$y_t = \alpha + \beta x_{t-1} + \gamma x_{t-2} + \delta x_{t-3} + v_t \tag{7.86}$$

In model (7.85), we have a fixed lag relation and in (7.86) a distributed lag model. As long as the x_{t-i}'s have predetermined values that are independent of the disturbances, the OLS estimates of the coefficients will be unbiased and consistent whether or not the disturbances are autocorrelated (see Section 6.2).

One difficulty in estimating the distributed lag models occurs when the number of time periods over which the lagged effect is distributed is large. A generalization of (7.86) would be

$$y_t = \alpha + \sum_{i=0}^{\infty} \beta_i x_{t-i} + w_t \tag{7.86a}$$

Notice that we are assuming that x_t, instead of x_{t-1}, is the first of the distributed lag terms, which is in line with the usual practice. Now, generally, if the lag index i becomes large in a model, it is not possible to estimate such a model, since the number of observations would be small in relation to the number of lagged terms, or since one cannot get accurate estimates because of the high collinearity among the lagged variables.

The estimation of models such as (7.86a) may not be entirely hopeless if some constraints can be placed on the lag coefficients. One very popular scheme is the assumption that the lag coefficients are a set of geometrically falling weights, say,

$$y_t = \alpha + \beta \sum_{i=0}^{\infty} \lambda^i x_{t-i} + u_t \tag{7.87}$$

The coefficients in this model cannot be estimated as the equation stands, but the use of Koyck's device, discussed previously, allows us to arrive at

$$y_t = \alpha(1 - \lambda) + \lambda y_{t-1} + \beta x_t + (u_t - \lambda u_{t-1}) \tag{7.88}$$

This model contains a lagged dependent variable y_{t-1} which is not independent of the disturbance term $(u_t - \lambda u_{t-1})$. The application of OLS to (7.88) no longer yields unbiased and consistent estimates; nor is the GLS procedure appropriate, although $(u_t - \lambda u_{t-1})$ is autocorrelated.

7.6.2. *Lagged Dependent Variable*

Models with lagged dependent variables are not always of the type shown in (7.88). Very often in economic analysis, a priori reasoning may suggest that the variation in a variable is partly or wholly explained by its own lagged value, say, lagged one period. For example, the famous habit persistence theory of the consumption function stipulates, in effect, that

$$c_t = \alpha + \beta c_{t-1} + \gamma y_t + u_t \tag{7.89}$$

where c_t is the current consumption expenditure, y_t is disposable income, and u_t has mean zero and constant variance with no autocorrelation. Models of the form (7.89) are also called autoregressive linear regression models. Notice that at time t, c_{t-1} is not independent of $u_{t-2}, u_{t-3}, \ldots$, since, for instance, c_{t-1} is determined in part by its lagged value c_{t-2} which is dependent on u_{t-2}. However, the contemporary value of u, u_t, is independent of c_{t-1}, so that by deduction c_t is independent of $u_{t+1}, u_{t+2}, \ldots$. In short, there is a partial dependence between the lagged dependent variable and the disturbances. Therefore, the OLS estimates of α, β, and γ in (7.89) are biased. But it can be shown that these OLS estimates are consistent, that their asymptotic variances exist, and that these variances are approximated by the usual OLS estimates of them. See Christ (1966, pp. 374–379) for an instructive proof of the consistency of the OLS estimator of β in a simple model $y_t = \beta y_{t-1} + u_t$. For an excellent survey article on the analysis of the various types of lag models, see Griliches (1967). For an ingenious technique for the analysis of distributed lag schemes, see Almon (1965).

CHAPTER 8

MULTIVARIATE REGRESSION

In the preceding chapters we have been concerned with the treatment of the dependence of a variable on another variable and on other variables. Our analysis was called (1), in the single independent variable case, simple regression analysis, and (2), in the case involving more than one independent variable, multiple regression analysis. Furthermore, in both cases we were interested in the behavior of one dependent variable. Consequently, we can call all of our preceding work univariate regression analysis. In this chapter, in contrast, we consider multivariate regression analysis. In essence, we discuss estimation and the related problems concerning the sets of regression equations, whether simple or multiple.

In the behavioral sciences, situations arise where we wish to consider the reactions of a set of related variables to a set of independent variables. In economics we often speak of endogenous variables and exogenous variables in the analysis of a problem, and at times, for forecasting and other purposes, one may wish to regress each of the endogenous variables on the same set of the exogenous variables included in the study. Again, for instance, in medicine, blood pressure and pulse rate may be the two variables that occur among several other independent variables, and for some purposes of analysis it may be better to think of blood pressure and pulse rate as individually dependent on the same set of independent variables. Of course, we might wish to see how dependent blood pressure is on pulse rate and the other independent variables, but this is a problem to be dealt with in the chapters on simultaneous equations.

Now, by definition, endogenous variables are jointly dependent on each other in a system and, consequently, it is often the case that interest centers on the behavioral dependence of an endogenous variable on some or all of the other endogenous variables and on some or all of the exogenous variables. The analysis of this type of relationship is in the realm of simultaneous equation analysis and will be postponed until Chapters 9 and 10. In this chapter, then, we confine our attention to the sets of regression equations where the dependent variables have no causal influence on one another. As our progress will make clear, our effort in this chapter will not only spell out some useful techniques in the analysis of equations which may seem totally unrelated but will also provide an important background for the study of simultaneous equations systems. We divide our work into (1) the multivariate linear regression model, and (2) the sets of linear regression equations. In the former we study a set of dependent variables, each having the same set of independent or explanatory variables. In the latter, we discuss a set of dependent variables that do not necessarily have the same independent variables.

8.1. Multivariate Linear Regression Model

In line with our discussions in Chapters 4 and 5, we deal here only with sets of linear regression equations whose independent variables are fixed. Since the problem of estimation in the multivariate linear regression model is a straightforward generalization of the univariate regression model in almost every respect, and since the direct uses of the multivariate model are not frequent in economics, we shall be brief in this section. Notice that the notations to be developed for the model and the basic concepts of the intercorrelated disturbances in the model are fundamental to our future work.

8.1.1. Notations and the Model

The notations on the models and their analysis that we shall discuss from now on become rather complex, so that it is advisable to take pains to learn the notations so as to quickly grasp the meaning of a matrix equation or a covariance matrix, and the like when it is encountered.

The extension of the single equation multiple regression model to the multi-equation situation is easily accomplished by assigning equational subscripts to the usual single equation model. That is, if the single equation model, as in (4.2), is

$$y = \beta_0 + \beta_1 X_1 + \beta_2 X_2 + \cdots + \beta_K X_K + u \tag{8.1a}$$

we assign subscript g to denote the individual equations as follows

$$y_g = \beta_{0g} + \beta_{1g} X_1 + \beta_{2g} X_2 + \cdots + \beta_{Kg} X_K + ug \tag{8.1b}$$

Say, $g = 1, 2, \ldots, G$, here y_g is the dependent variable of the gth equation, β_{1g} is the ith coefficient in the gth equation, and so on. Then, for a sample of size n, a full representation of the model and the sample is

$$\mathbf{y}_g = \mathbf{X}\boldsymbol{\beta}_g + \mathbf{u}_g \qquad g = 1, 2, \ldots, G \tag{8.1c}$$

where $\mathbf{y}_g$ is $n \times 1$, $\mathbf{X}$ is $n \times (K + 1)$, and $\mathbf{u}_g$ is $n \times 1$. To simplify notations we, henceforth, shall assume that we have the K independent variable instead of $(K + 1)$. Thus (8.1c) reduces to

$$\mathbf{y}_g = \mathbf{X}\boldsymbol{\beta}_g + \mathbf{u}_g \qquad g = 1, 2, \ldots, G \tag{8.2}$$

with change that $\mathbf{X}$ is now $n \times K$.

The assumptions of the model (8.2) are the ones of the fixed model discussed in Chapter 4. They are, for $g = 1, 2, \ldots, G$,

$$E\mathbf{u}_g = \mathbf{0} \tag{8.2a}$$

$$E\mathbf{u}_g\mathbf{u}_g' = \sigma^2\mathbf{I} = \sigma_{gg}\mathbf{I} \tag{8.2b}$$

$$\rho(\mathbf{X}) = K \tag{8.2c}$$

$$\mathbf{X} \text{ is fixed} \tag{8.2d}$$

One additional assumption that must be made for the model (8.2) is that there is a possible correlation among the disturbances of the different equations. We assume that for any two equations numbered g and g'

$$E\mathbf{u}_g\mathbf{u}_{g'} = \sigma_{gg'}\mathbf{I} \tag{8.2e}$$

meaning that the contemporaneous covariance of u_g and $u_{g'}$ is constant at $\sigma_{gg'}$ and that there are no lagged covariances among the u_g and $u_{g'}$ values at different time periods. To fix ideas, (8.2e) can be written as

$$E\begin{bmatrix} u_{g1}u_{g'1} & u_{g2}u_{g'2} & \cdots & u_{g1}u_{g'n} \\ u_{g2}u_{g'1} & u_{g2}u_{g'2} & \cdots & u_{g2}u_{g'n} \\ \vdots & \vdots & & \vdots \\ u_{gn}u_{g'1} & u_{gn}u_{g'2} & \cdots & u_{gn}u_{g'n} \end{bmatrix} = \begin{bmatrix} \sigma_{gg'} & 0 & 0 & \cdots & 0 \\ 0 & \sigma_{gg'} & 0 & \cdots & 0 \\ 0 & 0 & \sigma_{gg'} & \cdots & 0 \\ \vdots & \vdots & \vdots & & \vdots \\ 0 & 0 & 0 & \cdots & \sigma_{gg'} \end{bmatrix} = \sigma_{gg'}\mathbf{I} \tag{8.2f}$$

where $\mathbf{I}$ is a square matrix of order n. As is indicated from our earlier discussion, (8.2f) implies that the contemporaneous covariance of u_g and $u_{g'}$ is $\sigma_{gg'}$, for all time periods, and there is no serial correlation among the u_g and $u_{g'}$ variables at different periods of time.

8.1.2. Estimation

Because of the assumptions (8.2a to d) the application of the least-squares method to an individual equation will yield BLUE estimates of the coefficients of the equation. Furthermore, if u_g is normally distributed, the LS estimator will be a maximum likelihood estimator. But now the question is: Since an equation to be estimated is one of the G equations and since (8.2e) states that the disturbances are contemporaneously correlated, would it not be more efficient (relative to a given set of information) to estimate the G equations jointly?* Most of this subsection is devoted to answering this question.

To anticipate the result, we notice that there are at least three alternative methods for estimating the system (8.2) that satisfies the assumptions (8.2a to 8.2e), namely, equation-by-equation least squares, least squares applied to the entire set of equations simultaneously [see the system (8.3a)], and generalized least squares. We shall show that the last method will give the same estimates as does single equation least squares although, in fact, the three methods are equivalent under the assumptions given. It will be seen that the reasons for this result are two: (1) the disturbances are independent of all the independent variables, and (2) the observation matrix for the independent variables $\mathbf{X}$ is the same in all equations.

First we collect all the G equations and write them in a compact system

$$\begin{bmatrix} \mathbf{y}_1 \\ \mathbf{y}_2 \\ \cdot \\ \cdot \\ \cdot \\ \mathbf{y}_G \end{bmatrix} = \begin{bmatrix} \mathbf{X} & \mathbf{0} & \cdots & \mathbf{0} & \mathbf{0} \\ \mathbf{0} & \mathbf{X} & \cdots & \mathbf{0} & \mathbf{0} \\ \cdot & \cdot & \cdot & \cdot & \cdot \\ \cdot & \cdot & \cdot & \cdot & \cdot \\ \cdot & \cdot & \cdot & \cdot & \cdot \\ \mathbf{0} & \mathbf{0} & \cdots & \mathbf{0} & \mathbf{X} \end{bmatrix} \begin{bmatrix} \boldsymbol{\beta}_1 \\ \boldsymbol{\beta}_2 \\ \cdot \\ \cdot \\ \cdot \\ \boldsymbol{\beta}_G \end{bmatrix} + \begin{bmatrix} \mathbf{u}_1 \\ \mathbf{u}_2 \\ \cdot \\ \cdot \\ \cdot \\ \mathbf{u}_G \end{bmatrix} \tag{8.3}$$

or

$$\mathbf{y} = \mathbf{Z}\boldsymbol{\beta} + \mathbf{U} \tag{8.3a}$$

where $\mathbf{y}$ is $Gn \times 1$, $\mathbf{Z}$ is $Gn \times GK$, β is $GK \times 1$, and $\mathbf{U}$ is $Gn \times 1$. Written thus, it appears that the usual least-squares procedure is applicable, but on examination of the assumptions (8.2a to e) we have

$$E\mathbf{UU}' = \begin{bmatrix} \sigma_{11} & \sigma_{12} & \cdots & \sigma_{1G} \\ \sigma_{21} & \sigma_{22} & \cdots & \sigma_{2G} \\ \cdot & \cdot & \cdot & \cdot \\ \cdot & \cdot & \cdot & \cdot \\ \cdot & \cdot & \cdot & \cdot \\ \sigma_{G1} & \sigma_{G2} & \cdots & \sigma_{GG} \end{bmatrix} \otimes \mathbf{I} = \boldsymbol{\Sigma} \otimes \mathbf{I} = \boldsymbol{\Omega} \tag{8.4}$$

where $\otimes$ denotes Kronecker multiplication, that is, every element of the matrix immediately to the right of the first equality sign in (8.4) is to be multiplied by $\mathbf{I}$, which is of order n, so that the entire variance-covariance matrix of $\mathbf{U}$ is $Gn \times Gn$. Now, because the covariance matrix of $\mathbf{U}$ is not diagonal, our discussion in Section 6.2 tells us that the direct application

* By joint estimation we mean the estimation of all the equations in the model simultaneously in one process. The term 'joint' is used in contrast with the separate or the equation-by-equation method.

of least squares to (8.3a) will not yield the BLUE estimator of $\boldsymbol{\beta}$ there but, instead, that the use of the Aitken's generalized least-squares procedure will. Of course, as stated earlier, this latter procedure is really not the only answer for estimating the system (8.3a), and we shall see this shortly. In the meantime, we proceed with Aitken's procedure.

From Section 6.2, the BLUE estimator of $\boldsymbol{\beta}$ in (8.3a) is

$$\mathbf{b}^* = (\mathbf{Z}'\boldsymbol{\Omega}^{-1}\mathbf{Z})^{-1}\mathbf{Z}'\boldsymbol{\Omega}^{-1}\mathbf{y} \tag{8.5}$$

where $\boldsymbol{\Omega}$ is as defined in (8.4), and the variance-covariance of $\mathbf{b}^*$ is

$$\mathbf{V}(\mathbf{b}^*) = (\mathbf{Z}'\boldsymbol{\Omega}^{-1}\mathbf{Z})^{-1} \tag{8.6}$$

An investigation of the matrices in (8.5) shows that

$$(\mathbf{Z}'\boldsymbol{\Omega}^{-1}\mathbf{Z})^{-1} = [\mathbf{Z}'(\boldsymbol{\Sigma}^{-1} \otimes \mathbf{I})\mathbf{Z}]^{-1}$$

$$= \begin{pmatrix} \sigma_{11}(\mathbf{X}'\mathbf{X})^{-1} & \sigma_{12}(\mathbf{X}'\mathbf{X})^{-1} & \cdots & \sigma_{1G}(\mathbf{X}'\mathbf{X})^{-1} \\ \cdot & \cdot & \cdot & \cdot \\ \cdot & \cdot & \cdot & \cdot \\ \cdot & \cdot & \cdot & \cdot \\ \sigma_{G1}(\mathbf{X}'\mathbf{X})^{-1} & \sigma_{G2}(\mathbf{X}'\mathbf{X})^{-1} & \cdots & \sigma_{GG}(\mathbf{X}'\mathbf{X})^{-1} \end{pmatrix} \tag{8.7a}$$

since

$$\mathbf{Z}'\boldsymbol{\Omega}^{-1}\mathbf{Z}$$

$$= \begin{pmatrix} \mathbf{X}' & \mathbf{0} & \cdots & \mathbf{0} \\ \mathbf{0} & \mathbf{X}' & \cdots & \mathbf{0} \\ \cdot & \cdot & \cdot & \cdot \\ \cdot & \cdot & \cdot & \cdot \\ \cdot & \cdot & \cdot & \cdot \\ \mathbf{0} & \mathbf{0} & \cdots & \mathbf{X}' \end{pmatrix} \begin{pmatrix} \sigma^{11}\mathbf{I} & \sigma^{\ 2}\mathbf{I} & \cdots & \sigma^{1G}\mathbf{I} \\ \sigma^{21}\mathbf{I} & \sigma^{22}\mathbf{I} & \cdots & \sigma^{2G}\mathbf{I} \\ \cdot & \cdot & \cdot & \cdot \\ \cdot & \cdot & \cdot & \cdot \\ \cdot & \cdot & \cdot & \cdot \\ \sigma^{G1}\mathbf{I} & \sigma^{G2}\mathbf{I} & \cdots & \sigma^{GG}\mathbf{I} \end{pmatrix} \begin{pmatrix} \mathbf{X} & \mathbf{0} & \cdots & \mathbf{0} \\ \mathbf{0} & \mathbf{X} & \cdots & \mathbf{0} \\ \cdot & \cdot & \cdot & \cdot \\ \cdot & \cdot & \cdot & \cdot \\ \cdot & \cdot & \cdot & \cdot \\ \mathbf{0} & \mathbf{0} & \cdots & \mathbf{X} \end{pmatrix}$$

$$= \begin{pmatrix} \mathbf{X}'\sigma^{11} & \mathbf{X}'\sigma^{12} & \cdots & \mathbf{X}'\sigma^{1G} \\ \mathbf{X}'\sigma^{21} & \mathbf{X}'\sigma^{22} & \cdots & \mathbf{X}'\sigma^{2G} \\ \cdot & \cdot & \cdot & \cdot \\ \cdot & \cdot & \cdot & \cdot \\ \cdot & \cdot & \cdot & \cdot \\ \mathbf{X}'\sigma^{G1} & \mathbf{X}'\sigma^{G2} & \cdots & \mathbf{X}'\sigma^{GG} \end{pmatrix} \begin{pmatrix} \mathbf{X} & \mathbf{0} & \cdots & \mathbf{0} \\ \mathbf{0} & \mathbf{X} & \cdots & \mathbf{0} \\ \cdot & \cdot & \cdot & \cdot \\ \cdot & \cdot & \cdot & \cdot \\ \cdot & \cdot & \cdot & \cdot \\ \mathbf{0} & \mathbf{0} & \cdots & \mathbf{X} \end{pmatrix}$$

$$= \begin{pmatrix} \sigma^{11}\mathbf{X}'\mathbf{X} & \sigma^{12}\mathbf{X}'\mathbf{X} & \cdots & \sigma^{1G}\mathbf{X}'\mathbf{X} \\ \sigma^{21}\mathbf{X}'\mathbf{X} & \sigma^{22}\mathbf{X}'\mathbf{X} & \cdots & \sigma^{2G}\mathbf{X}'\mathbf{X} \\ \cdot & \cdot & \cdots & \cdot \\ \cdot & \cdot & \cdots & \cdot \\ \cdot & \cdot & \cdots & \cdot \\ \sigma^{G1}\mathbf{X}'\mathbf{X} & \sigma^{G2}\mathbf{X}'\mathbf{X} & \cdots & \sigma^{GG}\mathbf{X}'\mathbf{X} \end{pmatrix}$$

Furthermore,

$$(\mathbf{Z}'\boldsymbol{\Omega}^{-1}\mathbf{y}) = \mathbf{Z}'(\Sigma^{-1} \otimes \mathbf{I})\mathbf{y}$$

$$= \begin{pmatrix} \mathbf{X}' & \mathbf{0} & \cdots & \mathbf{0} \\ \mathbf{0} & \mathbf{X}' & \cdots & \mathbf{0} \\ \cdot & \cdot & \cdot & \cdot \\ \cdot & \cdot & \cdot & \cdot \\ \cdot & \cdot & \cdot & \cdot \\ \mathbf{0} & \mathbf{0} & \cdots & \mathbf{X}' \end{pmatrix} \begin{pmatrix} \sigma^{11}\mathbf{I} & \sigma^{12}\mathbf{I} & \cdots & \sigma^{1G} \\ \sigma^{21}\mathbf{I} & \sigma^{22}\mathbf{I} & \cdots & \sigma^{2G}\mathbf{I} \\ \cdot & \cdot & \cdot & \cdot \\ \cdot & \cdot & \cdot & \cdot \\ \cdot & \cdot & \cdot & \cdot \\ \sigma^{G1}\mathbf{I} & \sigma^{G2}\mathbf{I} & \cdots & \sigma^{GG}\mathbf{I} \end{pmatrix} \begin{pmatrix} \mathbf{y}_1 \\ \mathbf{y}_2 \\ \cdot \\ \cdot \\ \cdot \\ \mathbf{y}_G \end{pmatrix}$$

$$= \left[\begin{pmatrix} \sigma^{11}\mathbf{I} & \sigma^{12}\mathbf{I} & \cdots & \sigma^{1G}\mathbf{I} \\ \sigma^{21}\mathbf{I} & \sigma^{22}\mathbf{I} & \cdots & \sigma^{2G}\mathbf{I} \\ \cdot & \cdot & \cdot & \cdot \\ \cdot & \cdot & \cdot & \cdot \\ \cdot & \cdot & \cdot & \cdot \\ \sigma^{G1}\mathbf{I} & \sigma^{G2}\mathbf{I} & \cdots & \sigma^{GG}\mathbf{I} \end{pmatrix} \otimes \mathbf{X}' \right] \begin{pmatrix} \mathbf{y}_1 \\ \mathbf{y}_2 \\ \cdot \\ \cdot \\ \cdot \\ \mathbf{y}_G \end{pmatrix}$$

$$= \begin{pmatrix} \sum_{g=1}^{G} \sigma^{ig}\mathbf{X}'\mathbf{y} \\ \cdot \\ \cdot \\ \cdot \\ \sum_{g=1}^{G} \sigma^{Gg}\mathbf{X}'\mathbf{y} \end{pmatrix} \tag{8.7b}$$

In (8.7b) we denote the ijth element of the inverse of $\boldsymbol{\Sigma}$ by σ^{ij}. This practice will be followed elsewhere in this book. It is becoming clear from these developments that the estimator for a subvector of $\boldsymbol{\beta}$, say $\boldsymbol{\beta}_i$, corresponding to the ith equation is

$$\begin{aligned} \mathbf{b}_i^* &= \sum_{g=1}^{G} \sigma_{ig}(\mathbf{X}'\mathbf{X})^{-1} \sum_{j=1}^{G} \sigma^{gj}\mathbf{X}'\mathbf{y}_j \\ &= \sum_{g=1}^{G} \sum_{j=1}^{G} \sigma_{ig}\sigma^{gj}(\mathbf{X}'\mathbf{X})^{-1}\mathbf{X}'\mathbf{y}_j \\ &= (\mathbf{X}'\mathbf{X})^{-1}\mathbf{X}'\mathbf{y}_i \end{aligned} \tag{8.8}$$

This is because

$$\sum_{g=1}^{G} \sigma_{ig}\sigma^{gj} = \begin{cases} 1 & \text{if} \quad i = j \\ 0 & \text{if} \quad i \neq j \end{cases}$$

by virtue of the definition of the matrix $\boldsymbol{\Sigma}$ and its inverse $\boldsymbol{\Sigma}^{-1}$ (that is, $\boldsymbol{\Sigma}\boldsymbol{\Sigma}^{-1} = \mathbf{I}$). And the variance-covariance of the estimator in (8.8) is

$$\mathbf{V}(\mathbf{b}_i^*) = \sigma_{ii}(\mathbf{X}'\mathbf{X})^{-1} \tag{8.9}$$

as can be taken directly from (8.7a).

From (8.8) and (8.9), we are led to conclude that nothing can be gained in estimating the equations (8.2) jointly. The equation-by-equation least squares gives the desirable estimator.*

There is another situation where the joint estimation of the multivariate regression equations does not bring about any gains. This is the special case of structural independence among the G equations, or

$$E\mathbf{u}_g\mathbf{u}_{g'}' = \mathbf{0} \tag{8.2s}$$

with the covariance matrix corresponding to (8.4) being

$$E\mathbf{U}\mathbf{U}' = \begin{pmatrix} \sigma_{11} & 0 & \cdots & 0 \\ 0 & \sigma_{22} & \cdots & 0 \\ \cdot & \cdot & \cdot & \cdot \\ \cdot & \cdot & \cdot & \cdot \\ \cdot & \cdot & \cdot & \cdot \\ 0 & 0 & \cdots & \sigma_{GG} \end{pmatrix} \otimes \mathbf{I}$$

It is advisable to show that in the event that (8.2s) holds true, a similar proof that led to (8.8) and (8.9) can be carried out.

In a more general situation than (8.4) we may have the presence of heteroskedasticity and autocorrelation in the disturbances of an equation in the G equations. In such a general case the covariance matrix of the disturbance vector $\mathbf{U}$ corresponding to (8.4) would be

$$\boldsymbol{\Omega} = \begin{pmatrix} \boldsymbol{\Sigma}_{11} & \boldsymbol{\Sigma}_{12} & \cdots & \boldsymbol{\Sigma}_{1G} \\ \boldsymbol{\Sigma}_{21} & \boldsymbol{\Sigma}_{22} & \cdots & \boldsymbol{\Sigma}_{2G} \\ \cdot & \cdot & \cdot & \cdot \\ \cdot & \cdot & \cdot & \cdot \\ \cdot & \cdot & \cdot & \cdot \\ \boldsymbol{\Sigma}_{G1} & \boldsymbol{\Sigma}_{G2} & \cdots & \boldsymbol{\Sigma}_{GG} \end{pmatrix} \tag{8.4g}$$

where $\boldsymbol{\Sigma}_{ij}$, $i = j$, is an $n \times n$ matrix that denotes the temporal variances of the elements of $\mathbf{u}_i$ as well as their autocovariances, and for $i \neq j$, $\boldsymbol{\Sigma}_{ij}$ contains the contemporaneous and lagged covariances of the elements in the disturbance vectors $\mathbf{u}_i$ and $\mathbf{u}_j$. To illustrate these concepts in a two equation system, we write, for a sample size n,

$$\mathbf{y}_1 = \mathbf{X}\boldsymbol{\beta}_1 + \mathbf{u}_1 \qquad \text{and} \qquad \mathbf{y}_2 = \mathbf{X}\boldsymbol{\beta}_2 + \mathbf{u}_2$$

* A shorter proof of the equivalence of the "joint" estimator and the equation-by-equation OLS estimator is available through the use of these theorems: $(\mathbf{A} \otimes \mathbf{B})(\mathbf{C} \otimes \mathbf{D}) = \mathbf{AC} \otimes \mathbf{BD}$; $(\mathbf{A} \otimes \mathbf{B})^{-1} = \mathbf{A}^{-1} \otimes \mathbf{B}^{-1}$; and $(\mathbf{A} \otimes \mathbf{B})(\mathbf{A}^{-1} \otimes \mathbf{B}^{-1}) = \mathbf{AA}^{-1} \otimes \mathbf{BB}^{-1} = \mathbf{I}$. That is,

$$\mathbf{b}^* = (\mathbf{Z}'\boldsymbol{\Omega}^{-1}\mathbf{Z})^{-1}(\mathbf{Z}'\boldsymbol{\Omega}^{-1}\mathbf{y}) = [\mathbf{Z}'(\boldsymbol{\Sigma}^{-1} \otimes \mathbf{I})\mathbf{Z}]^{-1}[\mathbf{Z}'(\boldsymbol{\Sigma}^{-1} \otimes \mathbf{I})\mathbf{y}] = (\mathbf{Z}'\boldsymbol{\Sigma}^{-1} \otimes \mathbf{Z})^{-1}$$
$$\times (\mathbf{Z}'\boldsymbol{\Sigma}^{-1} \otimes \mathbf{y}) = [\boldsymbol{\Sigma} \otimes (\mathbf{Z}'\mathbf{Z})^{-1}][\boldsymbol{\Sigma}^{-1} \otimes \mathbf{Z}'\mathbf{y}] = \mathbf{I} \otimes (\mathbf{Z}'\mathbf{Z})^{-1}\mathbf{Z}'\mathbf{y} = (\mathbf{Z}'\mathbf{Z})^{-1}\mathbf{Z}'\mathbf{y}.$$

In the most general case

$$
E\mathbf{UU}' = E\begin{pmatrix} u_{11} \\ u_{21} \\ \cdot \\ \cdot \\ \cdot \\ u_{n1} \\ u_{12} \\ u_{22} \\ \cdot \\ \cdot \\ \cdot \\ u_{n2} \end{pmatrix} (u_{11}u_{21} \cdots u_{n1}u_{12}u_{22} \cdots u_{n2})
$$

$$
= \left(\begin{array}{cccc|cccc}
\sigma_{11,11} & \sigma_{11,21} & \cdots & \sigma_{11,n1} & \sigma_{11,12} & \sigma_{11,22} & \cdots & \sigma_{11,n2} \\
\sigma_{21,11} & \sigma_{21,21} & \cdots & \sigma_{21,n1} & \sigma_{21,12} & \sigma_{21,22} & \cdots & \sigma_{21,m2} \\
\cdot & \cdot & \cdot & \cdot & \cdot & \cdot & \cdot & \cdot \\
\cdot & \cdot & \cdot & \cdot & \cdot & \cdot & \cdot & \cdot \\
\cdot & \cdot & \cdot & \cdot & \cdot & \cdot & \cdot & \cdot \\
\sigma_{n1,11} & \sigma_{n1,21} & \cdots & \sigma_{n1,n1} & \sigma_{n1,12} & \sigma_{n1,22} & \cdots & \sigma_{n1,n2} \\
\hline
\sigma_{12,11} & \sigma_{12,21} & \cdots & \sigma_{12,n1} & \sigma_{12,12} & \sigma_{12,22} & \cdots & \sigma_{12,n2} \\
\sigma_{22,11} & \sigma_{22,21} & \cdots & \sigma_{22,n1} & \sigma_{22,12} & \sigma_{22,22} & \cdots & \sigma_{22,n2} \\
\cdot & \cdot & \cdot & \cdot & \cdot & \cdot & \cdot & \cdot \\
\cdot & \cdot & \cdot & \cdot & \cdot & \cdot & \cdot & \cdot \\
\cdot & \cdot & \cdot & \cdot & \cdot & \cdot & \cdot & \cdot \\
\sigma_{n2,11} & \sigma_{n2,21} & \cdots & \sigma_{n2,n1} & \sigma_{n2,12} & \sigma_{n2,22} & \cdots & \sigma_{n2,n2}
\end{array}\right)
$$

$$
= \left(\begin{array}{c|c} \boldsymbol{\Sigma}_{11} & \boldsymbol{\Sigma}_{12} \\ \hline \boldsymbol{\Sigma}_{21} & \boldsymbol{\Sigma}_{22} \end{array}\right) \tag{8.4h}
$$

Notice, for instance, that the diagonal elements of $\boldsymbol{\Sigma}_{11}$ are variances of the elements of $\mathbf{u}_1$ at different time periods, although the off-diagonal elements are autocovariances. $\boldsymbol{\Sigma}_{11}$ will correspond to the familiar Laurentian matrix mentioned in Section 5.4 if the second of each of the double subscripts is eliminated from the elements of $\boldsymbol{\Sigma}_{11}$.

If, indeed, the assumption (8.4g) holds true in model (8.3), then the LS method applied to the system (8.3) will no longer yield estimators with desirable properties. In particular, the LS estimators of the coefficients will be unbiased but will be inefficient. See, for example, Zellner (1961).

For the model described in (8.3) and accompanied by the assumptions

(8.2a to e), the estimation method can be equation-by-equation LS or the application of the generalized least-squares method. Both yield the same estimates. Also in the reference just cited, Zellner shows that the LS procedure applied to the system (8.3) directly, as if it were one equation, will yield the same estimates. In general, since the $\boldsymbol{\Omega}$ matrix in (8.4) is not known, it is consoling to know that the usual LS method, either applied equation-by-equation or by all equations at once, gives estimates with desirable properties.

8.1.3. Error of Forecast in Multivariate Regression Model

Although it is the case that the "joint" estimation procedures provide no gains over the equation-by-equation OLS estimation, there is one area in the application of the multivariate regression model where a joint approach is useful. If there is contemporaneous dependence among the disturbances of the component equations of the model, it is better to use a joint forecast region to accompany the point forecasts of the equations than to use individual forecast intervals. As will be shown later in Chapter 9, the multivariate regression model we have discussed thus far is equivalent to the so-called reduced form of a simultaneous equation system. The reduced form equations are frequently used for forecasting in econometric work; hence, the motivation for an analysis of the error of forecast in the multivariate regression models. We shall borrow considerable portions of the ideas and results contained in the present subsection from the joint work by Hooper and Zellner (1961).

Given the model in (8.3a), $\mathbf{y} = \mathbf{X}\boldsymbol{\beta} + \mathbf{U}$, and the covariance matrix, $E\mathbf{U}\mathbf{U}' = \boldsymbol{\Omega}$, we now consider observing $\mathbf{y}^F$, a $G \times 1$ column vector with the ith element corresponding to the ith equation, and $\mathbf{Z}^F$, a $G \times GK$ matrix for the forecast period F. Following (8.3), we notice that

$$\mathbf{Z}^F = \begin{bmatrix} \mathbf{X}^F & \mathbf{0} & \cdots & \mathbf{0} \\ \mathbf{0} & \mathbf{X}^F & \cdots & \mathbf{0} \\ \cdot & \cdot & \cdot & \cdot \\ \cdot & \cdot & \cdot & \cdot \\ \cdot & \cdot & \cdot & \cdot \\ \mathbf{0} & \mathbf{0} & \cdots & \mathbf{X}^F \end{bmatrix}$$

where $\mathbf{X}^F$ is a K-element row vector representing the observations on the K independent variables in the forecast period. (Recall that in (8.3) the matrix $\mathbf{X}$ is of order $n \times K$.) It then follows that, for the forecast period,

$$\mathbf{y}^F = \mathbf{Z}^F\boldsymbol{\beta} + \mathbf{u}^F \tag{8.10}$$

and for the least-squares estimate of $\boldsymbol{\beta}$, $\mathbf{b}$, we have the forecast value of $\mathbf{y}$

and the error of the forecast **f**, respectively, as follows

$$\hat{\mathbf{y}}^F = \mathbf{Z}^F\mathbf{b}$$

$$\mathbf{f} = \mathbf{y}^F - \hat{\mathbf{y}}^F = \mathbf{Z}^F(\boldsymbol{\beta} - \mathbf{b}) + \mathbf{u}^F \tag{8.10a}$$

Let us assume that $E\mathbf{u}^F = \mathbf{0}$, $E(\mathbf{u}^F)(\mathbf{u}^F)' = \boldsymbol{\Omega}$, and $E\mathbf{u}^F\mathbf{u}_i = \mathbf{0}$, for $i = 1, 2, \ldots, G$, so that the covariance matrix of the error of forecast is

$$\begin{aligned} E\mathbf{f}\mathbf{f}' &= E[\mathbf{Z}^F(\boldsymbol{\beta} - \mathbf{b}) + \mathbf{u}^F][\mathbf{Z}^F(\boldsymbol{\beta} - \mathbf{b}) + \mathbf{u}^F]' \\ &= \mathbf{Z}^F[\boldsymbol{\Omega} \otimes (\mathbf{X}'\mathbf{X})^{-1}](\mathbf{Z}^F)' + \boldsymbol{\Omega} \\ &= \boldsymbol{\Sigma}_{FF} \end{aligned} \tag{8.10b}$$

Since usually $\boldsymbol{\Sigma}_{FF}$ is not known because $\boldsymbol{\Omega}$ is often unknown, let us say that we estimate $\boldsymbol{\Omega}$ by

$$\hat{\boldsymbol{\Omega}} = \frac{1}{n - k}(\mathbf{Y}'\mathbf{Y} - \mathbf{P}'\mathbf{X}'\mathbf{X}\mathbf{P}) \tag{8.10c}$$

and $\boldsymbol{\Sigma}_{FF}$ by

$$\mathbf{S}_{FF} = \mathbf{Z}^F[\hat{\boldsymbol{\Omega}} \otimes (\mathbf{X}'\mathbf{X})^{-1}](\mathbf{Z}^F)' + \hat{\boldsymbol{\Omega}} \tag{8.10d}$$

where

$$\mathbf{Y} = (\mathbf{y}_1\, \mathbf{y}_2 \cdots \mathbf{y}_G)$$

and

$$\mathbf{P} = (\mathbf{b}_1\, \mathbf{b}_2 \cdots \mathbf{b}_G)$$

That is, **Y** is $n \times G$, **P** is $K \times G$, and $\mathbf{b}_i$ is the OLS estimate of the coefficients of the ith equation. Now from (8.10a) and (8.10d), we form the Hotellings T^2 statistic

$$T^2 = \mathbf{f}'\mathbf{S}_{FF}^{-1}\mathbf{f}$$

and it can be shown that*

$$F = \left[\frac{n - K - G + 1}{(n - K)G}\right]T^2$$

has the F distribution with G and $(n - K - G + 1)$ degrees of freedom if (8.10) and the distributional assumptions about $\mathbf{y}^F$ are, indeed, true. We then can define a set of points contained in the region satisfying the inequality

$$\left[\frac{n - K - G + 1}{(n - K)G}\right]T^2 \le F_\alpha \tag{8.10e}$$

where F_α is the value of F preselected at an arbitrary probability level α. Such a region is the joint forecast region for the G predicted values given by the G equations and may be interpreted in the same way as the interval for a single equation. Specifically, if repeated samples are taken, holding

* See, for example, Donald Morrison (1967, pp. 117–120) and the references cited there.

X and $\mathbf{X}^F$ constant, then $(1 - \alpha)$ percent of the time a region, as in (8.10e), will contain the true values of the forecasted variables—namely, $\mathbf{y}^F$.

An illustration given by Hooper and Zellner is a two-equation system

$$c_t = \pi_{11}X_{1t} + \pi_{12}X_{2t} + u_{1t}$$
$$y_t = \pi_{21}X_{1t} + \pi_{22}X_{2t} + u_{2t} \qquad (t = 1, \ldots, n)$$

where y_t = deflated disposable income per capita during t, and c_t = deflated consumer expenditure per capita during t. In our notation the least-squares estimates of $\begin{pmatrix}\pi_{21}\\ \pi_{12}\end{pmatrix}$ and $\begin{pmatrix}\pi_{21}\\ \pi_{22}\end{pmatrix}$ are, respectively, $\mathbf{b}_1$ and $\mathbf{b}_2$, so that $\mathbf{P} = (\mathbf{b}_1\, \mathbf{b}_2)$, and for the data Hooper and Zellner used

$$\mathbf{b}_1 = \begin{pmatrix}298.56\\ 1.499\end{pmatrix} \qquad \mathbf{b}_2 = \begin{pmatrix}285.79\\ 2.105\end{pmatrix}$$

the estimate of $\boldsymbol{\Omega}$ is

$$\hat{\boldsymbol{\Omega}} = \begin{bmatrix} 0.02507 & -0.02570 \\ -0.02570 & 0.03125 \end{bmatrix}^{-1} \tag{8.10f}$$

and the inverse of the moment matrix of the independent variables

$$(\mathbf{X}'\mathbf{X})^{-1} = \begin{bmatrix} 0.73294 & -0.0070133 \\ -0.0070133 & 0.00007498 \end{bmatrix} \tag{8.10g}$$

Suppose that $\mathbf{X}^F = (1 \quad 100)$, then given that $n = 13$, $K = 2$, and $G = 2$, we have the F value of 4.10 at the 5 percent level of significance and, following (8.10d) and (8.10e), we find that

$$\mathbf{S}_{FF} = 1.08007\ \hat{\boldsymbol{\Omega}} \tag{8.10h}$$

Suppose now that the predicted values of c_t and y_t are

$$\hat{c}^F = 448.454$$
$$\hat{y}^F = 496.287$$

Then we can use (8.10e), (8.10f), (8.10g), and (8.10h) to obtain

$$\begin{aligned} 9.472 = 0.02507(\hat{c}^F - c^F)^2 - 0.05139(\hat{c}^F - c^F)(\hat{y}^F - y^F) \\ + 0.03125(\hat{y}^F - y^F)^2 \end{aligned} \tag{8.10i}$$

This defines an ellipse in the parameter space of c^F and y^F, where the center of the ellipse is $(\hat{c}^F, \hat{y}^F)$, and the area covered by the ellipse is the forecast region. This region is shown in Figure 8.1. Notice that if we were to use

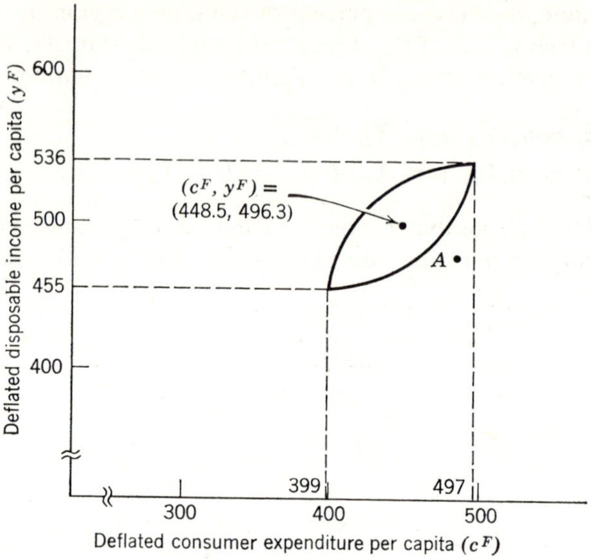

Figure 8.1

each single equation to obtain separate forecast intervals, points such as A can be in each of the forecast intervals but not in the joint forecast region. This observation is important in the appraisal of the forecasts given by a system of equations and indicates the usefulness of calculating joint forecast regions when making forecasts from several equations.

8.2 Sets of Regression Equations

The model discussed in Section 8.1 is quite restrictive in that in economic research we frequently encounter other types of multivariate response problems. When several dependent variables are each regressed on the same set of independent variables having the same observation matrix, we then have the multivariate regression model. In many problems it is quite possible that the observation matrix may be different for different dependent variables, although the independent variables may be the same. For instance, the consumption function for two metropolitan cities over a number of years may be analyzed by securing the local aggregates of both income and consumption expenditures for that period. Frequently, it is also the case that part of the observation matrix of the independent variables in one equation is different from another, although the rest of the matrix is the same. For instance, the total services expenditures in each of the two cities may be considered dependent on national income (the same observations for both cities) and on the respective cities' unemployment rate. In general, once we depart from the case of the same independent variables-same observation matrix, we face the area of analysis that we call the sets of regression equations.

There are three main reasons for wanting to consider a set of equations in the analysis of economic and other behavioral relationships: (1) the specification in a number of regression equations may be such that certain commonly correlated variables are not included in the equations to be estimated; (2) the dependent variables may be somehow correlated; and (3) when conditions in (1) and (2) exist, the disturbances of a set of equations would be correlated, at least, contemporaneously. In these cases it would be more efficient to apply a joint estimation procedure. The last point is not always obvious. Hence, the equations under analysis may be "seemingly unrelated."* We shall refer to a set of equations as a system of equations or merely as a system.

In the next subsection we give notations for a set of equations and describe several cases in which the analysis can be carried out in the context of a set of equations. Later, we shall discuss, in particular, two equations: (1) the efficient estimation or Aitken's generalized least squares used for seemingly unrelated regression equations, and (2) the setwise least squares utilized on a number of equations for saving degrees of freedom or for other reasons of computational or analytical convenience. In the setwise least squares no explicit consideration of the correlations among disturbances may be necessary.

8.2.1. Notations and Applications

As previously, let us suppose that there are G equations each having a set of independent variables and the corresponding observation matrix of the independent variables. Then we may write the set of equations as

$$\begin{aligned} \mathbf{y}_1 &= \mathbf{X}_1\boldsymbol{\beta}_1 + \mathbf{u}_1 \\ \mathbf{y}_2 &= \mathbf{X}_2\boldsymbol{\beta}_2 + \mathbf{u}_2 \\ &\cdot \\ &\cdot \\ &\cdot \\ \mathbf{y}_G &= \mathbf{X}_G\boldsymbol{\beta}_G + \mathbf{u}_G \end{aligned} \tag{8.11}$$

For convenience (although it is not necessary), let us also assume that the sample size for each of the equations is n. Thus, in (8.11), $\mathbf{y}_i$ is $n \times 1$ for

* See Arnold Zellner (1962, pp. 348–68). Theil (1970) refers to these equations as disturbance related.

$i = 1, 2, \ldots, G$; $\mathbf{X}_i$ is $n \times K_i$ where K_i can be different for different equations; $\boldsymbol{\beta}_i$ is $K_i \times 1$; and $\mathbf{u}_i$ is $n \times 1$ for $i = 1, 2, \ldots, G$.

The range of problems that fit the general framework of (8.11) is varied and numerous. We may be studying the behavior of a group of individuals for whom we have a panel data over a number of time periods, say T. Suppose that the behavior under study entails the consumption expenditures on durables d and services s. Now take individual A. The dependence of his consumption expenditures on income y, liquid asset holdings L, and the lagged consumption expenditures may be analyzed by the set of equations:

$$\begin{aligned} d_t &= \alpha_0 + \alpha_1 y_t + \alpha_2 L_t + \alpha_3 d_{t-1} + u_{1t} \\ s_t &= \beta_0 + \beta_1 y_t + \beta_2 L_t + \beta_3 s_{t-1} + u_{2t} \\ t &= 1, 2, \ldots, T \end{aligned} \tag{8.12}$$

Here it is quite possible that $Eu_{1t}u_{2t} \neq 0$ for $t = 1, 2, \ldots, T$. Also, the observation matrices for the independent variables in the two equations are not entirely the same.

It is also possible to consider a set of equations when we wish to analyze just the durables expenditures of a group of individuals, for example, 5. Thus

$$\begin{aligned} d_t^1 &= \alpha_0^1 + \alpha_1^1 y_t^1 + \alpha_2^1 L_t^1 + \alpha_3^1 d_{t-1}^1 + u_t^1 \\ d_t^2 &= \alpha_0^2 + \alpha_1^2 y_t^2 + \alpha_2^2 L_t^2 + \alpha_3^2 d_{t-1}^2 + u_t^2 \\ &\;\;\vdots \\ d_t^5 &= \alpha_0^5 + \alpha_1^5 y_t^5 + \alpha_2^5 L_t^5 + \alpha_3^5 d_{t-1}^5 + u_t^5 \\ t &= 1, 2, \ldots, T \end{aligned} \tag{8.13}$$

where superscripts refer to individuals. In this set we may have

$$Eu_t^i u_t^j \neq 0 \qquad i, j = 1, 2, \ldots, 5 \qquad t = 1, 2, \ldots, T \tag{8.13a}$$

as a result of, say, a possible fact that all individuals may be subject to some kind of common shock in the same time periods not accountable by the independent variables.

We may wish to study the apparel sales in four metropolitan areas of the United States. The sales in each locality may be dependent on the local consumers' income, climate, and so forth, and also all localities may be subject to a nationwide influence (for instance, wars). In this case the disturbance of the four equations that explain the apparel expenditures in the four areas may have a positive correlation.

There is no end to this enumeration, and the applicable situations may be spotted as they arise. We now consider the methods of the estimation of the set (8.11).

8.2.2. Efficient Estimation

Our effort in Subsection 8.1.2 showed that a multivariate regression model could be estimated equation-by-equation instead of by Aitken's joint procedure, even though the covariances of the disturbances were not zero. There, the crucial point was that the observation matrix of the independent variables was the same for every equation. Now in the set of equations (8.11) the observation matrix ($\mathbf{X}_i$) for one equation is assumed different from another and, in addition, the covariances of $\mathbf{u}_i$ and $\mathbf{u}_j$, $i, j = 1, 2, \ldots, G$, are not zero. Our task in this subsection is to clarify that under the conditions

$$\mathbf{X}_i \neq \mathbf{X}_j \qquad \text{for } i \neq j \tag{8.12a}$$

$$E\mathbf{u}_i\mathbf{u}_j' = \sigma_{ij}\mathbf{I} \qquad i, j = 1, 2, \ldots, G \tag{8.12b}$$

where $\sigma_{ij} \neq 0$, Aitken's generalized least squares applied to (8.11) yields estimators that are more efficient than the estimators given by the equation-by-equation LS procedure. The proof is similar to the one given in Section 8.1 and, therefore, will be brief.

We let the notations of (8.11) be

$$\mathbf{y} = \mathbf{X}\boldsymbol{\beta} + \mathbf{u} \tag{8.11a}$$

where

$$\mathbf{y} = \begin{pmatrix} \mathbf{y}_1 \\ \mathbf{y}_2 \\ \cdot \\ \cdot \\ \cdot \\ \mathbf{y}_G \end{pmatrix} \qquad \mathbf{X} = \begin{pmatrix} \mathbf{X}_1 & \mathbf{0} & \cdots & \mathbf{0} \\ \mathbf{0} & \mathbf{X}_2 & \cdots & \mathbf{0} \\ \cdot & \cdot & \cdot & \cdot \\ \cdot & \cdot & \cdot & \cdot \\ \cdot & \cdot & \cdot & \cdot \\ \mathbf{0} & \mathbf{0} & \cdots & \mathbf{X}_G \end{pmatrix}$$

$$\boldsymbol{\beta} = \begin{pmatrix} \boldsymbol{\beta}_1 \\ \boldsymbol{\beta}_2 \\ \cdot \\ \cdot \\ \cdot \\ \boldsymbol{\beta}_G \end{pmatrix} \qquad \mathbf{u} = \begin{pmatrix} \mathbf{u}_1 \\ \mathbf{u}_2 \\ \cdot \\ \cdot \\ \cdot \\ \mathbf{u}_G \end{pmatrix}$$

Furthermore, from (8.12b),

$$E\mathbf{uu}' = \begin{pmatrix} \sigma_{11}\mathbf{I} & \sigma_{12}\mathbf{I} & \cdots & \sigma_{1G}\mathbf{I} \\ \sigma_{21}\mathbf{I} & \sigma_{22}\mathbf{I} & \cdots & \sigma_{2G}\mathbf{I} \\ \cdot & \cdot & \cdot & \cdot \\ \cdot & \cdot & \cdot & \cdot \\ \cdot & \cdot & \cdot & \cdot \\ \sigma_{G1}\mathbf{I} & \sigma_{G2}\mathbf{I} & \cdots & \sigma_{GG}\mathbf{I} \end{pmatrix} = \begin{pmatrix} \sigma_{11} & \sigma_{12} & \cdots & \sigma_{1G} \\ \sigma_{21} & \sigma_{22} & \cdots & \sigma_{2G} \\ \cdot & \cdot & \cdot & \cdot \\ \cdot & \cdot & \cdot & \cdot \\ \cdot & \cdot & \cdot & \cdot \\ \sigma_{G1} & \sigma_{G2} & \cdots & \sigma_{GG} \end{pmatrix} \otimes \mathbf{I} = \mathbf{\Sigma} \otimes \mathbf{I} \tag{8.14}$$

Thus, now treating (8.11a) as one regression equation and noticing that $E\mathbf{uu}'$ is not diagonal, we apply Aitken's generalized least squares to (8.11a). Moving on, the variance-covariance matrix (8.14) implies that the transformation matrix on $\mathbf{y}$ and $\mathbf{X}$ and $\mathbf{u}$ is to be taken so that

$$\mathbf{T}'\mathbf{T} = (E\mathbf{uu}')^{-1} = (\mathbf{\Sigma} \otimes \mathbf{I})^{-1} \tag{8.15}$$

Thus for

$$\mathbf{Ty} = \mathbf{TX}\boldsymbol{\beta} + \mathbf{Tu} \tag{8.16}$$

the Aitken estimator for $\boldsymbol{\beta}$ is

$$\begin{aligned} \mathbf{b}^* &= (\mathbf{X}'\mathbf{T}'\mathbf{TX})^{-1}\mathbf{X}'\mathbf{T}'\mathbf{Ty} \\ &= [\mathbf{X}'(\mathbf{\Sigma} \otimes \mathbf{I})^{-1}\mathbf{X}]^{-1}[\mathbf{X}'(\mathbf{\Sigma} \otimes \mathbf{I})^{-1}\mathbf{y}] \\ &= \begin{bmatrix} \sigma^{11}\mathbf{X}_1'\mathbf{X}_1 & \sigma^{12}\mathbf{X}_1'\mathbf{X}_2 & \cdots & \sigma^{1G}\mathbf{X}_1'\mathbf{X}_G \\ \sigma^{21}\mathbf{X}_2'\mathbf{X}_1 & \sigma^{22}\mathbf{X}_2'\mathbf{X}_2 & \cdots & \sigma^{2G}\mathbf{X}_2'\mathbf{X}_G \\ \cdot & \cdot & \cdot & \cdot \\ \cdot & \cdot & \cdot & \cdot \\ \cdot & \cdot & \cdot & \cdot \\ \sigma^{G1}\mathbf{X}_G'\mathbf{X}_1 & \sigma^{G2}\mathbf{X}_G'\mathbf{X}_2 & \cdots & \sigma^{GG}\mathbf{X}_G'\mathbf{X}_G \end{bmatrix}^{-1} \begin{bmatrix} \sum_{g=1}^{G}\sigma^{ig}\mathbf{X}_1'\mathbf{y}_g \\ \cdot \\ \cdot \\ \cdot \\ \sum_{g=1}^{G}\sigma^{Gg}\mathbf{X}_G'\mathbf{y}_g \end{bmatrix} \end{aligned} \tag{8.17}$$

Notice the near identity of the last two matrices with the ones in (8.7a) and (8.7b) in Section 8.1. The variance-covariance of $\mathbf{b}^*$ is

$$\mathbf{V}(\mathbf{b}^*) = \begin{bmatrix} \sigma^{11}\mathbf{X}_1'\mathbf{X}_1 & \sigma^{12}\mathbf{X}_1'\mathbf{X}_2 & \cdots & \sigma^{1G}\mathbf{X}_1'\mathbf{X}_G \\ \sigma^{21}\mathbf{X}_2'\mathbf{X}_1 & \sigma^{22}\mathbf{X}_2'\mathbf{X}_2 & \cdots & \sigma^{1G}\mathbf{X}_2'\mathbf{X}_G \\ \cdot & \cdot & \cdot & \cdot \\ \cdot & \cdot & \cdot & \cdot \\ \cdot & \cdot & \cdot & \cdot \\ \sigma^{G1}\mathbf{X}_G'\mathbf{X}_1 & \sigma^{G2}\mathbf{X}_G'\mathbf{X}_2 & \cdots & \sigma^{GG}\mathbf{X}_G'\mathbf{X}_G \end{bmatrix}^{-1} \tag{8.18}$$

The estimator $\mathbf{b}^*$ has all the usual properties of the Aitken estimator discussed in Section 6.3. If the disturbances are normally distributed, the $\mathbf{b}^*$ is a maximum likelihood estimator. Now, in practice, the true variance-covariance matrix of $\mathbf{u}_i$ in (8.11) is not known, so that the matrix $\boldsymbol{\Sigma}$ in (8.15) must be estimated from a sample. Let us denote the $\boldsymbol{\Sigma}$ matrix estimated from the sample by $\mathbf{S} = \{s_{ij}\}$ and define

$$s_{ij} = \frac{1}{n - K_i} \hat{\mathbf{u}}_i{}'\hat{\mathbf{u}}_j \qquad i, j = 1, 2, \ldots, G \tag{8.15a}$$

and

$$\hat{\mathbf{u}}_i = \mathbf{y}_i - \mathbf{X}_i\mathbf{b}_i$$

where $\mathbf{b}_i$ is the usual LS estimator of $\boldsymbol{\beta}_i$ in (8.11), $i = 1, 2, \ldots, G$. Notice that K_i in (8.15a) is the number of the coefficients in the ith equation and that one can just as well use K_i. This choice is arbitrary but would not affect the asymptotic properties of the revised Aitken estimator $\mathbf{b}$ (also referred to as the Zellner estimator). Zellner has shown that

$$\mathbf{b} = [\mathbf{X}'(\mathbf{S} \otimes \mathbf{I})^{-1}\mathbf{X}]^{-1}[\mathbf{X}'(\mathbf{S} \otimes \mathbf{I})^{-1}\mathbf{y}] \tag{8.17a}$$

is consistent for $\mathbf{b}^*$ and $\sqrt{n}\,(\mathbf{b} - \boldsymbol{\beta})$ has the same asymptotic covariance matrix as that of $\sqrt{n}\,(\mathbf{b}^* - \boldsymbol{\beta})$. For the proof of the last two points and for related results, see Zellner (1962) and Zellner and Huang (1962).

8.2.3. What Do We Gain?

Since the use of the LS method for estimation equation-by-equation of the set (8.11) is much easier and less complicated than Aitken's method, we naturally ask what the gain is in the use of the latter. Of, course, we should observe that Aitken's estimator is BLUE for the system taken together, so that any other unbiased linear estimator, including the equation-by-equation LS estimator, will at best be as efficient. Here we exhibit the "size" of the gain in efficiency that accrues to Aitken's estimator relative to the single equation LS estimator. Let the single-equation LS estimator, say $\hat{\boldsymbol{\beta}}$,

$$\hat{\boldsymbol{\beta}} = (\mathbf{X}'\mathbf{D}^{-1}\mathbf{X})^{-1}\mathbf{X}'\mathbf{D}'\mathbf{y} \tag{8.19}$$

where

$$\mathbf{D} = \begin{bmatrix} \sigma_{11}\mathbf{I} & \mathbf{0} & \cdots & \mathbf{0} \\ \mathbf{0} & \sigma_{22}\mathbf{I} & \cdots & \mathbf{0} \\ \cdot & \cdot & \cdot & \cdot \\ \cdot & \cdot & \cdot & \cdot \\ \cdot & \cdot & \cdot & \cdot \\ \mathbf{0} & \mathbf{0} & \cdots & \sigma_{GG}\mathbf{I} \end{bmatrix} \tag{8.20}$$

In the last matrix, **I** is of order n. Since **D** is block-diagonal, it can be observed easily that, for instance, the first K_1 elements of $\hat{\boldsymbol{\beta}}$ above are really the LS estimates of the K_1 coefficients of the first equation in (8.11). That is,

$$\hat{\boldsymbol{\beta}}_1 = \left[\mathbf{X}_1'\left(\frac{1}{\sigma^{11}}\mathbf{I}\right)^{-1}\mathbf{X}_1\right]^{-1}\mathbf{X}_1'(\sigma_{11}\mathbf{I})\mathbf{y}_1$$

$$= (\mathbf{X}_1'\mathbf{X}_1)^{-1}\mathbf{X}_1'\mathbf{y}_1$$

The gain in efficiency is looked at in the generalized variance sense. Consequently, we take the ratio of the generalized variance of the Aitken estimator to that of the LS single-equation estimator and let the ratio be denoted by α as follows:

$$\alpha = \frac{|[\mathbf{X}'(\boldsymbol{\Sigma} \otimes \mathbf{I})^{-1}\mathbf{X}]^{-1}|}{|[\mathbf{X}'\mathbf{D}^{-1}\mathbf{X}]^{-1}|} \tag{8.21}$$

If we can show that, in general, α is less than 1, then we can determine that there is a gain in efficiency and how much the gain is in the Aitken procedure. Writing (8.21) in another form, we have

$$\frac{1}{\alpha} = |\mathbf{X}'(\boldsymbol{\Sigma} \otimes \mathbf{I})^{-1}\mathbf{X}|\,|(\mathbf{X}'\mathbf{D}^{-1}\mathbf{X})^{-1}|$$

$$= |\mathbf{X}'(\boldsymbol{\Sigma} \otimes \mathbf{I})^{-1}\mathbf{X}|\,|(\mathbf{X}_1'\mathbf{X}_1)^{-1}\sigma_{11}|\,|(\mathbf{X}_2'\mathbf{X}_2)^{-1}\sigma_{22}|\cdots|(\mathbf{X}_G'\mathbf{X}_G)^{-1}\sigma_{GG}| \tag{8.21a}$$

Also, from Bellman (1960, p. 127), we notice that

$$|\mathbf{X}'(\boldsymbol{\Sigma} \otimes \mathbf{I})^{-1}\mathbf{X}| \le |\mathbf{X}_1'\mathbf{X}_1\sigma^{11}|\,|\mathbf{X}_2'\mathbf{X}_2\sigma^{22}|\cdots|\mathbf{X}_G'\mathbf{X}_G\sigma^{GG}| \tag{8.21b}$$

with the equality holding true with (a) $\sigma_{ij} = 0$ for $i \neq j$, *and/or* (b) $\mathbf{X}_i\mathbf{X}_j = 0$ for $i \neq j$. It is important to observe that if either one of (a) or (b) holds true, the equality holds true. On combining (8.21a) and (8.21b), we have

$$\frac{1}{\alpha} \le |\mathbf{I}_1\sigma_{11}\sigma^{11}|\cdot|\mathbf{I}_2\sigma_{22}\sigma^{22}|\cdots|\mathbf{I}_G\sigma_{GG}\sigma^{GG}| \tag{8.21c}$$

where $\mathbf{I}_i$, $i = 1, 2, \ldots, G$, is a unit matrix of order K_i. From the last expression it follows that $\alpha \le 1$. Now, when $\sigma_{ij} = 0$ for $i \neq j$, the Aitken estimator in (8.14) reduces to the single-equation LS estimator, so that $\alpha = 1$. Thus there is no gain in efficiency if the disturbances of the equations have zero covariances.

Recall that in the discussion of the multivariate regression model the observation matrix for the independent variables was the same for every equation and, as such, there was no need for the Aitken procedure, even though the covariances of the disturbance terms can be zero. The same reasoning applies to the role of the $\mathbf{X}_i$ matrices with only a minor change. Here, what we want to study is the effect that the difference in $\mathbf{X}_i$ matrices would have on the gains that the 'more different' $\mathbf{X}_i$ is from $\mathbf{X}_j$ the greater the gain of efficiency in the Aitken procedure if $\sigma_{ij} \neq 0$ for $i \neq j$. To partially confirm this, we write (8.21c) as

$$(\sigma_{11}\sigma^{11})^{-K_1}(\sigma_{22}\sigma^{22})^{-K_2} \cdots (\sigma_{GG}\sigma^{GG})^{-K_G} \leq \alpha \leq 1 \tag{8.22}$$

and observe that if $\mathbf{X}_i'\mathbf{X}_j = 0$ and $\sigma_{ij} \neq 0$ for $i \neq j$, α will be equal to the lower bound shown in (8.22).* This α represents the maximal gain possible. For a two-equation system with $\mathbf{X}_1'\mathbf{X}_2 = 0$, the left-hand side of (8.22) is

$$(\sigma_{11}\sigma^{11})^{-K_1}(\sigma_{22}\sigma^{22})^{-K_2} = (1 - \rho^2)^{K_1+K_2} \tag{8.22a}$$

(show this as an exercise by direct inversion of the 2×2 $\mathbf{\Sigma}$ matrix where we take $\sigma_{12} = \rho\sqrt{\sigma_{11}\sigma_{12}}$) in which ρ is the correlation coefficient of the two disturbance terms. And the expression in the right-hand side gives the amount of gain in efficiency by using the Aitken method. Of course, in general, $\mathbf{X}_i'\mathbf{X}_j \neq 0$ for $i \neq j$. It develops that the reduction in the possible maximal gain in efficiency is determined by the canonical correlation between $\mathbf{X}_i$ and $\mathbf{X}_j$, and the correlation between the disturbances u_i and u_j. For the details, see Zellner and Huang (1962) and the references.

We now exhibit the results from the LS and the Aitken estimation of three consumption-expenditure equations for a partial illustration of the preceding discussion.

ILLUSTRATION. In a study of consumer purchases of goods and services conducted by the author the total expenditure is decomposed into three categories of expenditures: durable goods, nondurable goods, and services. Each of these types of expenditure in real terms is hypothesized to be a linear function of real disposable personal income Y, own relative price, own expenditure lagged one period. Annual data for the period 1936 to 1963 with the exception of the war years 1941 to 1946 is used for direct LS

* When $\sigma_{ij} = 0$ for $i \neq j$, the Aitken estimator of $\boldsymbol{\beta}$ or $\mathbf{b}^*$ is the single equation LS estimator and $\alpha = 1$.

estimation with the following result:

$$C_d = -31.87 + 0.195\,Y - 0.809P_d - 0.0154(C_d)_{-1}$$
$$(64.42)\quad (0.0251)\quad (0.577)\quad (0.133)$$
$$R^2 = 0.980 \tag{8.23a}$$

$$C_{nd} = -44.91 + 0.198\,Y + 1.68P_{nd} + 0.404(C_{nd})_{-1}$$
$$(48.47)\quad (0.0234)\quad (0.544)\quad (0.0715)$$
$$R^2 = 0.994 \tag{8.23b}$$

$$C_s = -73.79 + 0.108\,Y + 0.563P_s + 0.733(C_s)_{-1}$$
$$(27.03)\quad (0.0205)\quad (0.269)\quad (0.0633)$$
$$R^2 = 0.997 \tag{8.23c}$$

In these three equations, C_d, C_{nd}, and C_s are, respectively, real durable goods, nondurable, and services expenditures, and P_d, P_{nd}, and P_s are the respective relative prices. Then the residuals of these equations are estimated to yield the following residual correlation matrix

	$\hat{u}_d$	$\hat{u}_{nd}$	$\hat{u}_s$
$\hat{u}_d$	1.00	−0.114	0.152
$\hat{u}_{nd}$		1.00	−0.119
$\hat{u}_s$			1.00

where $\hat{u}_d$, $\hat{u}_{nd}$, and $\hat{u}_s$ denote, respectively, the LS estimates of the disturbances in the C_d, C_{nd}, and C_s equations.

The sample covariance matrix corresponding to this matrix is

$$\begin{bmatrix} 113.91 & -12.79 & 10.89 \\ -12.79 & 111.12 & -8.42 \\ 10.89 & -8.42 & 45.08 \end{bmatrix} \tag{8.23d}$$

and its elements are as defined in (8.15a). The covariance matrix (8.23d) then serves as an estimate of $\boldsymbol{\Sigma}$ in (8.17) so as to give the approximate Aitken estimator as described in (8.17a). The resulting estimates of the expenditure equations are

$$C_d = -52.96 + 0.196\,Y - 0.612P_d - 0.0149(C_d)_{-1} \tag{8.24a}$$
$$(64.04)\quad (0.0248)\quad (0.573)\quad (0.132)$$

$$C_{nd} = -50.13 + 0.210\,Y + 1.815P_{nd} + 0.366(C_{nd})_{-1} \tag{8.24b}$$
$$(48.30)\quad (0.0232)\quad (0.541)\quad (0.0708)$$

$$C_s = -77.73 - 0.112\,Y + 0.606P_s + 0.721(C_s)_{-1} \tag{8.24c}$$
$$(26.81)\quad (0.0202)\quad (0.267)\quad (0.0623)$$

The gain in efficiency is observed in the smaller standard errors in (8.24a to c). Notice that this is for illustration only. The statistical insignificance of the correlations among the disturbances would lead one to doubt the usefulness of the joint estimation in an empirical situation.

8.3. Setwise Least Squares

Because of analytical and/or statistical reasons, it is sometimes useful to apply a technique that we shall call setwise least squares. By setwise LS we shall mean a technique of estimating what is basically a set of equations jointly as one regression equation. The technique is particularly useful when analyzing a number of individuals or cross sections over time. Since the technique is applicable in a variety of situations in empirical work, we shall discuss two cases of analysis where setwise LS is applied. Notice that setwise LS is primarily a technique suitable for specific situations and may or may not derive from the efficient estimation we have discussed in the preceding sections.

8.3.1. Production Function of a Fishing Industry

In a study of the production function of the Pacific halibut industry, a seven-year panel data (1958 to 1964) was collected for ten halibut fishing boats; the data for two of these years are shown in Table 5.4. The basic production function for each boat to be estimated is

$$\ln q = A + \lambda t + \alpha \ln k + \beta \ln L + \sigma \ln C + u \tag{8.25}$$

where t is time ($1958 = 1$, $1959 = 2$, etc.) and the other variables are as defined in Subsection 5.2.2. Since we have only seven observations for each boat, the fitting of (8.25) with this sample size will leave us with very few degrees of freedom. However, if it is hypothesized that the coefficients α, β, and σ are the same for all the boats, although only the time coefficient is different for each boat (that is, the degree of technical progress may differ among boats), we can rewrite (8.25) as

$$\begin{aligned} \ln q = A' + \lambda_1 t_1 + \lambda_2 t_2 + \cdots + \lambda_{10} t_{10} \\ + \alpha \ln k + \beta \ln L + \sigma \ln C + u' \end{aligned} \tag{8.25a}$$

where the t_i and λ_i, $i = 1, 2, \ldots, 10$, are, respectively, the t variable and its coefficient of the ith boat, and the latter equation can be estimated with the entire sample of 70 observations. Then the degrees of freedom will be (70, 14). To carry out the calculation, each t_i becomes an observation column of 70 elements with values $1, 2, \ldots, 7$ for each of the years 1958, 1959, . . . , 1964, respectively, located in the ith seven rows of the observation matrix and with 63 zeros in the rest of the column. To see this we write below an abbreviated version of the **X** matrix that illustrates how the data for calculation may be set up.

$$
\begin{array}{c}
\\
\mathbf{X} =
\end{array}
\begin{array}{c}
\begin{array}{cccccccc}
A' & t_1 & t_2 & \cdots & t_{10} & \ln K & \ln L & \ln C
\end{array}\\
\left[\begin{array}{cccccccc}
1 & 1 & 0 & & 0 & & & \\
1 & 2 & 0 & & 0 & \cdot & \cdot & \cdot \\
1 & 3 & 0 & & 0 & \cdot & \cdot & \cdot \\
1 & 4 & 0 & & 0 & \cdot & \cdot & \cdot \\
1 & 5 & 0 & & 0 & \cdot & \cdot & \cdot \\
1 & 6 & 0 & & 0 & 7.3085 & 6.7742 & 2.7147 \\
1 & 7 & 0 & & 0 & 7.4782 & 6.9632 & 2.55953 \\
1 & 0 & 1 & \cdots & 0 & & & \\
1 & 0 & 2 & & 0 & \cdot & \cdot & \cdot \\
1 & 0 & 3 & & 0 & \cdot & \cdot & \cdot \\
1 & 0 & 4 & & 0 & \cdot & \cdot & \cdot \\
1 & 0 & 5 & & 0 & \cdot & \cdot & \cdot \\
1 & 0 & 6 & & 0 & 6.9713 & 5.9269 & 2.7143 \\
1 & 0 & 7 & & 0 & 6.8406 & 5.8171 & 2.9460 \\
\cdot & \cdot & \cdot & & \cdot & \cdot & \cdot & \cdot \\
\cdot & \cdot & \cdot & & \cdot & \cdot & \cdot & \cdot \\
\cdot & \cdot & \cdot & & \cdot & \cdot & \cdot & \cdot \\
1 & 0 & 0 & \cdots & 1 & \cdot & \cdot & \cdot \\
1 & 0 & 0 & & 2 & \cdot & \cdot & \cdot \\
1 & 0 & 0 & & 3 & \cdot & \cdot & \cdot \\
1 & 0 & 0 & & 4 & \cdot & \cdot & \cdot \\
1 & 0 & 0 & & 5 & \cdot & \cdot & \cdot \\
1 & 0 & 0 & & 6 & 7.0783 & 6.3026 & 3.1380 \\
1 & 0 & 0 & & 7 & 7.1776 & 6.4739 & 3.0168
\end{array}\right]
\end{array}
\qquad (8.26)
$$

Notice that part of the data shown in Table 5.4 appears here. The first seven rows of **X** contain values of the variables indicated at the top of **X** for Boat Number 1 whose $\ln K$, $\ln L$, and $\ln C$ values for the years 1963 and 1964 appear in the last three columns of the sixth and seventh rows. Presumably, the $\ln K$, $\ln L$, and $\ln C$ values for 1958 to 1962 appear in place of the dots shown in the last three columns. Identify the similar observations for Boats Numbers 2 and 10, in order to determine how the data are placed in **X**. The LS estimate of (8.25a) is

$$\begin{aligned}
\ln q = {} & \underset{(0.0157)}{0.212} + \underset{(0.0145)}{0.0151 t_1} + \underset{(0.0174)}{0.0574 t_2} + \underset{(0.219)}{0.0959 t_3} + \underset{(0.0141)}{0.06611 t_4} \\
& - \underset{(0.0179)}{0.0261 t_5} + \underset{(0.0168)}{0.0447 t_6} + \underset{(0.0142)}{0.0166 t_7} + \underset{(0.0147)}{0.0292 t_8} \\
& + \underset{(0.0144)}{0.0217 t_9} + \underset{(0.0134)}{0.0311 t_{10}} + \underset{(0.143)}{0.5431 \ln K} + \underset{(0.149)}{0.4161 \ln L} \\
& + \underset{(0.0754)}{0.331 \ln C}
\end{aligned} \tag{8.27}$$

$$R^2 = 0.910$$

8.3.2. A Multi-Cross-Section Study of Automobile Buying

In contrast to the preceding example, sometimes a researcher may have data in the form of successive cross sections. Data like those provided by the University of Michigan's Survey of Consumer Finances are a case in point. We observe that the successive cross sections differ from panel data in that the latter represent observations on the same sample units for a number of time periods, whereas in the successive cross sections sample units and sizes may differ from one cross section to another.

To study households' buying patterns with respect to automobiles, we might take a cross section of households in a given year, make note of those households that are buying and are not buying cars, gather information about the economic and demographic characteristics of these households, and possibly consider what the probability is for any household to purchase a car given its economic and demographic status. With respect to purchase behavior, it is reasonable to believe that beside the households' economic and demographic circumstances there are certain time-series market variables that may be relevant for analysis. Now, market variables such as the wholesale price of cars and the rate of unemployment are the same for all households in a given year (cross section) but will vary over time in a number of cross sections. Thus, if there are several successive cross-sections data, it is possible to incorporate the market variables into a relation explaining purchase.

To illustrate this point, we consider a linear probability function that relates purchasing a new car to certain economic and demographic variables as well as to some market variables. The relation is in the nature of the ones discussed in Section 7.3. The data to be used are obtained from the 1956, 1958, 1959, 1960, and 1961 Surveys of Consumer Finances,* and the relation

* The author is grateful to the University of Michigan's Survey Research Center and Social Systems Research Institute at the University of Wisconsin for making these data available.

of interest is

$$P_n = \alpha_0 + \alpha_1 Y + \alpha_2 Y^2 + \alpha_3 S_1 + \alpha_4 S_2 + \alpha_5 S_3 + \alpha_6 S_4 + \alpha_7 M + \alpha_8 N + \alpha_9 L_2 \alpha_{10} L_3 + \alpha_{11} R + \alpha_{12} S_x + \alpha_{13} H_1 + \alpha_{14} E_3 + \alpha_{15} A + \alpha_{16} A^2 + \alpha_{17} \Delta U + \alpha_{18} \Delta WPI + \alpha_{19} \Delta I + u \quad (8.28)$$

where the variables that have not been explained in Table 5.3 are defined as follows:

p_n = 1 if the spending unit bought a new car during the year, 0 otherwise (we consider only the purchasers of cars).

S_1 = 1 if the first car SU owned at the beginning of the year was bought new and was 3 years old or less, 0 otherwise.

S_2 = 1 if the first car SU owned at the beginning of the year was bought new and was more than 3 years old, 0 otherwise.

S_3 = 1 if the first car SU owned at the beginning of the year was bought used and was 5 years old or less, 0 otherwise.

S_4 = 1 if the first car SU owned at the beginning of the year was bought used and was more than 5 years old, 0 otherwise. (The residual class, S_5, includes SU's now owning any car at the beginning of the year.)

L_2 = 1 if SU is located in suburban areas of metropolitan cities, 0 otherwise.

L_3 = 1 if SU is located in rural areas or areas adjacent to suburbs, 0 otherwise. (The residual class, L_1, includes SU's located in areas not specified above, metropolitan cities.)

R = 1 if head of SU is nonwhite, 0 otherwise.

S_x = 1 if head of SU is female, 0 otherwise.

H_6 = 1 if SU owns home, 0 otherwise.

E_1 = 1 if head of SU received 13 or more years of education, 0 otherwise.

A = Age of head of SU, in number of years.

A^2 = Age squared, in 100's of years.

ΔU = Average change in the quarterly rates of unemployment during the year.

ΔWPI = Average quarterly change in the wholesale price of new automobiles.

ΔI = Average quarterly change in the car inventory in the hands of dealers.

Table 8.1 P_n: Probability of Purchasing a New Car

Variable	1955	1957	1958	1959	1960	Setwise
Constant	−0.0828	−0.1822	−0.1605	0.0864	−0.1538	−0.0814
Y	0.0546[a]	0.0296[a]	0.0443[a]	0.0030	0.0555[a]	0.0252[a]
Y^2	−0.0008[a]	−0.0004	−0.0010	−0.0001	−0.0014[a]	−0.0003[a]
S_1	0.4098[a]	0.4022[a]	0.3873[a]	0.5374	0.3056[a]	0.4347[a]
S_2	0.4059[a]	0.3060[a]	0.3840[a]	0.4168[a]	0.2414[a]	0.3681[a]
S_3	0.1035[a]	0.2353[a]	0.1632[a]	0.1452	0.0956	0.1446[a]
S_4	0.0204	0.0289	0.0360	0.0438	−0.0465	0.0161
M	0.0537	0.0019	−0.1349[a]	0.1342[a]	−0.0571	0.0231
N	−0.0346[a]	−0.0397[a]	−0.0432[a]	0.0177	−0.0419[a]	−0.0365[a]
L_2	0.0190	−0.0241	0.0373	−0.0861[a]	−0.0480	−0.0250
L_3	0.0078	−0.0218	0.0370	−0.0400	−0.0190	−0.0355[a]
R	0.1093	−0.0241	−0.0211	−0.0752	0.0166	−0.0246
S_x	0.0475	0.1442[a]	0.0584	0.1235[a]	0.0056	0.05034
H	0.0458	0.1175[a]	0.0192	0.0328	−0.0265	0.0279[a]
E_1	0.0075	0.0061	0.0784[a]	0.1273[a]	0.0831[a]	0.0694[a]
A	0.0042	0.0146[a]	0.0098	0.0034	0.0157	0.0146[a]
A^2	−0.0001	−0.0001	−0.0002	−0.00002	−0.0001	−0.0002[a]
ΔU	—	—	—	—	—	−0.00018[a]
ΔWPI	—	—	—	—	—	−0.00005[a]
ΔI	—	—	—	—	—	−0.00008[a]
$\bar{R}^2$	0.314	0.278	0.272	0.275	0.265	0.278
S_e	0.4127	0.4094	0.3939	0.4112	0.4141	0.4116
n	840	716	650	689	541	3436

[a] Coefficient is significant at the 5 percent level by conventional test.
$\bar{R}^2$ is the coefficient of determination corrected for degrees of freedom.

The results of the LS calculations relative to (8.28) are shown in Table 8.1. We first estimated the relation (8.28) without the market variables for each of the cross sections and then pooled the cross sections in order to allow the effect of ΔU, ΔWPI, and ΔI to be estimated. Observe that the value of a market variable in a given year is same the for all spending units during that year so that, for example, if $\Delta U = 25$ percent in 1955, then all of the 840 spending units will have 25 as the observation for the variable ΔU. In this manner we are able to find an estimate of a relation incorporating both cross-section and time-series information.

CHAPTER 9

INTRODUCTION TO SIMULTANEOUS EQUATIONS

9.1. Introduction

A combination of the types of regression models discussed in Chapters 6, 7, and 8 provides the formal ingredients of the simultaneous equation systems in economic analysis In the opening paragraphs of Chapter 8, we referred to a set of dependent variables, which may be related or interdependent, that depended on a set of independent variables; and in the multivariate regression model each of the dependent variables was regressed on the same set of independent variables. In a simultaneous equation system the restriction of the multivariate regression model that one and only one dependent variable of the related set of dependent variables appears in a regression equation is relaxed, whereby allowing the explicit specification of the dependence relation of one dependent variable on one or more of the other jointly dependent variables. When this type of specification is permitted, we have a regression equation that has similar statistical properties as does the stochastic independent variables equation discussed in Chapter 7. That is, in an equation of a simultaneous equation system we generally assume the statistical dependence between the jointly dependent variables appearing in the right-hand side of the equation and the disturbance term of that equation.

We illustrate the preceding discussion with a simple Keynesian model. The basic point in this example is that the consumption expenditure is a function of income, but income, by definition, is the result of adding the consumption expenditure and the investment expenditure. Thus the statistical form of the model is

$$C = \alpha + \beta Y + u \tag{9.1a}$$

$$Y = C + I \tag{9.1b}$$

where I is the investment expenditure and is assumed to be exogenous (in the sense that it is not determined by C or Y, but by an outside variable or variables). $Eu = 0$; $Eu^2 = \sigma^2$; and $EIu = 0$. Two features common to most simultaneous equation systems, in the system (9.1a to b) can be observed: (1) the mutual dependence of C and Y, and (2) the dependence between the disturbance and the explanatory variable Y. This latter point can be seen as follows. Solving the system of C and Y in terms of the exogenous variable I, we have

$$C = \frac{\alpha}{1-\beta} + \frac{\beta}{1-\beta} I + \frac{1}{1-\beta} u \tag{9.2a}$$

$$Y = \frac{\alpha}{1-\beta} + \frac{1}{1-\beta} I + \frac{1}{1-\beta} u \tag{9.2b}$$

The multiplication of (9.2b) by u yields

$$Yu = \frac{\alpha}{1-\beta} u + \frac{1}{1-\beta} Iu + \frac{1}{1-\beta} u^2$$

and the expectation of the last expression is

$$EYu = \frac{1}{1-\beta} \sigma^2$$

It follows then that Y and u are not uncorrelated. The last point recalls our earlier discussion in Section 7.5 where we indicated that in the event that the disturbance term is correlated with one or more of the explanatory variables in a regression equation it is no longer correct to apply the least-squares method for estimation of the equation since, by doing so, one obtains biased and inconsistent estimates.

In a system of simultaneous equations the kind of problems discussed in the preceding paragraph are generally the rule. However, there are systems, for example, recursive ones, where these problems are not present, although in the minority they may be. Therefore, our objective in this chapter is (1) to sort out a few of the types of simultaneous equation systems, (2) to consider some of the taxonomy of the variables and equations entering these systems, and (3) to develop the concept of identification that is fundamental in simultaneous equation systems. The problems of the estimation of some of the systems are considered in the next chapter.

Before proceeding, we observe that the basic objectives of Chapters 9 and 10 are to familiarize the student with the rudiments of simultaneous equation systems at a relatively elementary level and to give him some of the more frequently used tools in economic research. As such, these chapters are not intended to be a full-fledged treatment of the subject.

9.2. Terminology and Notations

As is true in any branch of specialized learning, there has evolved an extensive body of technical terms in the field of econometric analysis. Thus we now must clarify some of the concepts and jargon basic to our later development of this analysis. In the process we shall also develop some notations.

9.2.1. The Structural Form

Our discussion of a model and a structure in the first paragraphs of Chapter 5, leads us to state that a model embodies the general form of a structure. Thus, a model is represented by a structural form in econometrics. In simultaneous equation analysis, of course, a model usually contains more than one behavioral relation, so that the structural form may consist of more than one equation. A structural form, then, is a general expression of the ways in which a group of variables are related. In its most general state a simultaneous equation model has the structural form

$$\mathbf{y}'\mathbf{B} + \mathbf{x}'\mathbf{\Gamma} + \mathbf{u}' = \mathbf{0}' \tag{9.4}$$

where $\mathbf{y}$ is, say, a $G \times 1$ column vector of G variables to be explained by the model; $\mathbf{x}$ is, say, a $K \times 1$ column vector of K variables; $\mathbf{B}$ is a $G \times G$ matrix of coefficients with, at least, the diagonal elements not equal to zero; $\mathbf{\Gamma}$ is a $K \times G$ matrix of coefficients; and $\mathbf{u}$ is a $G \times 1$ vector of disturbance term. An alternative way to write (9.4) is

$$\begin{aligned}
\beta_{11}y_1 + \beta_{21}y_2 + \cdots + \beta_{G1}y_G + \gamma_{11}x_1 + \cdots + \gamma_{K1}x_K + u_1 &= 0 \\
\beta_{12}y_1 + \beta_{22}y_2 + \cdots + \beta_{G2}y_G + \gamma_{12}x_1 + \cdots + \gamma_{K2}x_K + u_2 &= 0 \\
&\;\vdots \\
\beta_{1G}y_1 + \beta_{2G}y_2 + \cdots + \beta_{GG}y_G + \gamma_{1G}x_1 + \cdots + \gamma_{KG}x_K + u_G &= 0
\end{aligned} \tag{9.4a}$$

Notice that in (9.4) the preceding equations are lined up in the manner of a horizontal string. Furthermore, in (9.4),

$$\mathbf{B} = \begin{pmatrix} \beta_{11} & \beta_{12} & \cdots & \beta_{1G} \\ \beta_{21} & \beta_{22} & \cdots & \beta_{2G} \\ \cdot & \cdot & \cdot & \cdot \\ \cdot & \cdot & \cdot & \cdot \\ \cdot & \cdot & \cdot & \cdot \\ \beta_{G1} & \beta_{G2} & \cdots & \beta_{GG} \end{pmatrix}$$

$$\mathbf{\Gamma} = \begin{pmatrix} \gamma_{11} & \gamma_{12} & \cdots & \gamma_{1G} \\ \gamma_{21} & \gamma_{22} & \cdots & \gamma_{2G} \\ \cdot & \cdot & \cdot & \cdot \\ \cdot & \cdot & \cdot & \cdot \\ \cdot & \cdot & \cdot & \cdot \\ \gamma_{K1} & \gamma_{K2} & \cdots & \gamma_{KG} \end{pmatrix}$$

so that the gth column of $\mathbf{B}$, the gth column of $\mathbf{\Gamma}$, and the gth element of $\mathbf{u}'$ in (9.4), when fully multiplied out, corresponds to the gth equation in (9.4a). The sample and the model can be written as

$$\mathbf{YB} + \mathbf{X\Gamma} + \mathbf{U} = 0 \tag{9.4n}$$

where $\mathbf{Y}$ is $n \times G$, $\mathbf{X}$ is $n \times K$, and $\mathbf{U}$ is $n \times G$, with $\mathbf{B}$ and $\mathbf{\Gamma}$ remaining unchanged.

Now consider (9.4a). If all the β_{ij}, γ_{ij}, are known, and if the marginal and joint distributions of $u_1, u_2, \ldots, u_G$ are known, the problems of statistical estimation do not arise. But the more frequent reason for considering an econometric model is that a hypothesis must be tested and, for this purpose, the hypothesis must be cast in a model suitable for statistical estimation. Now if one is to estimate the equations in (9.4a), assuming that the adequate data necessary for estimation are available, and that no knowledge exists on the distribution of $u_1, u_2, \ldots, u_G$, he immediately encounters the problem that statistically one equation cannot be distinguished from any other in the system. This motivates a discussion of prior specification (we say this for pedagogical convenience, and it may be advisable to think of other motives later).

9.2.2. *Prior Specification*

Prior specification is a result of explicit and/or implicit theorizing and constitutes specific statements regarding the variables, their probability distributions, the parameters, and the forms of the equations of a model.

ENDOGENOUS AND EXOGENOUS VARIABLES. Variables that are considered to be jointly dependent on each other while being affected by some other variables are called endogenous variables. These variables are the object of the explanation sought by a simultaneous equation system. Some endogenous variables affect the jointly dependent variables with a lag; they are called lagged endogenous variables. The variables that enter the system, affect the endogenous variables, but not being affected by them are called exogenous variables. Lagged endogenous variables and exogenous variables taken together are called predetermined variables. They are "pre" in the sense of prior in time or exterior to the current, jointly dependent variables.

As an example, C and Y in (9.1a) and (9.1b) are endogenous and I is exogenous. This, of course, is entirely from prior theoretical considerations that say, in part, that the investment is determined by autonomous factors, such as population and technology. It is quite possible that economic theory, in some other context of analysis, might specify that investment is endogenous as in the case when the growth of income is considered in the light of the multiplier and the accelerator. At times, the distinction between endogenous and exogenous variables can be quite hazy, but all depends on the theoretical consideration of the analyst.

Usually, endogenous variables are represented by y_i, and predetermined and exogenous variables by x_i, as in (9.4a). In a general situation, y_i and x_i can all be random variables and can have joint probability distributions. In the usual practice we often regard x_i as given, and we consider only y_i, along with the disturbance terms u_i, to have a joint distribution. Endogenous variables are correlated with the disturbances, at least, contemporaneously, but predetermined variables are not.

STRUCTURAL EQUATIONS. For the reason stated at the end of Subsection 9.2.1 and for theoretical considerations, a structural equation can be specified to include endogenous and predetermined variables of the system. And by virtue of the specified inclusion, the variables that are not included are naturally excluded.

If there are G endogenous variables, there must be G equations to explain these variables. A model is complete when there are as many structural equations as there are endogenous variables, so that the endogenous variables can be solved for in terms of the predetermined variables.

A part of the job of specifying a structural equation or relation is to be explicit on the form of the parameters and the variables entering the equation and on the probability distribution of the disturbance term. As for the form, we usually deal with linear equations in the sense that the relation is linear in the parameters and also in the variables (some exceptions to the latter occur but, sometimes, can be handled as if the linearity in the variables is preserved). Observe that the structural form (9.4a) is linear in both the parameters and the variables.

Structural relations or equations may be classified into four types: (1) behavioral, such as demand and supply equations, (2) technological, such as production function, (3) institutional, such as income tax-income relation, and (4) identity, mere definitional or accounting.

Our discussion in Section 7.2 leads us to classify structural equations and the resulting models into yet three types: (1) shock model, a model in which the variables of the structural equations are free from error of measurement so that the disturbances provide random "shocks" to the equations, (2) error model, a model in which the equations are exact but the variables are subject to measurement errors, and (3) error and shock model, a combination of (1) and (2). Most of the results in econometrics today concern the first type only.

9.3. Identifying a Structural Equation

In this section we make an effort to clarify in some detail the identification problem referred to in Section 9.1.2. We discuss situations in which the identification of parameters in a structural equation is not possible and then move on to consider how this identification can be achieved. In a more formal and advanced treatment of identification, it is advisable to consult Malinvaud (1966, Chapter 18) or Fisher (1966).

9.3.1. Supply or Demand?

A simple and yet instructive illustration of the identification problem is the following. Economic theory states that the quantity supplied of the commodity depends on the price of the commodity and that the quantity demanded of the commodity also depends on the price. Thus an economist engaged in the estimation of statistical relationships for the market might secure time-series observations on the quantity that cleared the market and the corresponding price and might use this set of data for estimating the model.

$$D = \alpha_0 + \alpha_1 p + u_1 \tag{9.21a}$$

$$S = \beta_0 + \beta_1 p + u_2 \tag{9.21b}$$

$$D = S \tag{9.21c}$$

where D and S are, respectively, quantities demanded and supplied. Now without any prior knowledge on the distributions of u_1, and u_2 and/or any restrictions on the coefficients α's and β's, statistical estimation can yield only one empirical relation, since there would be only one price and one quantity series. This would leave one in the predicament of not being able to say whether the estimated relation is the demand or the supply schedule.

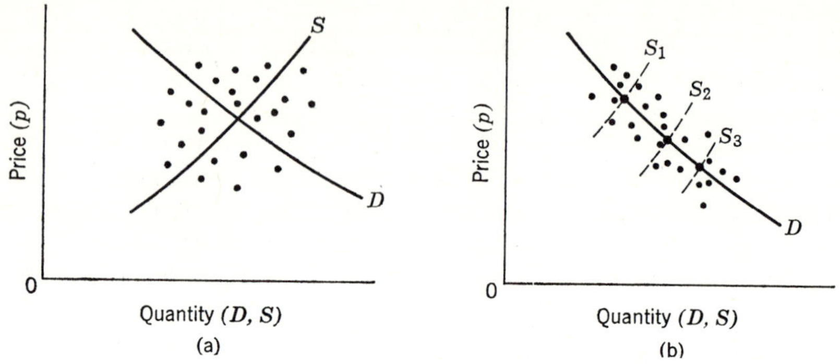

Figure 9.1*a* and *b*

Now suppose that additional knowledge is injected into the model, say, the assumptions that $Eu_1 = Eu_2 = 0$ and that $\sigma^2(u_2) > \sigma^2(u_1)$. Then the variability of S conditional on the price is greater than that of D, and we would expect that in the intrinsic sense the supply schedule shifts more randomly than the demand schedule. Thus it may be that the intersections of these shifts are observed as the points in a scatter diagram from which a statistical relation close to the demand curve (9.21c) can be obtained. Such an estimated demand curve is designated by D in Figure 9.1*b*.

An extreme but instructive case of the general situation just discussed in the preceding paragraph is the estimation of the schedule for an agricultural commodity using annual data. Since the annual quantity of farm produce can be subject to major changes because of weather and other unpredictable conditions and since in any one year the supply is fixed (assuming that no storage facility is available), the annual observed sales and price data would most likely give rise to the scattered points in Figure 9.2*b*. The real situation

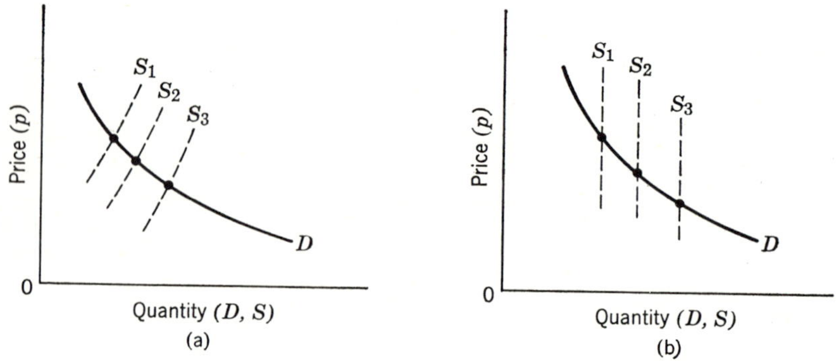

Figure 9.2*a* and *b*

causing the phenomenon in Figure 9.2*b*, actually, may be the function drawn in Figure 9.2*a*. In this example the true supply relation may be assumed to be a function of weather (say, W) in addition to the price variable, so that the system (9.21a to c) may be written as

$$D = \alpha_0 + \alpha_1 P + u_1 \tag{9.22a}$$

$$S = \beta_0 + \beta_1 P + \gamma W + v_2 \tag{9.22b}$$

$$D = S \tag{9.22c}$$

(where the variability of γW and v_2 taken together may be assumed to be greater than the variability of u_1). Thus, the fitting of a quantity-price relation with the data appearing in Figure 9.2*b* would allow the unique estimation of the demand relation (9.21a). Hence, the demand function is identified. (We shall discuss the meaning of identification more formally later.)

But is the supply relation in (9.22b) identified? No, it is not. And the reason is that the supply relation is "observationally equivalent" to another structure that can be implied by the system. That is, given (9.22c), one can multiply (9.22a) by a constant, add the result to both sides of (9.22b), and come up with a relation, say,

$$S + kD = \beta_0 + \beta_1 P + \gamma W + v_2 + \alpha_0 k + \alpha_1 kp + ku_1$$

or

$$S = \frac{1}{1 + k} [(\beta_0 + \alpha_0 k) + (\beta_1 + \alpha_1 k)p + \gamma W + (ku_1 + v_2)$$

or

$$S = \theta_0 + \theta_1 p + \gamma W + v_2' \tag{9.23b}$$

where $\theta_0 = (\beta_0 + \alpha_0 k)/(1 + k)$ and $v_2' = (ku_1 + v_2)$, for example. Thus the statistical estimation of the relation such as (9.23b) that is based on the observations on S, p, and W does not give estimates that can be uniquely associated with those of the parameters in (9.22b). Somewhat formally, in the parameter space consisting of values that are admissible for any linear relation $S = L$ (constant, p, W) $+ v_2'$, we have no way of identifying the set of values that belong uniquely to (9.22b). The model (9.22a to c) is not identified but the structural relation (9.22a) is.

Suppose now that the disposable personal income (Y) should appear as an explanatory variable in the demand relation so the true system becomes

$$D = \alpha_0 + \alpha_1 p + \alpha_2 Y + u_1' \tag{9.24a}$$

$$S = \beta_0 + \beta_1 p + \beta_2 W + u_2' \tag{9.24b}$$

$$D = S \tag{9.24c}$$

Here it can be determined easily that the relations (9.24a) and (9.24b) are not observationally equivalent. We notice that if two relations are observationally equivalent, this fact cannot be altered by any manipulation of the sample information or by an increase in sample size. Only the prior specification of relations entering the model can break the observational equivalence between any two relations. Notice, too, that whether a system is identified is not a matter of cleverness in specification but a matter of what the true model is.

9.3.2. *Reduced Form and Identification*

In the preceding section we indicated that observationally equivalent relations cannot be identified if there is no further information regarding the parameters and variables in the relations. Although the problem of identification arises for structural relations, no such problem is present for the reduced form equations that correspond to these structural equations. This statement is elaborated on.

Given a structural form

$$\mathbf{y}'\mathbf{B} + \mathbf{x}'\boldsymbol{\Gamma} + \mathbf{u}' = \mathbf{0}' \tag{9.25}$$

where $\mathbf{y}$ and $\mathbf{x}$ are column vectors of endogenous and predetermined variables, the reduced form corresponding to it is

$$\mathbf{y}' = -\mathbf{x}'\boldsymbol{\Gamma}\mathbf{B}^{-1} - \mathbf{u}'\mathbf{B}^{-1} = \mathbf{x}\boldsymbol{\Pi} + \mathbf{V}$$

To illustrate the above, let us take a system with two endogenous variables (y_1, y_2) and three exogenous variables (x_1, x_2, x_3). The structural form of this system is

$$\begin{aligned}(y_1 \quad y_2)\begin{pmatrix}\beta_{11} & \beta_{12}\\ \beta_{21} & \beta_{22}\end{pmatrix} + (x_1 \quad x_2 \quad x_3)\begin{pmatrix}\gamma_{11} & \gamma_{12}\\ \gamma_{21} & \gamma_{22}\\ \gamma_{31} & \gamma_{32}\end{pmatrix}\\ = (0 \quad 0)\end{aligned} \tag{9.25a}$$

Or, equivalently,

$$\begin{aligned}\beta_{11}y_1 + \beta_{21}y_2 + \gamma_{11}x_1 + \gamma_{21}x_2 + \gamma_{31}x_3 + u_1 = 0\\ \beta_{12}y_1 + \beta_{22}y_2 + \gamma_{12}x_1 + \gamma_{22}x_2 + \gamma_{32}x_3 + u_2 = 0\end{aligned} \tag{9.26a}$$

The reduced form for (9.25a) is

$$\begin{aligned}(y_1 \quad y_2) &= -(x_1 \quad x_2 \quad x_3)\begin{pmatrix}\gamma_{11} & \gamma_{12}\\ \gamma_{21} & \gamma_{22}\\ \gamma_{31} & \gamma_{32}\end{pmatrix}\begin{pmatrix}\beta_{11} & \beta_{12}\\ \beta_{21} & \beta_{22}\end{pmatrix}^{-1}\\ &\quad - (u_1 \quad u_2)\begin{pmatrix}\beta_{11} & \beta_{12}\\ \beta_{21} & \beta_{22}\end{pmatrix}^{-1}\\ &= (x_1 \quad x_2 \quad x_3)\begin{pmatrix}\pi_{11} & \pi_{12}\\ \pi_{21} & \pi_{22}\\ \pi_{31} & \pi_{32}\end{pmatrix} + (v_1 \quad v_2)\end{aligned} \tag{9.25b}$$

In the ordinary notation the reduced form equations are

$$
\begin{aligned}
y_1 &= \pi_{11}x_1 + \pi_{21}x_2 + \pi_{31}x_3 + v_1 \\
y_2 &= \pi_{12}x_1 + \pi_{22}x_2 + \pi_{32}x_3 + v_2
\end{aligned} \tag{9.26b}
$$

Since all the endogenous variables are now written in terms of the predetermined variables, it is easy to see that the two equations in (9.26b) are not observationally equivalent. To state this another way: given the exogenous variables and the probability distributions of v_1 and v_2, the probability laws that generate the observations of y_1 and on y_2 are not the same. Thus the coefficient of all the reduced form equations are identifiable. A more rigorous approach in considering the identification of a system or structure like (9.25a) is to view the system as one of the many structures that are characterized by different values of **B** and **Γ** and by different joint probability densities of **u** and to determine if the structure is unique among the many, given the predetermined variables and the joint density of **u**. Furthermore, since the reduced form codes all the information that can be recovered from the sample observations, observationally equivalent structures will have the same reduced form.

It stands to reason, therefore, that if the reduced form equations are always identified, the identification of the structural equations can be had by establishing the unique correspondence between the reduced form coefficients and the structural coefficients. Thus, in general, one way to achieve the identification of the structural coefficients **B** and **Γ** in (9.25) is to ensure that these coefficients can be uniquely associated with the reduced form coefficients **Π**. That is to say, we would require that

$$
-\mathbf{\Gamma B}^{-1} = \mathbf{\Pi} \tag{9.27}
$$

But this is not generally possible because for a known **Π** we shall have too many unknowns in **Γ** and **B**. Consider the system (9.25a), for instance. The **B** matrix is 2 × 2, **Γ** is 3 × 2, and **Π** is 3 × 2. And denoting the ijth elements of $\mathbf{B}^{-1}$ by β^{ij}, we have

$$
\begin{pmatrix} \gamma_{11} & \gamma_{12} \\ \gamma_{21} & \gamma_{22} \\ \gamma_{31} & \gamma_{32} \end{pmatrix} \begin{pmatrix} \beta^{11} & \beta^{12} \\ \beta^{21} & \beta^{22} \end{pmatrix} = \begin{pmatrix} \pi_{11} & \pi_{12} \\ \pi_{21} & \pi_{22} \\ \pi_{31} & \pi_{32} \end{pmatrix} \tag{9.27a}
$$

or what is an equivalent system of equations

$$
\begin{aligned}
-\beta^{11}\gamma_{11} - \beta^{21}\gamma_{12} &= \pi_{11} \\
-\beta^{11}\gamma_{21} - \beta^{21}\gamma_{22} &= \pi_{21} \\
-\beta^{11}\gamma_{31} - \beta^{21}\gamma_{32} &= \pi_{31} \\
-\beta^{12}\gamma_{11} - \beta^{22}\gamma_{12} &= \pi_{12} \\
-\beta^{12}\gamma_{21} - \beta^{22}\gamma_{22} &= \pi_{22} \\
-\beta^{12}\gamma_{31} - \beta^{22}\gamma_{32} &= \pi_{32}
\end{aligned} \tag{9.27b}
$$

Here, we have, for π_{ij} known, a system of equations in ten unknowns. The situation is worse if you realize that each β^{ij} is a function of β_{ij}. Thus, in general, the unique association between the elements of $\mathbf{\Pi}$ and those in $\mathbf{\Gamma}$ and $\mathbf{B}$ in (9.27) is not possible. This difficulty can be avoided if, in the specification of the system, some of the elements of $\mathbf{\Gamma}$ and $\mathbf{B}$ are reduced to zero. That is, if some of the endogenous and predetermined variables are excluded from the structural equations. This is the case of having the identification through zero restriction. Of course, there are other ways in which identification can be obtained. For instance, the other restrictions on the coefficients of the structural equations, such as specifying that one coefficient is some multiple of another, have the same effect as excluding one variable. Also, as discussed in Section 9.3.1, the identification can be had by knowing something about the distribution of the disturbances. Zero restriction, however, is the most frequently used mode of having identification. We next consider the examination of conditions for identifiability under zero restriction.

9.3.3. Rank and Order Conditions

We discuss the conditions for the identifiability of a structural relation under zero restriction in some generality. Consider any equation in a system such as (9.25). Let us assume that there are all together G endogenous and K predetermined variables in the system. We further assume that in the equation of interest we include G^* of the G endogenous variables and K^* of the K predetermined variables, so that we shall have

$$G^* + G^{**} = G$$

and

$$K^* + K^{**} = K$$

Clearly, G^{**} and K^{**} are the numbers of endogenous and predetermined variables, respectively, excluded from the equation of interest. It follows then that a general expression for any equation in the system can be written as

$$\mathbf{y}'_* \boldsymbol{\beta}_* + \mathbf{y}'_{**}\boldsymbol{\beta}_{**} + \mathbf{x}'_* \boldsymbol{\gamma}_* + \mathbf{x}'_{**}\boldsymbol{\gamma}_{**} + \mathbf{u}' = \mathbf{0}'$$

or, in matrix notation,

$$(\mathbf{y}'_* \; \mathbf{y}'_{**})\begin{pmatrix} \boldsymbol{\beta}_* \\ \boldsymbol{\beta}_{**} \end{pmatrix} + (\mathbf{x}'_* \; \mathbf{x}'_{**})\begin{pmatrix} \boldsymbol{\gamma}_* \\ \boldsymbol{\gamma}_{**} \end{pmatrix} + \mathbf{u}' = \mathbf{0}' \tag{9.28}$$

Notice that the partition of vectors of the endogenous and predetermined variables follows the convention that $\mathbf{y}_*$ are the G^* endogenous variables included in the equation and that $\mathbf{y}_{**}$ are the G^{**} excluded endogenous variables. Similarly, this is true for the included K^* predetermined variables ($\mathbf{x}_*$) and for the excluded K^{**} predetermined variables ($\mathbf{x}_{**}$). To fix ideas about the equation under discussion, we observe that $\begin{pmatrix} \boldsymbol{\beta}_* \\ \boldsymbol{\beta}_{**} \end{pmatrix}$ and $\begin{pmatrix} \boldsymbol{\gamma}_* \\ \boldsymbol{\gamma}_{**} \end{pmatrix}$ are columns, respectively, of $\mathbf{B}$ and $\boldsymbol{\Gamma}$, although (9.28) may also be written as

$$(\mathbf{y}'_* \ \mathbf{y}'_{**}) \begin{pmatrix} \beta_1 \\ \beta_2 \\ \cdot \\ \cdot \\ \cdot \\ \beta_{G^*} \\ 0 \\ 0 \\ \cdot \\ \cdot \\ \cdot \\ 0 \end{pmatrix} + (\mathbf{x}'_* \ \mathbf{x}'_{**}) \begin{pmatrix} \gamma_1 \\ \gamma_2 \\ \cdot \\ \cdot \\ \cdot \\ \gamma_{K^*} \\ 0 \\ 0 \\ \cdot \\ \cdot \\ \cdot \\ 0 \end{pmatrix} + \mathbf{u}' = \mathbf{0}' \tag{9.28a}$$

Now the reduced form of (9.28) is

$$(\mathbf{y}'_* \ \mathbf{y}'_{**}) = (\mathbf{x}'_* \ \mathbf{x}'_{**}) \begin{pmatrix} \boldsymbol{\Pi}_{*,*} & \boldsymbol{\Pi}_{*,*} \\ \boldsymbol{\Pi}_{**,*} & \boldsymbol{\Pi}_{**,**} \end{pmatrix} + (\mathbf{v}'_* \ \mathbf{v}'_{**}) \tag{9.28b}$$

Here $\boldsymbol{\Pi}_{*,*}$ is $K^* \times G^*$, $\boldsymbol{\Pi}_{**,*}$ is $K^{**} \times G^*$, $\boldsymbol{\Pi}_{*,**}$ is $K^* \times G^{**}$ and $\boldsymbol{\Pi}_{**,**}$ is $K^{**} \times G^{**}$, so that the matrix of the reduced form coefficients is K by G. Now suppose that we take the second and the third terms of the matrix equation (9.28a) to the right-hand side of the equation and that we multiply (9.28b) by the vector $\begin{pmatrix} \boldsymbol{\beta}_* \\ \boldsymbol{\beta}_{**} \end{pmatrix}$. In the order mentioned, we have

$$(\mathbf{y}'_* \ \mathbf{y}'_{**}) \begin{pmatrix} \boldsymbol{\beta}_* \\ \boldsymbol{\beta}_{**} \end{pmatrix} = (\mathbf{x}'_* \ \mathbf{x}'_{**}) \begin{pmatrix} -\boldsymbol{\gamma}_* \\ -\boldsymbol{\gamma}_{**} \end{pmatrix} - \mathbf{u}' \tag{9.28c}$$

and

$$(\mathbf{y}'_* \ \mathbf{y}'_{**}) \begin{pmatrix} \boldsymbol{\beta}_* \\ \boldsymbol{\beta}_* \end{pmatrix} = (\mathbf{x}'_* \ \mathbf{x}'_{**}) \begin{pmatrix} \boldsymbol{\Pi}_{*,*} & \boldsymbol{\Pi}_{*,**} \\ \boldsymbol{\Pi}_{**,*} & \boldsymbol{\Pi}_{**,**} \end{pmatrix} \begin{pmatrix} \boldsymbol{\beta}_* \\ \boldsymbol{\beta}_{**} \end{pmatrix} + (\mathbf{v}'_* \ \mathbf{v}'_{**}) \begin{pmatrix} \boldsymbol{\beta}_* \\ \boldsymbol{\beta}_{**} \end{pmatrix} \tag{9.28d}$$

Consider the correspondence of the coefficient matrices associated with $(\mathbf{x}'_* \quad \mathbf{x}'_{**})$. Immediately, we have

$$\begin{pmatrix} -\boldsymbol{\gamma}_* \\ -\boldsymbol{\gamma}_{**} \end{pmatrix} = \begin{pmatrix} \boldsymbol{\Pi}_{*,*} & \boldsymbol{\Pi}_{*,**} \\ \boldsymbol{\Pi}_{**,*} & \boldsymbol{\Pi}_{**,**} \end{pmatrix} \begin{pmatrix} \boldsymbol{\beta}_* \\ \boldsymbol{\beta}_{**} \end{pmatrix} \tag{9.29}$$

since $\boldsymbol{\gamma}_{**} = \mathbf{0}$ and $\boldsymbol{\beta}_{**} = \mathbf{0}$. Thus

$$\boldsymbol{\Pi}_{*,*} \boldsymbol{\beta}_* = -\boldsymbol{\gamma}_* \tag{9.29a}$$

$$\boldsymbol{\Pi}_{**,*} \boldsymbol{\beta}_* = \mathbf{0} \tag{9.29b}$$

where (9.29a) is a system of K^* equations in G^* unknowns (assuming for the moment that $\boldsymbol{\gamma}_*$ is known) and where (9.29b) is a system of K^{**} homogeneous equations in G^* unknowns. The latter system of equations provides a starting point for the identifiability conditions.

We know from matrix algebra that a system of n homogeneous linear equations in m unknowns can have a nontrivial solution if the rank of the coefficient matrix has a rank less than m and that the system can have a solution unique up to a factor of proportionality if the rank of the coefficient matrix is equal to $m - 1$. By applying these rules to (9.29b), we observe that for $\boldsymbol{\beta}_*$ to have a nontrivial solution we must have

$$\rho(\boldsymbol{\Pi}_{**.*}) < G^* \tag{9.30a}$$

and that to have a nontrivial solution unique up to a factor of proportionality we should require that

$$\rho(\boldsymbol{\Pi}_{**.*}) = G^* - 1 \tag{9.30b}$$

Now, all along, our objective has been the unique determination of $\boldsymbol{\beta}_*$ and $\boldsymbol{\gamma}_*$ in (9.29a) and (9.29b), knowing that the elements of $\boldsymbol{\Pi}$ are identified. But the condition (9.30b) by itself cannot contribute toward a unique determination of $\boldsymbol{\beta}_*$ in (9.29b). The solution of this problem lies in the so-called normalization rule which we now elaborate.

In the general structural specification of a system such as (9.25) the $\mathbf{B}$ matrix consists of elements that have arbitrary but fixed values. Thus in an illustrative system like (9.26a) β_{11} and β_{22} can have any value. A rule that states that the endogenous variable to be explained in an operation should assume the value of 1 is called the normalization rule. For some purposes in econometric analysis this rule is quite useful. An application of this rule to (9.26a) would allow us to write the system as

$$\begin{aligned} &y_1 + \frac{\beta_{21}}{\beta_{11}} y_2 + \frac{\gamma_{11}}{\beta_{11}} x_1 + \text{other terms} \\ &y_2 + \frac{\beta_{12}}{\beta_{22}} y_1 + \frac{\gamma_{12}}{\beta_{22}} x_1 + \text{other terms} \end{aligned} \tag{9.26b}$$

We observe that the application of the normalization rule to the solution of $\boldsymbol{\beta}_*$ in (9.29b) with the proviso that (9.30b) is satisfied will make for a unique determination of $\boldsymbol{\beta}_*$, since $\boldsymbol{\beta}_*$ has a solution that is unique up to a factor of proportionality on account of (9.30b). This result, in turn, makes possible a unique determination of $\boldsymbol{\gamma}_*$ in (9.29a), since, we note again, $\boldsymbol{\Pi}_{*.*}$ is identified. Consequently, we can state that the coefficients of a structural equation are identified if, and only if, $\rho(\boldsymbol{\Pi}_{**.*}) = G^* - 1$; if the rank is different from $G^* - 1$, the parameters are not identified and, hence, is the equation. An offshoot of this condition (called rank condition for identifiability) is the so-called order condition frequently used in practice. That is, the necessary condition for $\rho(\mathbf{\Pi}_{**,*}) = G^* - 1$ to hold true is that the smaller of the dimensionality of $\mathbf{\Pi}_{**,*}$ be not less than $G^* - 1$. Since the columns of $\mathbf{\Pi}_{**,*}$ are G^* in number so as to be conformable with $\boldsymbol{\beta}_*$ for multiplication and, since the normalization is being applied, we in effect have only $G^* - 1$ independent columns in $\mathbf{\Pi}_{**,*}$. It follows then that the number of rows of $\mathbf{\Pi}_{**,*}$ is precisely the number of the predetermined variables excluded from the equation being considered for estimation. Thus we say that the order condition for the identifiability is that the number of predetermined variables excluded from the equation be no less than $G^* - 1$. Suppose now that the last condition is satisfied, then two very important mutually exclusive cases must be distinguished: (1) if the number of excluded predetermined variables (K^{**}) is equal to $G^* - 1$, we call the equation "just-identified"; (2) if $K^{**} > G^* - 1$, the equation is overidentified, in the sense that there is more than enough information to recover the structure from the reduced form. The meaning of these cases is explained in detail in Subsection 10.2.1. Notice again that the order condition is necessary but not sufficient for identifiability.

To illustrate the use of the order condition, let us return to (9.22a) and (9.22b). The entire system consists of two endogenous variables and one exogenous variable. In the demand relation $G^* = 2$ and $K^{**} = 1$, so that $K^{**} = G^* - 1$ and we have just identification under order condition. Looking at the supply relation, we have $G^* = 2$ and $K^{**} = 0$ and, therefore, $K^{**} < G^* - 1$. Hence, the supply relation is not identified or underidentified.

9.3.4. Identification and Estimation

Our preceding discussions make it clear that the identification is definitely (make no mistake about it) prior to estimation. Of course, this statement is applicable only where there is a genuine interest in the determination and the estimation of the behavioral coefficients, since in the underidentified case there is no meaning in structural estimation. Recall, however, that the reduced form coefficients are always identified. Thus, in some instances, particularly if forecasting is of interest, one may estimate the reduced form equation corresponding to the endogenous variable, the structural equation for which is underidentified.

In the strict sense, then, structural estimation is meaningful only when the equation is identified. And depending on whether the equation is over- or just- identified, alternative methods of estimation are to be chosen. This and the various subjects of estimation are discussed in the next chapter.

CHAPTER 10

SIMULTANEOUS EQUATIONS—SOME ESTIMATION PROCEDURES

In this chapter our primary concern is to obtain a basic understanding of some of the estimation procedures useful for structural equations in a simultaneous system. Attention also is given to the application of the estimated equations or systems.

There are a considerable number of estimation procedures in use, and they fall into two classes: (1) single equation methods (or equation-by-equation), and (2) system methods. The first group of methods estimates one structural equation at a time, and the second group of procedures estimates all the structural equations in a system in one process. Both groups of methods exploit the fact that a structural equation is part of a large system, but in the first group this exploitation is done with less formality. This point will become clear later. In any case, in the first group belong direct least squares, indirect least squares, two-stage least squares, limited-information single-equation method, and other *k*-class estimators, and in the second group fall full information maximum likelihood, three-stage least squares, linearized maximum likelihood, and Wold's generalized interdependent systems procedure. Most of these procedures are described in detail in Christ (1966), Goldberger (1964), Chow (1964), and Theil (1970). Since it is not the purpose of this book to engage in a full discourse of econometric methods, we shall only discuss some of the procedures indicated above—the ones associated with the use of the least-squares method. Mainly because of their simplicity, the alternative least-squares procedures are observed to be the more frequently used procedures of the ones available for simultaneous equation estimation.

We divide this chapter into four sections. They are indirect least squares, (ILS), two-stage least squares (2SLS), three-stage least squares (3SLS), and applications.

10.1. Indirect Least Squares

In direct least squares (ILS) is a procedure that is useful in the estimation of a just identified structural equation in a simultaneous system but, as will be seen shortly, is relatively cumbersome in comparison with the other related and equivalent procedures. We present ILS mainly for pedagogical reasons. To discuss ILS, we must begin with the reduced-form equations and their estimation. To anticipate the result, the name indirect least squares comes from the fact that the procedure consists of finding the reduced-form estimates first and then of solving for the structural parameters in terms of the reduced-form estimates.

10.1.1. Reduced Form Estimation

To begin with the right perspective, notice that here we are interested in the estimation of the structural parameters of one of the relations in a simultaneous equations system. Let the equation to be estimated, in general terms and following the notations in Subsection 9.3.3, be

$$(\mathbf{y}'_* \ \mathbf{y}'_{**})\begin{pmatrix}\boldsymbol{\beta}_*\\ \boldsymbol{\beta}_{**}\end{pmatrix} + (\mathbf{x}'_* \ \mathbf{x}'_{**})\begin{pmatrix}\boldsymbol{\gamma}_*\\ \boldsymbol{\gamma}_{**}\end{pmatrix} + \mathbf{u}' = \mathbf{0}' \tag{10.1}$$

where $*$ denotes the variables and the associated parameters appearing in the equation and $**$ those not appearing in the equation. Recall (9.28) and (9.29a) for further clarification. Presumably, (10.1) is taken from the structural form (9.25) which has the reduced form $\mathbf{y}' = \mathbf{x}'\boldsymbol{\Pi} + \mathbf{v}$. Now the partition of the reduced-form coefficient matrix $\boldsymbol{\Pi}$ so as to conform to the ways in which endogenous and predetermined variables are included in and excluded from (10.1) gives another way of writing the reduced form as follows:

$$(\mathbf{y}'_* \ \mathbf{y}'_{**}) = (\mathbf{x}'_* \ \mathbf{x}'_{**})\begin{pmatrix}\boldsymbol{\Pi}_{*,*} & \boldsymbol{\Pi}_{*,**}\\ \boldsymbol{\Pi}_{**,*} & \boldsymbol{\Pi}_{**,**}\end{pmatrix} + (\mathbf{v}'_* + \mathbf{v}'_{**}) \tag{10.2}$$

By continuing and reviewing our discussion in Subsection 9.3.3, we observe that, if equation (10.1) is identified, the estimates of $\boldsymbol{\beta}_*$ and $\boldsymbol{\gamma}_*$ can be deduced from the estimates of the reduced form coefficients through the following relations, after an appropriate normalization.

$$\boldsymbol{\Pi}_{*,*}\boldsymbol{\beta}_* = -\boldsymbol{\gamma}_* \tag{10.3a}$$

$$\boldsymbol{\Pi}_{**,*}\boldsymbol{\beta}_* = \mathbf{0} \tag{10.3b}$$

These discussions motivate reduced-form estimation. Notice that by "identified" we mean both just- and overidentified cases. Indirect least squares builds on the condition for a just-identified equation, but in the meantime we consider some general results from reduced-form estimation.

Our treatment of the multivariate regression model in Chapter 8 makes it abundantly clear that the usual least-squares method is eminently suitable for the estimation of the reduced-form equations. Now by writing (10.2) for a sample of size n, we have

$$(\mathbf{Y}_* \, \mathbf{Y}_{**}) = (\mathbf{X}_* \, \mathbf{X}_{**})\begin{pmatrix} \mathbf{\Pi}_{*,*} & \mathbf{\Pi}_{*,**} \\ \mathbf{\Pi}_{**,*} & \mathbf{\Pi}_{**,**} \end{pmatrix} + (\mathbf{V}_* \, \mathbf{V}_{**}) \tag{10.4a}$$

or compactly

$$\mathbf{Y} = \mathbf{X\Pi} + \mathbf{V} \tag{10.4b}$$

where $\mathbf{Y}_*$ is $n \times G^*$, $\mathbf{Y}_{**}$ is $n \times G^{**}$, $\mathbf{X}$ is $n \times K^*$, $\mathbf{X}_{**}$ is $n \times K^{**}$, and the dimensionalities of the partitioned submatrices of $\mathbf{\Pi}$ can be deduced from the ones given for $\mathbf{Y}_*$, $\mathbf{Y}_{**}$, $\mathbf{X}_*$, and $\mathbf{X}_{**}$. For instance, $\mathbf{\Pi}_{**,*}$ is $K^{**} \times G^*$. The LS estimate of the reduced-form coefficient matrix is, then,

$$\hat{\mathbf{\Pi}} = (\mathbf{X}'\mathbf{X})^{-1}\mathbf{X}'\mathbf{Y} \tag{10.5}$$

and the partitioning of $\hat{\mathbf{\Pi}}$ to correspond to the structural relation (10.1) is

$$\begin{pmatrix} \hat{\mathbf{\Pi}}_{*,*} & \hat{\mathbf{\Pi}}_{*,**} \\ \hat{\mathbf{\Pi}}_{**,*} & \hat{\mathbf{\Pi}}_{**,**} \end{pmatrix} = \begin{pmatrix} \mathbf{X}_*'\mathbf{X}_* & \mathbf{X}_*'\mathbf{X}_{**} \\ \mathbf{X}_{**}'\mathbf{X}_* & \mathbf{X}_{**}'\mathbf{X}_{**} \end{pmatrix}^{-1} \begin{pmatrix} \mathbf{X}_*'\mathbf{Y}_* & \mathbf{X}_*'\mathbf{Y}_{**} \\ \mathbf{X}_{**}'\mathbf{Y}_* & \mathbf{X}_{**}'\mathbf{Y}_{**} \end{pmatrix} \tag{10.5a}$$

In (10.5) the ith column of $\hat{\mathbf{\Pi}}$ would be

$$\hat{\mathbf{\Pi}}_i = (\mathbf{X}'\mathbf{X})^{-1}\mathbf{X}'\mathbf{Y}_i \tag{10.5b}$$

where $\hat{\mathbf{\Pi}}_i$ is $K \times 1$ and consists of the estimates of the coefficients in the ith reduced-form equation and $\mathbf{Y}_i$ is $n \times 1$, the latter being the observation on the ith endogenous variable. The last consideration leads us to observe that the LS estimate of $\mathbf{\Pi}$ is formed of the separate (or equation-by-equation) LS estimates of the coefficients of the reduced-form equations. From this viewpoint and from the discussions of Section 8.1, we readily see that either equation-by-equation or joint estimation of (10.4b) will yield the same estimates of $\hat{\mathbf{\Pi}}$ and of the related covariance matrix and, in addition, these estimates are consistent. Since consistency is invariant under rational operations, consistent estimates of $\boldsymbol{\beta}_*$ and $\boldsymbol{\gamma}_*$ in (10.1) can be obtained through the relations (10.3a) and (10.3b). Summarizing then, we need to solve

$$\hat{\mathbf{\Pi}}_{*,*}\boldsymbol{\beta}_* = -\boldsymbol{\gamma}_* \tag{10.5c}$$

$$\hat{\mathbf{\Pi}}_{**,*}\boldsymbol{\beta}_* = \mathbf{0} \tag{10.5d}$$

first for $\boldsymbol{\beta}_*$ in (10.5d) and, then, after normalization, for $\boldsymbol{\gamma}_*$ in (10.5c). Now, if the number of rows of $\mathbf{\Pi}_{**,*}$ is $(G^* - 1)$, there can be a unique solution for $\boldsymbol{\beta}_*$ (that is, the number of equations is equal to the number of unknowns). But, if the number of rows of $\boldsymbol{\Pi}_{**,*}$ is greater than $(G^* - 1)$, there will be more equations than unknowns. In the last case, there can be choices as to which $(G^* - 1)$ rows of $\boldsymbol{\Pi}_{**,*}$ for solving $\boldsymbol{\beta}_*$ and, as such, the solutions of $\boldsymbol{\beta}_*$ are not necessarily identical. The uniqueness with which the consistent estimates of the reduced-form equations can be transformed into consistent estimates of the parameters in a structural equation is the topic of the next sections.

There are several questions of practical importance in estimating a system like (10.4b). For instance, the K predetermined variables may be so collinear that the LS estimation cannot be carried out. We shall have more to say about this sort of problem when we discuss two-stage least squares.

10.1.2. The ILS Procedure

Thus far in discussing the reduced-form estimation we have said only that the structural equation of interest was identified, meaning that the rank of $\boldsymbol{\Pi}_{**,*}$ in (10.3b) was less than or equal to $G^* - 1$. In the just-identified case we know that, after normalization, $\boldsymbol{\beta}_*$ in (10.3b) can be uniquely solved for in terms of the matrix $\boldsymbol{\Pi}_{**,*}$. Thus, the unique and consistent estimate of $\boldsymbol{\beta}_*$ can be had, given the consistent estimate of $\boldsymbol{\Pi}_{**,*}$. The use of this idea in estimating the structural equation is the ILS procedure.

Thus, to find consistent estimates of the elements of $\boldsymbol{\beta}_*$ and $\boldsymbol{\gamma}_*$ in (10.1), we go through the following process.

1. Notice clearly the endogenous and predetermined variables in the system that are included in the equation to be estimated.
2. Obtain the LS estimate of the reduced form coefficient matrix, using all the predetermined variables as independent variables as in (10.5).
3. Pull out from the estimated $\boldsymbol{\Pi}$ matrix the elements that correspond to the included endogenous variables and to the excluded predetermined variables in the manner of (10.5a).
4. Use the $\hat{\boldsymbol{\Pi}}_{**,*}$ in place of $\boldsymbol{\Pi}_{**,*}$ in (10.3b) with the coefficient of the endogenous variable to be explained by the equation assigned the value -1 in the $\boldsymbol{\beta}_*$ vector, and solve

$$\hat{\boldsymbol{\Pi}}_{**,*}\begin{pmatrix} -1 \\ \beta_2 \\ \cdot \\ \cdot \\ \cdot \\ \beta_{G^*} \end{pmatrix} = \mathbf{0} \qquad (10.6a)$$

for $\beta_2, \beta_3, \ldots, \beta_{G^*}$ in terms of the elements of $\hat{\boldsymbol{\Pi}}_{**,*}$.

5. Pull out the elements of $\hat{\mathbf{\Pi}}$ corresponding to the included endogenous variables and included predetermined variables to form $\hat{\mathbf{\Pi}}_{*,*}$.
6. Use the last matrix along with the estimated value of $\boldsymbol{\beta}_*$ vector and solve, as in (10.3a),

$$\hat{\mathbf{\Pi}}_{*,*}\begin{pmatrix} -1 \\ \hat{\beta}_2 \\ \cdot \\ \cdot \\ \cdot \\ \hat{\beta}_{G^*} \end{pmatrix} = -\boldsymbol{\gamma}_* \tag{10.6b}$$

for the elements of $\boldsymbol{\gamma}_*$. This completes the ILS procedure of estimating the structural parameters $\boldsymbol{\beta}_*$ and $\boldsymbol{\gamma}_*$ in (10.1).

This procedure becomes rather cumbersome if the size of $\mathbf{\Pi}$ is large and the numbers K^{**} and/or G^* are large. Also having to solve (10.6a to b), when these numbers are large, is not pleasant. It turns out that in the just-identified equation the ILS procedure gives the same estimates as the estimates obtained by 2SLS or by the instrumental variable technique. This proof is available in Goldberger (1964, pp. 329–334).

We now illustrate the preceding discussion with a two-equation example. Consider the system

$$\begin{aligned} y_1 + \beta_1 y_2 + \gamma_{11}x_1 + \gamma_{12}x_2 + u_1 &= 0 \\ \beta_2 y_1 + \quad y_2 + \qquad\qquad\qquad \gamma_{23}x_3 + u_2 &= 0 \end{aligned} \tag{10.6c}$$

where y_1 and y_2 are endogenous and x_1, x_2, and x_3 are exogenous. Notice that the first equation is just-identified and the second equation is over-identified. Also, the coefficients have been normalized. Suppose that we estimate the coefficients of the unrestricted reduced form for (10.6c)

$$(y_1 \quad y_2) = (x_1 \quad x_2 \quad x_3)\begin{pmatrix} \pi_{11} & \pi_{21} \\ \pi_{12} & \pi_{22} \\ \pi_{13} & \pi_{23} \end{pmatrix} + (v_1 \quad v_2) \tag{10.6e}$$

by

$$\hat{\mathbf{\Pi}} = \begin{pmatrix} \hat{\mathbf{\Pi}}_{11} & \hat{\mathbf{\Pi}}_{21} \\ \hat{\mathbf{\Pi}}_{12} & \hat{\mathbf{\Pi}}_{22} \\ \hat{\mathbf{\Pi}}_{13} & \hat{\mathbf{\Pi}}_{23} \end{pmatrix} = \begin{pmatrix} p_{11} & p_{21} \\ p_{12} & p_{22} \\ p_{13} & p_{23} \end{pmatrix} = (\mathbf{X}'\mathbf{X})^{-1}\mathbf{X}'\mathbf{Y} \tag{10.6f}$$

where $\mathbf{X} = (\mathbf{x}_1 \quad \mathbf{x}_2 \quad \mathbf{x}_3)$ and $\mathbf{Y} = (\mathbf{y}_1 \quad \mathbf{y}_2)$ and $\mathbf{x}_i$ and $\mathbf{y}_i$ are the observation vectors. Now let us take the second equation in (10.6c) and try to apply the ILS procedure. This involves first partitioning the matrix in (10.6f), so as to reflect the included endogenous variables and the excluded exogenous variables. Thus, by (10.5a),

$$\hat{\mathbf{\Pi}}_{**,*} = \begin{pmatrix} p_{12} & p_{22} \\ p_{13} & p_{23} \end{pmatrix}$$

Utilizing (10.5d), we observe that

$$\hat{\mathbf{\Pi}}_{**,*}\boldsymbol{\beta}_{*} = \begin{pmatrix} p_{12} & p_{22} \\ p_{13} & p_{23} \end{pmatrix}\begin{pmatrix} \beta_2 \\ -1 \end{pmatrix} = \begin{pmatrix} 0 \\ 0 \end{pmatrix}$$

and solving for β_2, we have

$$\beta_2 = p_{22}/p_{12} = p_{23}/p_{13} \tag{10.6g}$$

We then conclude that for the overidentified equation we cannot obtain the unique estimates of the coefficients by ILS. In contrast, take the first equation where y_1 and y_2 are the included endogenous variables and x_3 is the excluded exogenous variable. And, by (10.5d),

$$\hat{\mathbf{\Pi}}_{**,*} = (p_{13} \quad p_{23})$$

and

$$\hat{\mathbf{\Pi}}_{**,*}\boldsymbol{\beta}_{*} = (p_{13} \quad p_{23})\begin{pmatrix} -1 \\ \beta_1 \end{pmatrix} = 0$$

Therefore

$$\beta_1 = p_{23}/p_{13}$$

Furthermore, by (10.5c),

$$\hat{\mathbf{\Pi}}_{*,*}\boldsymbol{\beta}_{*} = \begin{pmatrix} p_{11} & p_{21} \\ p_{12} & p_{22} \end{pmatrix}\begin{pmatrix} -1 \\ \beta_1 \end{pmatrix} = -\begin{pmatrix} \gamma_{11} \\ \gamma_{12} \end{pmatrix}$$

It then follows that

$$-p_{11} + \beta_1 p_{21} = -\gamma_{11}$$

$$-p_{12} + \beta_1 p_{22} = -\gamma_{12}$$

so that γ_{11} and γ_{12} can be uniquely determined, given the solution for β_1. Thus, ILS is appropriate for the estimation of the equation that is just-identified.

One of the estimation procedures for handling an overidentified equation is the method of two-stage least squares. Before launching our formal discussion of this method in the next section, let us indicate how this method might be applied for the estimation of the second equation above. The first-stage least squares entails regressing y_1 in the second equation on x_1, x_2, and x_3 and obtaining the estimated y_1, $\hat{y}_1$,

$$\hat{\mathbf{y}}_1 = (\mathbf{X}'\mathbf{X})^{-1}\mathbf{X}'\mathbf{y}_1$$

The second-stage least squares calls for regression of y_2 on $\hat{y}_1$ and x_3 as follows

$$y_2 = -\beta_2\hat{y}_1 - \gamma_{23}x_3 - u_2$$

whereby one can obtain the estimates of β_2 and γ_{23} without the difficulty that involves (10.6g), as in the case of using ILS.

10.2. Two-Stage Least Squares (2SLS)

We indicated in the preceding section that indirect least squares was appropriate for the estimation of a just-identified structural equation. If a structural equation is overidentified, it becomes necessary to use some other procedures of estimation. These procedures include two-stage least squares (2SLS) and limited information single-equation estimation.

In this section we first consider the meaning of overidentification in relation to indirect least squares, and then discuss the 2SLS procedure, its interpretation, and some related computational problems.

10.2.1. The Meaning of Overidentification

Consider the matrix equation (10.5d) in the preceding section. The number of unknown elements of $\boldsymbol{\beta}^*$ to be solved for in terms of the elements of $\hat{\boldsymbol{\Pi}}_{**.*}$ is $(G^* - 1)$, given that one of the elements of $\boldsymbol{\beta}^*$ has been set at -1. Now the only way in which (10.5d) can have a nontrivial unique solution is for the rank of $\hat{\boldsymbol{\Pi}}_{**.*}$ to be equal to $(G^* - 1)$. Furthermore, overidentification means that in (10.5d) there are more equations than unknowns. And it becomes necessary to pick among the K^{**} (which is greater than $G^* - 1$ by assumption) equations a subset consisting of $G^* - 1$ equations. It is possible, then, that there is more than one choice of a subset of $G^* - 1$ equations. That is, some of the rows of $\boldsymbol{\Pi}_{**.*}$ in

$$\boldsymbol{\Pi}_{**.*}\boldsymbol{\beta}_* = \mathbf{0} \qquad (10.7)$$

will be thrown away. Another way of saying this is that some specifying assumptions about the model as reflected in the $\boldsymbol{\Pi}_{**.*}$ matrix are thrown out. As will be clear soon, 2SLS uses all the information in (10.7) in its estimation process.

10.2.2. The Procedure of 2SLS

This procedure for simultaneous equation estimation is independently due to Theil (1958) and to Basmann (1957). The basic idea is (1) to find the first-stage LS estimated values of the explanatory endogenous variables in the equation, and (2) to use these estimated values in place of the observed values of these endogenous variables for the second-stage LS estimation.

More specifically, take any structural equation

$$\mathbf{y} = \mathbf{Y}_1\boldsymbol{\beta}_1 + \mathbf{X}_1\boldsymbol{\gamma}_1 + \mathbf{v} \tag{10.8}$$

where $\boldsymbol{\beta}_1$ and $\boldsymbol{\gamma}_1$ are structural coefficients corresponding to the ones shown in (10.5c) and (10.5d) as follows: $\boldsymbol{\beta}_1' = -(\beta_2 \quad \beta_3 \quad \cdots \quad \beta_{G^*})$ and $\boldsymbol{\gamma}_1' = -(\gamma_1 \quad \gamma_2 \quad \cdots \quad \gamma_{K^*})$. We use the LS method to find $\hat{\mathbf{Y}}_1$ by reference to the unrestricted reduced-form equations for $y_2, y_3, \ldots, y_{G^*}$. That is, the reduced-form equations are

$$\mathbf{Y}_1 = \mathbf{X}\boldsymbol{\Pi}_1 + \mathbf{V} \tag{10.11}$$

where $\mathbf{Y}_1$ is $n \times (G^* - 1)$ and $\boldsymbol{\Pi}_1$ is $K \times (G^* - 1)$. These equations are unrestricted in the sense that the zero restrictions in the $\mathbf{B}$ and $\boldsymbol{\Gamma}$ matrices in the structural form are not reflected in $\boldsymbol{\Pi}_1$ (recall that we take $G^* - 1$ reduced-form equations out of the entire G equations in $\mathbf{Y} = \mathbf{X}\boldsymbol{\Pi} + \mathbf{V} = -\mathbf{X}\boldsymbol{\Gamma}\mathbf{B}^{-1} - \mathbf{U}\mathbf{B}^{-1}$). From (10.11) we obtain

$$\hat{\mathbf{Y}}_1 = \mathbf{X}(\mathbf{X}'\mathbf{X})^{-1}\mathbf{X}'\mathbf{Y}_1 \tag{10.11a}$$

since the LS $\hat{\boldsymbol{\Pi}}_1 = (\mathbf{X}'\mathbf{X})^{-1}\mathbf{X}'\mathbf{Y}_1$; then we replace $\mathbf{Y}_1$ in (10.8) by $\hat{\mathbf{Y}}_1$ just found and get LS estimator of $\boldsymbol{\beta}_1$ and $\boldsymbol{\gamma}_1$ there. That is, we obtain the second-stage LS estimates from

$$\mathbf{Y} = \hat{\mathbf{Y}}_1\boldsymbol{\beta}_1 + \mathbf{X}_1\boldsymbol{\gamma}_1 + \mathbf{w} \tag{10.12}$$

by applying the usual LS method. The 2SLS estimates of $\boldsymbol{\beta}_1$ and $\boldsymbol{\gamma}_1$ are shown to be consistent by Christ (1966, pp. 438–440). Briefly the idea is this. From (10.11)

$$\hat{\mathbf{V}}_1 = \mathbf{Y}_1 - \hat{\mathbf{Y}}_1 \tag{10.12a}$$

where $\hat{\mathbf{V}}$ is $n \times (G^* - 1)$ and contains the LS estimated residuals of the unrestricted reduced form disturbances, we obtain, by substitution into (10.8),

$$\mathbf{y} = \hat{\mathbf{Y}}_1\boldsymbol{\beta}_1 + \mathbf{X}_1\boldsymbol{\gamma}_1 + [\mathbf{v} + \hat{\mathbf{V}}\boldsymbol{\beta}_1] \tag{10.13}$$

The proof of consistency proceeds by considering the independence of $(\mathbf{v} + \hat{\mathbf{V}}\boldsymbol{\beta}_1)$ from $\hat{\mathbf{Y}}_1$ and $\mathbf{X}_1$ in the limit.

The consistency of the 2SLS estimator can also be shown by observing that the estimation procedure is an application of the instrumental variable technique.† Also, as will be apparent in the next section, the 2SLS estimation can be interpreted as an application of Aitken's generalized least-squares method. These ways of interpreting 2SLS are interesting but are not our immediate concern here. We next consider some computational problems of 2SLS.

† In addition to Goldberger (1964, pp. 329–344), also, see Klein (1955, pp. 147–153).

10.2.3. Computational Notes

Our discussion in the preceding subsection outlined what the 2SLS procedure is. The computational formula is developed below.

Take (10.12) and consider direct LS estimation. Then the 2SLS estimate of $\begin{pmatrix} \boldsymbol{\beta}_1 \\ \boldsymbol{\gamma}_1 \end{pmatrix}$ is

$$\begin{pmatrix} \mathbf{b}_1 \\ \mathbf{c}_1 \end{pmatrix} = \begin{pmatrix} \hat{\mathbf{Y}}_1'\hat{\mathbf{Y}}_1 & \hat{\mathbf{Y}}_1'\mathbf{X}_1 \\ \mathbf{X}_1'\hat{\mathbf{Y}}_1 & \mathbf{X}_1'\mathbf{X}_1 \end{pmatrix}^{-1} \begin{pmatrix} \hat{\mathbf{Y}}_1'\mathbf{y} \\ \mathbf{X}_1'\mathbf{y} \end{pmatrix} \tag{10.14}$$

or, by rewriting, we obtain

$$\begin{pmatrix} \hat{\mathbf{Y}}_1'\hat{\mathbf{Y}}_1 & \hat{\mathbf{Y}}_1'\mathbf{X}_1 \\ \mathbf{X}_1'\hat{\mathbf{Y}}_1 & \mathbf{X}_1'\mathbf{X}_1 \end{pmatrix} \begin{pmatrix} \mathbf{b}_1 \\ \mathbf{c}_1 \end{pmatrix} = \begin{pmatrix} \hat{\mathbf{Y}}_1'\mathbf{y} \\ \mathbf{X}_1'\mathbf{y} \end{pmatrix} \tag{10.14a}$$

a set of normal equations. Now it can be shown that

$$\hat{\mathbf{Y}}_1'\hat{\mathbf{Y}}_1 = \hat{\mathbf{Y}}_1'\mathbf{Y}_1 = \mathbf{Y}_1'\hat{\mathbf{Y}}_1 \tag{10.15a}$$

$$\mathbf{X}_1'\hat{\mathbf{Y}}_1 = \mathbf{X}_1'\mathbf{Y}_1 \tag{10.15b}$$

That is, (1) the cross-products matrix of the LS estimated $\hat{\mathbf{Y}}_1$ with itself is the same as the cross-product matrix of the LS estimate of $\mathbf{Y}_1$ with $\mathbf{Y}_1$ itself, and (2) the cross-product matrix of $\mathbf{X}_1$ with the LS estimated $\hat{\mathbf{Y}}_1$ is the same as $\mathbf{X}_1$ times the raw $\mathbf{Y}_1$—results due to the LS properties involved in these quantities. Show this as an exercise, or see Goldberger (1964, p. 332). Of course, $\hat{\mathbf{Y}}_1'\hat{\mathbf{Y}}_1 \neq \mathbf{Y}_1'\mathbf{Y}_1$. By using (10.15a) and (10.15b), we can reduce (10.14a) to

$$\begin{pmatrix} \mathbf{Y}_1'\hat{\mathbf{Y}}_1 & \mathbf{Y}_1'\mathbf{X}_1 \\ \mathbf{X}_1'\mathbf{Y}_1 & \mathbf{X}_1'\mathbf{X}_1 \end{pmatrix}^{-1} \begin{pmatrix} \mathbf{b}_1 \\ \mathbf{c}_1 \end{pmatrix} = \begin{pmatrix} \hat{\mathbf{Y}}_1'\mathbf{Y} \\ \mathbf{X}_1'\mathbf{Y} \end{pmatrix} \tag{10.16}$$

and further to

$$\begin{pmatrix} \mathbf{Y}_1'\mathbf{X}(\mathbf{X}'\mathbf{X})^{-1}\mathbf{X}'\mathbf{Y}_1 & \mathbf{Y}_1'\mathbf{X}_1 \\ \mathbf{X}_1'\mathbf{Y}_1 & \mathbf{X}_1'\mathbf{X}_1 \end{pmatrix} \begin{pmatrix} \mathbf{b}_1 \\ \mathbf{c}_1 \end{pmatrix} = \begin{pmatrix} \mathbf{Y}_1'\mathbf{X}(\mathbf{X}'\mathbf{X})^{-1}\mathbf{X}'\mathbf{Y} \\ \mathbf{X}_1'\mathbf{Y} \end{pmatrix} \tag{10.17}$$

noticing that the LS estimator of $\mathbf{\Pi}_1$ in (10.11) is

$$\mathbf{P}_1 = (\mathbf{X}'\mathbf{X})^{-1}\mathbf{X}'\mathbf{Y}_1 \tag{10.11b}$$

Thus (10.17) is an expression involving only the sample observations. Although this formula is useful for the analysis and for the computation of a relatively small equation we observe that computer programs that handle the computation according to the two-stage procedure as described in Subsection 10.2.2 are available. One such computer program has been written by Straud, Zellner, and Chau (1963), with subsequent revision by Thornber and Zellner, which not only does the 2SLS estimation but also gives Zellner's estimate (see Chapter 8) and the 3SLS estimate.

A consistent estimate of the asymptotic variance covariance matrix of the 2SLS estimator $\begin{pmatrix} \mathbf{b}_1 \\ \mathbf{c}_1 \end{pmatrix}$ is

$$\mathbf{Var}\begin{pmatrix} \mathbf{b}_1 \\ \mathbf{c}_1 \end{pmatrix} = s^2 \begin{pmatrix} \hat{\mathbf{Y}}_1'\hat{\mathbf{Y}}_1 & \hat{\mathbf{Y}}_1'\mathbf{X}_1 \\ \mathbf{X}_1'\hat{\mathbf{Y}}_1 & \mathbf{X}_1'\mathbf{X}_1 \end{pmatrix}^{-1} \tag{10.18}$$

with

$$s^2 = \frac{1}{n - (G^* - 1) - K^*} (\mathbf{y} - \mathbf{Y}_1\mathbf{b}_1 - \mathbf{X}_1\mathbf{c}_1)'(\mathbf{y} - \mathbf{Y}_1\mathbf{b}_1 - \mathbf{X}_1\mathbf{c}_1) \tag{10.19}$$

Notice that $\hat{\mathbf{Y}}_1$ was used to find the consistent estimates of $\boldsymbol{\beta}_1$ and $\boldsymbol{\gamma}_1$, $\mathbf{b}_1$, and $\mathbf{c}_1$, respectively, but that in computing the residuals for the structural equation the raw observed $\mathbf{Y}_1$, not $\hat{\mathbf{Y}}_1$, is used.

Another problem that frequently arises in the first-stage least squares is the question of choosing among unrestricted and restricted reduced-form equations and the somewhat arbitrary dropping of some predetermined variables from these equations. First, take the unrestricted reduced-form equations in (10.11) where each of the $G^* - 1$ endogenous variables is a function of K predetermined variables. Now in econometric models sometimes it is the case that the time-series data used do not have a sufficient number of observations (say, sample size is less than or about the same as the number K), so that the LS estimation of the reduced-form equation would not be possible. Even if there are enough observations, it may be the case that the predetermined variables are highly correlated so as to yield unreliable estimates.

When the problems of insufficient observations and/or high collinearity are present we can resort to purely instrumental variable technique for simultaneous equation estimation. By the instrumental variable technique in the present context we mean that judgment must be exercised to pick a subset of the K predetermined variables and that we must use the subset as independent variables for finding the LS estimates of the $G^* - 1$ endogenous variables and then must employ these estimates as the instruments for the endogenous variables present in the equation for the second-stage LS estimation. In choosing the subset for the first-stage LS, a few points of empirical interest can be noted: (1) if several variables are observed to be highly correlated, pick only one of them and use it as one of the independent variables; (2) theory and/or other knowledge might indicate that some exogenous or lagged endogenous variables are more significant explanators of the endogenous variables in question; and (3) it is theoretically crucial and computationally efficient to use the same set of predetermined variables for each endogenous variable.

Of course, if high collinearity is the only problem, the principal component analysis can be applied to the K predetermined variables prior to the first-stage least squares, and then we can use a handful of the resulting principal components for the first-stage LS estimation. Another alternative would be to apply nonlinear estimation procedure (which we discussed in Section 7.4) with explicit restriction on the coefficients representing the identifying specification of the $\mathbf{B}$ and $\mathbf{\Gamma}$ matrices.

These problems of estimation in obtaining the "reduced form" estimates are not peculiar to 2SLS; they are present also in the limited information single-equation estimation as well as in the three-stage least squares.

We now consider this latter estimation procedure.

10.3. Three-Stage Least Squares

Recall our earlier statement that 3SLS is a system procedure as compared with single equation procedures. By a system procedure or method, we mean that, when estimating a system of simultaneous equations, the method provides that all the identifiability restrictions of the equations of the system are brought to bear in the estimation process. Let us clarify this statement by an example.

10.3.1. Single Equation and Systems Methods Illustrated

$$D = \alpha_0 + \alpha_1 i + \alpha_2 Y + \alpha_3\left(\frac{R}{P}\right) + \alpha_4\left(\frac{M}{A}\right)_{-1} + \alpha_5 L + v_1 \qquad (10.20a)$$

$$S = \beta_0 + \beta_1 i + \beta_2 i^L_{-2} + \beta_3 R^P_{-3} + \beta_4 S^A + v_2 \qquad (10.20b)$$

$$\Delta i = \gamma_0 + \gamma_1(S - D) + \gamma_2\, \Delta i^L_{-1} + v_3 \qquad (10.20c)$$

where the variables are defined as follows.

D = Demand flow of nonfarm residential mortage loans in millions of dollars.

S = Supply flow of nonfarm residential mortgage loans in millions of dollars.

i = Average market mortgage yield.

Y = Disposable personal income in billions of dollars.

R = Rent component of the consumer price index (1957–1959 = 100).

P = Construction cost index for residence by Boeckh and Associated.

M = Amount of 1- to 4-family mortgage debt outstanding at the beginning of the quarter, in billions of dollars.

A = Amount of financial assets held by the household sector at the beginning of the quarter, in billions of dollars.

R^P = Average reserve position of the member banks, or free reserves in millions of dollars during the quarter.

i^L = Market yield of long-term capital market instruments (new issues) other than home mortgages.

S^A = Net increase in savings at savings and loan associations, mutual savings banks, commercial banks, and life insurance companies, in millions of dollars.

The endogenous variables of the system (10.20a to c) are $D\ (= y_1)$, $S\ (= y_2)$, and $i\ (= y_3)$. The predetermined variables are $Y\ (= x_1)$, $R/P = x_2$, $(M/A)_{-1}$ $(= x_3)$, $\Delta L\ (= x_4)$, $i^L_{-1}\ (= x_5)$, $i^L_{-2}\ (= x_6)$, $R^P_{-3}\ (= x_7)$, $S^A\ (= x_8)$, $i_{-1}\ (= x_9)$, and the constant $(= x_{10})$. All of the equations are overidentified so that the 2SLS or 3SLS estimation is appropriate. (Investigate the order condition of identifiability for each equation.)

Suppose that we decide to use 2SLS, a single equation method, to estimate the coefficients in (10.20a). The procedure requires that y_3, which appears in (10.20a), be regressed on $x_1, x_2, \ldots, x_{10}$ to provide the first-stage estimate of y_3, or $\hat{y}_3$. Then least squares is run on

$$y_1 = \alpha_0 + \alpha_1 \hat{y}_3 + \alpha_2 x_1 + \alpha_3 x_2 + \alpha_4 x_3 + \alpha_5 x_4 + v_1$$

to obtain the 2SLS estimates, and similarly for the other two structural equations. Notice that here in estimating any one equation no account is made of how the other two equations are specified (that is, the exclusion and inclusion of certain variables). In a system method the prior specification as to what variables are included in and excluded from *each of the* equations of the system is explicitly accounted for in the estimation procedure. Thus under the system method the equation (10.20a) is estimated simultaneously with equations (10.20b and c) while an account is made of the fact that in equation (10.20b) the variables y_2, x_1, x_2, x_3, x_4, x_5, and x_9 do not appear and in equation (10.20c) the variables x_1, x_2, x_3, x_4, x_7, and x_8 do not appear. It stands to reason, then, that if there are errors in the specification of the equations in the system, a system method would be using erroneous information in the estimation.

10.3.2. The 3SLS Procedure

We interpret this estimation procedure as involving an application of Aitken's generalized least squares. To observe this, let us start from the 2SLS estimation with a minor change in notation to allow for the fact that we are now dealing with all of the equations in a system.*

* It is assumed that all of the equations to be estimated are identified and that identities have been properly eliminated.

First, we want to determine that the 2SLS is an application of Aitken's generalized least squares to the structural estimation. Recall that the structural equation to be estimated is

$$\begin{aligned} \mathbf{y} &= \mathbf{Y}_1\boldsymbol{\beta}_1 + \mathbf{X}_1\boldsymbol{\gamma}_1 + \mathbf{v} \\ &= (\mathbf{Y}_1 \quad \mathbf{X}_1)\begin{pmatrix} \boldsymbol{\beta}_1 \\ \boldsymbol{\gamma}_1 \end{pmatrix} + \mathbf{v} \\ &= \mathbf{Z}\boldsymbol{\theta} + \mathbf{v} \end{aligned} \tag{10.8}$$

where $\mathbf{Z} = (\mathbf{Y}_1 \quad \mathbf{X}_1)$, and $\boldsymbol{\theta} = \begin{pmatrix} \boldsymbol{\beta}_1 \\ \boldsymbol{\gamma}_1 \end{pmatrix}$. Suppose now that we choose to transform the equation by multiplying it through by the observation matrix of all of the predetermined variables, namely, $\mathbf{X} = (\mathbf{X}_1 \quad \mathbf{X}_2)$. Then

$$\mathbf{X}'\mathbf{y} = \mathbf{X}'\mathbf{Z}\boldsymbol{\theta} + \mathbf{X}'\mathbf{v} \tag{10.21a}$$

so that the covariance of the new disturbance vector $\mathbf{X}'\mathbf{v}$ in (10.21a) is

$$E\mathbf{X}'\mathbf{v}\mathbf{v}'\mathbf{X} = E\mathbf{X}'\sigma^2\mathbf{I}\mathbf{X} = \sigma^2\mathbf{X}'\mathbf{X} \tag{10.21b}$$

which is a nonscalar matrix, and the application of Aitken's GLS is called for. Since the part of the covariance matrix corresponding to $\boldsymbol{\Omega}$ in (6.2b) is $\mathbf{X}'\mathbf{X}$ and since $\boldsymbol{\Omega}^{-1} = (\mathbf{X}'\mathbf{X})^{-1}$, the pure Aitken's estimator of $\boldsymbol{\theta}$, $\boldsymbol{\theta}^*$ by (6.8) is

$$\boldsymbol{\theta}^* = [\mathbf{Z}'\mathbf{X}(\mathbf{X}'\mathbf{X})^{-1}\mathbf{X}'\mathbf{Z}]^{-1}[\mathbf{Z}'\mathbf{X}(\mathbf{X}'\mathbf{X})^{-1}\mathbf{X}'\mathbf{y}] \tag{10.21c}$$

Returning to the notation that $\mathbf{Z} = (\mathbf{Y}_1 \quad \mathbf{X}_1)$, we observe that

$$\begin{aligned} \boldsymbol{\theta}^* &= \left[\begin{pmatrix} \mathbf{Y}_1' \\ \mathbf{X}_1' \end{pmatrix}\mathbf{X}(\mathbf{X}'\mathbf{X})^{-1}\mathbf{X}'(\mathbf{Y}_1 \quad \mathbf{X}_1)\right]^{-1}\begin{bmatrix} \mathbf{Y}_1'\mathbf{X}(\mathbf{X}'\mathbf{X})^{-1}\mathbf{X}'\mathbf{y} \\ \mathbf{X}_1'\mathbf{X}(\mathbf{X}'\mathbf{X})^{-1}\mathbf{X}'\mathbf{y} \end{bmatrix} \\ &= \begin{bmatrix} \mathbf{Y}'\mathbf{X}(\mathbf{X}'\mathbf{X})^{-1}\mathbf{X}'\mathbf{Y}_1 & \mathbf{Y}_1'\mathbf{X}(\mathbf{X}'\mathbf{X})^{-1}\mathbf{X}'\mathbf{X}_1 \\ \mathbf{X}_1'\mathbf{X}(\mathbf{X}'\mathbf{X})^{-1}\mathbf{X}'\mathbf{Y}_1 & \mathbf{X}_1'\mathbf{X}(\mathbf{X}'\mathbf{X})^{-1}\mathbf{X}'\mathbf{X}_1 \end{bmatrix}^{-1}\begin{bmatrix} \mathbf{Y}_1'\mathbf{X}(\mathbf{X}'\mathbf{X})^{-1}\mathbf{X}'\mathbf{y} \\ \mathbf{X}_1'\mathbf{X}(\mathbf{X}'\mathbf{X})^{-1}\mathbf{X}'\mathbf{y} \end{bmatrix} \\ &= \begin{bmatrix} \mathbf{Y}_1'\mathbf{X}(\mathbf{X}'\mathbf{X})^{-1}\mathbf{X}'\mathbf{Y}_1 & \mathbf{Y}_1'\mathbf{X}_1 \\ \mathbf{X}_1'\mathbf{Y}_1 & \mathbf{X}_1'\mathbf{Y}_1 \end{bmatrix}^{-1}\begin{bmatrix} \mathbf{Y}_1'\mathbf{X}(\mathbf{X}'\mathbf{X})^{-1}\mathbf{X}'\mathbf{y} \\ \mathbf{X}_1'\mathbf{y} \end{bmatrix} \end{aligned} \tag{10.21d}$$

which is identical with the 2SLS estimator obtainable from (10.17). In arriving at the last line of (10.21d), we use the result

$$\begin{aligned} \mathbf{X}_1'\mathbf{X}(\mathbf{X}'\mathbf{X})^{-1}\mathbf{X}' &= (\mathbf{X}_1'\mathbf{X}_1 \quad \mathbf{X}_1'\mathbf{X}_2)\begin{pmatrix} \mathbf{X}_1'\mathbf{X}_1 & \mathbf{X}_1'\mathbf{X}_2 \\ \mathbf{X}_2'\mathbf{X}_1 & \mathbf{X}_2'\mathbf{X}_2 \end{pmatrix}^{-1}\begin{pmatrix} \mathbf{X}_1' \\ \mathbf{X}_2' \end{pmatrix} \\ &= (\mathbf{I} \quad \mathbf{0})\begin{pmatrix} \mathbf{X}_1' \\ \mathbf{X}_2' \end{pmatrix} \end{aligned}$$

See Goldberger (1964, p. 332).

The Aitken-GLS interpretation of the 2SLS estimation can be extended to the system method called 3SLS. Suppose that the equation in (10.8) is one of the M equations of a system, say the ith, we can write the system as

$$\mathbf{y}_i = \mathbf{Z}_i\boldsymbol{\theta}_i + \mathbf{v}_i \qquad i = 1, 2, \ldots, M \tag{10.22}$$

To fix ideas, let us observe that $\mathbf{y}_i$ is $n \times 1$, $\mathbf{Z}_i$ is $n \times (G_i{}^* - 1 + K_i{}^*)$ where $G_i{}^*$ and $K_i{}^*$ denote, respectively, the numbers of the endogenous and predetermined variables appearing in the ith equation, and $\boldsymbol{\theta}_i$ is $(G_i{}^* - 1 + K_i{}^*) \times 1$. Since all the M equations are to be estimated simultaneously, let us write the entire set of equations in a large matrix equations as follows:

$$\begin{pmatrix} \mathbf{y}_1 \\ \mathbf{y}_2 \\ \cdot \\ \cdot \\ \cdot \\ \mathbf{y}_M \end{pmatrix} = \begin{pmatrix} \mathbf{Z}_1 & & & \\ & \mathbf{Z}_2 & & \bigcirc \\ & & \ddots & \\ & \bigcirc & & \mathbf{Z}_M \end{pmatrix} \begin{pmatrix} \boldsymbol{\theta}_1 \\ \boldsymbol{\theta}_2 \\ \cdot \\ \cdot \\ \cdot \\ \boldsymbol{\theta}_M \end{pmatrix} + \begin{pmatrix} \mathbf{v}_1 \\ \mathbf{v}_2 \\ \cdot \\ \cdot \\ \cdot \\ \mathbf{v}_M \end{pmatrix} \tag{10.22a}$$

or

$$\mathbf{q} = \mathbf{C}\boldsymbol{\lambda} + \mathbf{w} \tag{10.22b}$$

with the understanding that

$$\mathbf{q} = \begin{pmatrix} \mathbf{y}_1 \\ \mathbf{y}_2 \\ \cdot \\ \cdot \\ \cdot \\ \mathbf{y}_M \end{pmatrix} \qquad \mathbf{C} = \begin{pmatrix} \mathbf{Z}_1 & & & \\ & \mathbf{Z}_2 & & \bigcirc \\ & & \ddots & \\ & \bigcirc & & \mathbf{Z}_M \end{pmatrix}$$

$$\boldsymbol{\lambda} = \begin{pmatrix} \boldsymbol{\theta}_1 \\ \boldsymbol{\theta}_2 \\ \cdot \\ \cdot \\ \cdot \\ \boldsymbol{\theta}_M \end{pmatrix} \qquad \mathbf{w} = \begin{pmatrix} \mathbf{v}_1 \\ \mathbf{v}_2 \\ \cdot \\ \cdot \\ \cdot \\ \mathbf{v}_M \end{pmatrix}$$

Thus in form (10.22b) is similar to the linear regression model we discussed in Chapters 4 and 5, where we had $\mathbf{y} = \mathbf{X}\boldsymbol{\beta} + \mathbf{u}$, and is a system analog of (10.21a).

Consider now the correlation among the disturbances of the M equations. We assume that the contemporaneous covariances among $\mathbf{v}_i$ are nonzero but that lagged auto and serial correlations are absent. Thus, for any time period t, $E\mathbf{ww}' = \mathbf{\Lambda}$, a $Mn \times Mn$ matrix. If we denote $E\mathbf{v}_i\mathbf{v}_j' = \sigma_{ij}\mathbf{I}$, then

$$\mathbf{\Lambda} = \begin{pmatrix} \sigma_{11} & \sigma_{12} & \cdots & \sigma_{1M} \\ \sigma_{21} & \sigma_{22} & \cdots & \sigma_{2M} \\ \cdot & \cdot & \cdot & \cdot \\ \cdot & \cdot & \cdot & \cdot \\ \cdot & \cdot & \cdot & \cdot \\ \sigma_{M1} & \sigma_{M2} & \cdots & \sigma_{MM} \end{pmatrix} \otimes \mathbf{I} = \mathbf{\Sigma} \otimes \mathbf{I} \tag{10.23}$$

Since the off-diagonal elements of $\mathbf{\Sigma}$ are nonzero and it is assumed that no two $\mathbf{Z}_i$ are identical, the application of Aitken's procedure is appropriate and desirable in the estimation of (10.22b). The question is: What transformation matrix should we choose?

To answer this question, let us premultiply (10.22b) by,

$$\mathbf{H}' = \begin{pmatrix} \mathbf{X}' & & & \bigcirc \\ & \mathbf{X}' & & \\ & & \ddots & \\ \bigcirc & & & \mathbf{X}' \end{pmatrix} \tag{10.24}$$

where $\mathbf{H}$ is $Mn \times MK$, with $\mathbf{X}$ being $n \times K$. Thus

$$\mathbf{H}'\mathbf{q} = \mathbf{H}'\mathbf{C}\boldsymbol{\lambda} + \mathbf{H}'\mathbf{w} \tag{10.25}$$

so that for

$$E(\mathbf{H}'\mathbf{ww}'\mathbf{H}) = \mathbf{\Omega} \tag{10.26}$$

Aitken's GLS estimator of $\boldsymbol{\lambda}$ in (10.25) would be

$$\hat{\boldsymbol{\lambda}} = [\mathbf{C}'\mathbf{H}\Omega^{-1}\mathbf{H}'\mathbf{C}]^{-1}\mathbf{C}'\mathbf{H}\Omega^{-1}\mathbf{H}'\mathbf{q} \tag{10.27}$$

Since in (10.25) and (10.26)

$$\mathbf{H}'\mathbf{q} = \begin{pmatrix} \mathbf{X}'\mathbf{y}_1 \\ \mathbf{X}'\mathbf{y}_2 \\ \cdot \\ \cdot \\ \cdot \\ \mathbf{X}'\mathbf{y}_M \end{pmatrix} \qquad \mathbf{H}'\mathbf{C} = \begin{pmatrix} \mathbf{X}'\mathbf{Z}_1 & & & \bigcirc \\ & \mathbf{X}'\mathbf{Z}_2 & & \\ & & \ddots & \\ \bigcirc & & & \mathbf{X}'\mathbf{Z}_M \end{pmatrix}$$

$$\mathbf{H}'\mathbf{w} = \begin{pmatrix} \mathbf{X}'\mathbf{v}_1 \\ \mathbf{X}'\mathbf{v}_2 \\ \cdot \\ \cdot \\ \cdot \\ \mathbf{X}'\mathbf{v}_M \end{pmatrix}$$

$$E(\mathbf{H}'\mathbf{w}\mathbf{w}'\mathbf{H}) = E\begin{pmatrix}\mathbf{X}'\mathbf{v}_1\\ \mathbf{X}'\mathbf{v}_2\\ \cdot\\ \cdot\\ \cdot\\ \mathbf{X}'\mathbf{v}_M\end{pmatrix}(\mathbf{v}_1'\mathbf{X}\cdots\mathbf{v}_M'\mathbf{X})$$

$$= \begin{pmatrix}\sigma_{11}\mathbf{X}'\mathbf{X} & \sigma_{12}\mathbf{X}'\mathbf{X} & \cdots & \sigma_{1M}\mathbf{X}'\mathbf{X}\\ \cdot & \cdot & \cdot & \cdot\\ \cdot & \cdot & \cdot & \cdot\\ \cdot & \cdot & \cdot & \cdot\\ \sigma_{M1}\mathbf{X}'\mathbf{X} & \sigma_{M2}\mathbf{X}'\mathbf{X} & \cdots & \sigma_{MM}\mathbf{X}'\mathbf{X}\end{pmatrix}$$

$$= \mathbf{\Sigma}\otimes\mathbf{X}'\mathbf{X}$$

and since

$$[E(\mathbf{H}'\mathbf{w}\mathbf{w}'\mathbf{H}]^{-1} = \mathbf{\Sigma}^{-1}\otimes(\mathbf{X}'\mathbf{X})^{-1}$$

the estimator (10.27) can be rewritten as follows:

$$\hat{\boldsymbol{\lambda}} = \begin{bmatrix}\sigma^{11}\mathbf{Z}_1'\mathbf{X}(\mathbf{X}'\mathbf{X})^{-1}\mathbf{X}'\mathbf{Z}_1 & \cdots & \sigma^{1M}\mathbf{Z}_1'\mathbf{X}(\mathbf{X}'\mathbf{X})^{-1}\mathbf{X}'\mathbf{Z}_M\\ \sigma^{21}\mathbf{Z}_2'\mathbf{X}(\mathbf{X}'\mathbf{X})^{-1}\mathbf{X}'\mathbf{Z}_1 & \cdots & \sigma^{2M}\mathbf{Z}_2'\mathbf{X}(\mathbf{X}'\mathbf{X})^{-1}\mathbf{X}'\mathbf{Z}_M\\ \cdot & \cdot & \cdot\\ \cdot & \cdot & \cdot\\ \cdot & \cdot & \cdot\\ \sigma^{M1}\mathbf{Z}_M'\mathbf{X}(\mathbf{X}'\mathbf{X})^{-1}\mathbf{X}'\mathbf{Z}_1 & \cdots & \sigma^{MM}\mathbf{Z}_M'\mathbf{X}(\mathbf{X}'\mathbf{X})^{-1}\mathbf{X}'\mathbf{Z}_M\end{bmatrix}^{-1}$$

$$\times\begin{bmatrix}\sum_{i=1}^{M}\sigma^{1i}\mathbf{Z}_1'\mathbf{X}(\mathbf{X}'\mathbf{X})^{-1}\mathbf{X}\,\mathbf{y}_i\\ \sum_{i=1}^{M}\sigma^{2i}\mathbf{Z}_2'\mathbf{X}(\mathbf{X}'\mathbf{X})^{-1}\mathbf{X}'\mathbf{y}_i\\ \cdot\\ \cdot\\ \cdot\\ \sum_{i=1}^{M}\sigma^{Mi}\mathbf{Z}_M'\mathbf{X}(\mathbf{X}\,\mathbf{X})^{-1}\mathbf{X}'\mathbf{y}_i\end{bmatrix} \qquad (10.27a)$$

It may be desirable to compare this expression with the ones in (8.7a) and (8.7b).

Of course, in practice, $\mathbf{\Sigma}$ matrix is rarely known prior to estimation, thus, making it necessary to estimate $\mathbf{\Sigma}$ from the sample. Therefore, for practical purposes, we recapitulate and supplement the preceding discussion as follows:

First Stage. Find the LS estimates of the values of all the endogenous variables through the reduced-form equations where each y_i is regressed on all of the K predetermined variables (see also subsection 10.2.3).

Second Stage. Using the first stage estimates, apply the LS to each structural equation to find the 2SLS estimates of the latter's coefficients. Obtain the 2SLS residuals of each of the structural equations and form an estimate of the $\mathbf{\Sigma}$ matrix as follows. Let the typical element of $\mathbf{\Sigma}$ be σ_{ij} and the 2SLS residuals of the ith structural equation be $\hat{\mathbf{v}}_i$. Then the estimate of σ_{ij} is

$$\hat{\sigma}_{ij} = \frac{1}{n} \hat{\mathbf{v}}_i'\hat{\mathbf{v}}_j \tag{10.28}$$

Third Stage. We form $\hat{\mathbf{\Sigma}}$ according to (10.28), and then $\hat{\mathbf{\Sigma}}^{-1}$ is obtained. The elements of the inverse of the estimated covariance matrix are used to derive $\hat{\boldsymbol{\lambda}}$ as shown in (10.27a) with σ^{ij} there replaced by $\hat{\sigma}^{ij}$.

The asymptotic covariance matrix of the 3SLS estimator of $\boldsymbol{\lambda}$ is

$$\mathbf{V}(\hat{\boldsymbol{\lambda}}) = \begin{bmatrix} \sigma^{11}\mathbf{Z}_1'(\mathbf{X}'\mathbf{X})^{-1}\mathbf{X}'\mathbf{Z}_1 & \cdots & \sigma^{1M}\mathbf{Z}_1'\mathbf{X}(\mathbf{X}'\mathbf{X})^{-1}\mathbf{X}'\mathbf{Z}_M \\ \sigma^{21}\mathbf{Z}_2'(\mathbf{X}'\mathbf{X})^{-1}\mathbf{X}'\mathbf{Z}_1 & \cdots & \sigma^{2M}\mathbf{Z}_2'\mathbf{X}(\mathbf{X}'\mathbf{X})^{-1}\mathbf{X}'\mathbf{Z}_M \\ \cdot & \cdot & \cdot \\ \cdot & \cdot & \cdot \\ \cdot & \cdot & \cdot \\ \sigma^{M1}\mathbf{Z}_M'(\mathbf{X}'\mathbf{X})^{-1}\mathbf{X}'\mathbf{Z}_1 & \cdots & \sigma^{MM}\mathbf{Z}_M'\mathbf{X}(\mathbf{X}'\mathbf{X})^{-1}\mathbf{X}'\mathbf{Z}_M \end{bmatrix}^{-1} \tag{10.29}$$

This can be estimated from the sample by replacing σ^{ij} by $\hat{\sigma}^{ij}$.

With respect to the properties of the last estimator as well as to the computational designs for obtaining $\hat{\boldsymbol{\lambda}}$, see the work of Zellner and Theil (1962), who developed the 3SLS procedure. Here it is sufficient to say that $\hat{\boldsymbol{\lambda}}$ estimated with the aid of $\mathbf{\Sigma}$ are consistent and are more efficient than the corresponding 2SLS estimators, for instance, the ones in (10.17). Computer programs in FORTRAN are available for computing both the 2SLS and 3SLS estimates and the related statistics. See Straud, Zellner, and Chau (1963).

10.3.3. A Comparison of 2SLS and 3SLS Estimates

The Aitken procedure clearly indicates that if the $\mathbf{\Sigma}$ matrix is known, the 3SLS estimators gain some efficiency over the corresponding 2SLS estimators. Of course, in practice it is difficult to have $\mathbf{\Sigma}$ or it is not known so that the estimates of $\mathbf{\Sigma}$ are used in finding the 3SLS estimates of the equation coefficients. In this subsection we exhibit a small model estimated by the 2SLS and 3SLS procedures.

The model is a slight modification of (10.20a to c) and is restated below:

$$D = \alpha_0 + \alpha_1 i + \alpha_2 Y + \alpha_3\left(\frac{R}{P}\right) + \alpha_4\left(\frac{M}{A}\right)_{-1} + \alpha_5\,\Delta L + v_1 \tag{10.31}$$

$$S = \beta_0 + \beta_1 i + \beta_2 i^L_{-2} + \beta_3 R^P_{-3} + \beta_4 S^A + v_2 \tag{10.32}$$

$$\Delta i = \gamma_0 + \gamma_1(S - D) + \gamma_2\,\Delta i^L_{-1} + v_3 \tag{10.33}$$

Here the Δ denotes the first-difference operation. For instance, $\Delta i = i_t - i_{t-1}$,

Table 10.1 Mortgage Credit Market Alternative Least-Squares Estimates of the Model

Equation	Coefficient of	Coefficient Estimates by 2SLS	Coefficient Estimates by 3SLS
(10.31)D	Constant	−5900.0	−10956.0
		(9825.6)	(9207.0)
	i	−2893.6	−3098.3
		(463.4)	(434.0)
	Y	87.14	90.40
		(7.9)	(7.1)
	R/P	89.96	140.45
		(88.4)	(82.4)
	M/F_{-1}	−488.90	−482.43
		(257.2)	(223.9)
	ΔL	1006.5	1112.2
		(391.8)	(337.5)
(10.32)S	Constant	−122760.0	−178520.0
		(7878.8)	(72730.0)
	i	39042.0	56547.0
		(25124.0)	(23224.0)
	i^L_{-2}	−21983.0	−31834.0
		(15430.0)	(14295.0)
	R^P_{-3}	2.344	2.542
		(1.48)	(1.28)
	S^A	0.2198	−0.3934
		(0.420)	(0.354)
(10.33)Δi	Constant	0.02161	0.02238
		(0.0181)	(0.0182)
	$(S - D)$	−0.00001591	−0.00002818
		(0.0000165)	(0.0000173)
	Δi^L_{-1}	0.2435	0.2106
		(0.0983)	(0.0920)

where t is a time subscript; also, the writing of the time subscript in general has been eliminated from the equations.

We provide the estimates obtained by the two alternative procedures in Table 10.1. For this estimation we used United States quarterly data for the period 1953 to 1963. The data, the specification of the model, and other details about the model and the variables are found in Huang (1966, 1967). The figures in the parentheses below the estimated coefficients are standard errors. Notice that for a coefficient the standard error of its 2SLS estimate is, with few exceptions, larger than its 3SLS counterpart. It is of interest to observe that the simple correlation matrix of the residuals calculated with the 2SLS estimates of the coefficients is

	v_1	v_2	v_3
v_1	1.000	−0.503	0.311
v_2		1.000	−0.417
v_3			1.000

Zellner and Theil show that, if the model is just-identified, there is no gain in the 3SLS over the 2SLS. We also indicate in passing that a specification error in one equation will be carried into the estimates of the other equations if the 3SLS is used; the 2SLS will be free of this problem.

10.4. Applications

Having estimated the structural equations of the model, we might ask about the ways in which an estimated model can be used. Estimated structural relationships do have many uses. For example, the parameter estimates in a structural relation may be examined for consistency with prior specification or theoretical expectation. Furthermore, the hypotheses about the parameters may be formerly tested by the use of the estimated results. More often than not, however, a researcher working with simultaneous equation systems is interested in obtaining the solution or the derived reduced form of an estimated system and in making certain uses that are made of solved reduced forms. The ones that we shall discuss are (1) forecasting, (2) simulation, and (3) the model's dynamic properties.

10.4.1. Derived Reduced Form

Recall that in our discussion of the 2SLS we used the reduced-form equations for the endogenous variables appearing in the right-hand side of a structural equation and that these equations contained on their right-hand side all of the predetermined variables of the system. We called such a reduced form unrestricted in the sense that the identifying zero restrictions in the structural equations are not reflected in the reduced form. That is, given the specification of the **B** and **Γ** matrices in a structural model

$$\mathbf{YB} + \mathbf{X\Gamma} + \mathbf{U} = \mathbf{0} \qquad (10.36a)$$

the restricted reduced form is

$$\begin{aligned} \mathbf{Y} &= -\mathbf{X\Gamma B}^{-1} - \mathbf{UB}^{-1} \qquad (10.36b) \\ &= \mathbf{X\Pi} + \mathbf{V} \end{aligned}$$

whereas the unrestricted reduced form would be

$$\mathbf{Y} = \mathbf{X\Delta} + \mathbf{W} \qquad (10.36c)$$

where the elements of **Δ** are LS estimated without regard to the restriction $\mathbf{\Delta} = -\mathbf{\Gamma B}^{-1}$. It is along the line of the restricted reduced form that we discuss the derived reduced form.

For our discussion in this section only let us write out a system, say, consisting of M equations, as follows

$$\begin{aligned} y_1 &= \boldsymbol{\beta}_1\mathbf{Y} + \boldsymbol{\gamma}_1\mathbf{X} + u_1 \\ y_2 &= \boldsymbol{\beta}_2\mathbf{Y} + \boldsymbol{\gamma}_2\mathbf{X} + u_2 \\ &\vdots \\ y_M &= \boldsymbol{\beta}_M\mathbf{Y} + \boldsymbol{\gamma}_M\mathbf{X} + u_M \end{aligned} \tag{10.38}$$

where $\mathbf{Y}$ is an M-dimensional column vector of endogenous variables, or $\mathbf{Y}' = (y_1 \ y_2 \ \cdots \ y_M)$, $\boldsymbol{\beta}_i$ is a $1 \times M$ row vector of coefficients in the ith equation, with, at least, the ith element of the vector equal to zero, and $\mathbf{X}$ is a column vector of predetermined variables, say, of dimensionality $K \times 1$, so that $\boldsymbol{\gamma}_i$ is a $1 \times K$ row vector. In matrix notation this system becomes, for time period t,

$$\mathbf{Y}_t = \mathbf{B}\mathbf{Y}_t + \boldsymbol{\Gamma}\mathbf{X}_t + \mathbf{u}_t \tag{10.38a}$$

where $\mathbf{B}$ is $M \times M$ whose ith row is $\boldsymbol{\beta}_i$ and $\boldsymbol{\Gamma}$ is $M \times K$ whose ith row is $\boldsymbol{\gamma}_i$. Solving (10.38a) for $\mathbf{Y}_t$ in terms of predetermined variables gives

$$\mathbf{Y}_t = (\mathbf{I} - \mathbf{B})^{-1}\boldsymbol{\Gamma}\mathbf{X}_t + (\mathbf{I} - \mathbf{B})^{-1}\mathbf{u}_t \tag{10.38b}$$

which is equivalent to the reduced form in, say, (10.4b). Now given the estimates of $\mathbf{B}$ and $\boldsymbol{\Gamma}$, or $\hat{\mathbf{B}}$ and $\hat{\boldsymbol{\Gamma}}$, respectively, we define the derived reduced form of (10.38a) by

$$\mathbf{Y}_t = (\mathbf{I} - \hat{\mathbf{B}})^{-1}\hat{\boldsymbol{\Gamma}}\mathbf{X}_t + (\mathbf{I} - \hat{\mathbf{B}})^{-1}\hat{\mathbf{u}}_t \tag{10.39}$$

where $\hat{\mathbf{u}}_t = \mathbf{Y}_t - (\hat{\mathbf{B}}\mathbf{Y}_t + \hat{\boldsymbol{\Gamma}}\mathbf{X}_t)$. We shall call the following expected derived reduced form

$$\hat{Y}_t = (\mathbf{I} - \hat{\mathbf{B}})^{-1}\hat{\boldsymbol{\Gamma}}\mathbf{X}_t \tag{10.40}$$

where the residuals are ignored.

Example 1. In terms of the system shown in (10.31 to 10.33) and in Table 10.1, the 3SLS estimates of the model gives matrices for the derived reduced form as follows:

$$(\mathbf{I} - \hat{\mathbf{B}}) = \begin{array}{c} \begin{array}{ccc} (D) & (S) & (\Delta i) \end{array} \\ \begin{bmatrix} 1 & 0 & 3098.3 \\ 0 & 1 & -56547.0 \\ -0.00002818 & 0.00002818 & 1 \end{bmatrix} \end{array} \tag{10.40a}$$

$$(\mathbf{I} - \hat{\mathbf{B}})^{-1} = \begin{array}{c} \begin{array}{ccc} (D) & (S) & (i) \end{array} \\ \begin{bmatrix} 0.97370 & 0.02630 & -1163.3 \\ 0.59826 & 0.40174 & 21230.0 \\ 0.00001058 & -0.00001058 & 0.37544 \end{bmatrix} \end{array} \tag{10.40b}$$

Now the $\hat{\mathbf{\Gamma}}$ matrix as formed by the estimated coefficients shown in Table 10.1 is

	(Y)	(R/P)	$(M/A)_{-1}$	(ΔL)	(i^L_{-1})	(i^L_{-2})	(R^P_{-3})	(S^A)	(i_{-1})	(Constant)
(D)	90.40	140.45	−482.43	1112.2	0	0	0	0	0	−10956.0
(S)	0	0	0	0	0	−31834.0	2.542	−0.3934	0	−178520.0
(i)	0	0	0	0	0.2106	−0.2106	0	0	1	0.02238

Premultiplying this matrix by $(\mathbf{I} - \hat{\mathbf{B}})^{-1}$ then yields the desired derived reduced form:

$$(\mathbf{I} - \hat{\mathbf{B}})^{-1}\hat{\mathbf{\Gamma}} =$$

	(Y)	(R/P)	$(M/A)_{-1}$	(ΔL)	(i^L_{-1})	(i^L_{-2})	(R^P_{-3})	(S^A)	(i_{-1})	(Constant)
(D)	88.02	136.8	−469.7	1083.0	−245.0	−1082.2	0.06685	−0.01035	−1163.3	−15388.9
(S)	54.08	84.03	−288.6	665.4	4471.0	−17260.0	1.021	−0.1580	21230.0	−77798.0
(i)	0.0009564	0.001486	−0.005104	0.01177	0.07907	−0.4159	−0.00002689	−0.000004161	0.3754	−1.9962

These coefficients give the one-shot effects of the predetermined variables on the endogenous variables *after* the interdependency specified in the system is taken into account. See the discussion of the impact multipliers in Subsection 10.4.4.

10.4.2. Forecasting

One important reason for the preceding discussion of the derived reduced form is that forecasts cannot be made from structural equations (except, of course, when the equations are already in reduced form). Recall that usually a structural equation contains more than one endogenous variable and that the variables are jointly dependent. A method must be devised to forecast the values of the jointly dependent variables so as to reflect the joint dependence among the endogenous variables and their dependence on the exogenous variables. Such a method is by the use of the derived reduced form.

Thus for a given set of the future values of the predetermined variables, say $\mathbf{X}_F$, the forecast values of the endogenous variables will be

$$\hat{\mathbf{Y}}_F = (\mathbf{I} - \hat{\mathbf{B}})^{-1}\hat{\mathbf{\Gamma}}\mathbf{X}_F \tag{10.41}$$

and the errors of the forecast are

$$\begin{aligned} \mathbf{e}_F &= \mathbf{Y}_F - \hat{\mathbf{Y}}_F \\ &= (\mathbf{I} - \mathbf{B})^{-1}\mathbf{\Gamma}\mathbf{X}_F + (\mathbf{I} - \mathbf{B})^{-1}\mathbf{u}_F - (\mathbf{I} - \hat{\mathbf{B}})^{-1}\hat{\mathbf{\Gamma}}\mathbf{X}_F \\ &= [(\mathbf{I} - \mathbf{B})^{-1}\mathbf{\Gamma} - (\mathbf{I} - \hat{\mathbf{B}})^{-1}\hat{\mathbf{\Gamma}}]\mathbf{X}_F + (\mathbf{I} - \mathbf{B})^{-1}\mathbf{u}_F \end{aligned} \tag{10.41e}$$

Thus, the errors of forecast, as in our earlier discussion of forecast by the LS estimators, consists of the sampling error in the estimator of the **B** and **Γ** matrices and of the variability of the disturbances terms. As to how the variance-covariance matrix of $\mathbf{e}_F$ may be estimated, see Christ (1966, pp. 568–571) and Goldberger, Nagar, and Odeh (1961).

It will be well to notice that there are two types of forecast depending on the nature of the values of $\mathbf{X}_F$ used. *Ex ante* forecasting means that some or all of the $\mathbf{X}_F$ values are unknown and are estimated before these values are used in (10.41). On the other hand, *ex post* forecasting means that the values of $\mathbf{X}_F$ are known for the forecast period. Thus, if forecasting is used as a means of testing how good a model is (or how good the estimated equations of the model are), effort must be made to obtain *ex post* forecasts since these forecasts will not contain "impurities" created by the errors in estimating or forecasting the values of $\mathbf{X}_F$.

10.4.3. Simulation

Frequently for policy and other decision-making purposes an estimated econometric model is used for simulation. The basic objective is to study the behavior pattern of the jointly dependent variables for certain assumed changes in the values of the predetermined variables. For instance, one might be interested in investigating what would happen to the variables D, S, and Δi in (10.20a to c) for a certain time period, if for that period variables such as L and R^P in the model are given some assumed changes. Such variables as can be manipulated to reflect a policy or a decision-making process are called instrument variables, while the dependent variables are sometimes called target variables. For detail on the terminology and methodology, refer to literature sources. For a good introduction, see Orcutt (1960).

Simulation is usually of two types, exact and stochastic. Continuing the discussion relating to (10.39) and (10.40), we call a simulation exact if the estimates of $\mathbf{B}$ and $\mathbf{\Gamma}$ alone are used for projecting the values of the dependent variables. It is exact in the sense that the estimated coefficients are assumed given for the analysis and no account is made of the probability distributions of the disturbances (or shocks in the equations). Simulation which explicitly uses known or assumed probability laws about the errors in the equations is called stochastic.

10.4.4. Model's Dynamic Properties and Multipliers

The standard references on this topic are Goldberger (1959), Baumol (1959), and Theil and Boot (1962). For the present discussion we borrow mostly from the work of Goldberger and from that of Theil and Boot.

Recall that the Keynesian multiplier describes a relationship between and the increment in consumption expenditures and the eventual total increase in national income. The circumstances in which the multiplier process is assumed to take place are as follows: (1) one-period, or once for all, injection of increased consumption expenditure (ΔC) is made, (2) the increment ΔC has its effect felt throughout the economic system until the next equilibrium level of the system is reached and, (3) one takes an inventory of the total increment in income (ΔY) at the equilibrium point and finds that $\Delta Y = k\,\Delta C$, with k being the multiplier defined as $k = 1/(1 - \beta)$ in the consumption function $C = \alpha + \beta Y$. This analysis is strictly in the realm of comparative statics.

If we have in an econometric model lagged endogenous variables the model will be called dynamic, and for such a model it is possible to study the time paths and changes over time of the endogenous variables in the system *as the result* of giving certain changes to one or more exogenous variables in the system. The changes can be a one-shot occurrence; or they can be changes to certain levels, and such levels are maintained for a number of time periods. No doubt, the reader has noticed the similarity of the present discussion to that of simulation, but he will soon see that our interest is in simulation as well as in learning, through the estimated model, the inherent stability or instability of the system, and sometimes the net changes in the endogenous variables after a certain time period or after an indefinite number of periods *as the result* of changing one or more exogeneous variables by one measurement unit or by fixing the variables at certain levels. In simulation, it is not necessary that lagged endogenous variables exist. Of particular interest here is a few types of multipliers, their meaning, and the inherent stability or instability of the system.

To consider these problems, we write lagged endogenous variables explicitly into (10.38b) by separating the components of $\mathbf{X}_t$ into two vectors of the same dimension as $\mathbf{X}_t$. For simplicity, we assume that we have only endogenous variables lagged one time period, so that $\mathbf{X}_t$ separates into $\mathbf{Y}_{t-1}$ and $\mathbf{Z}_t$, both being $K \times 1$. By reassigning symbols to the coefficient matrices we have, in the form of (10.38b), or an expected derived reduced form,

$$\mathbf{Y}_t = \mathbf{A}\mathbf{Y}_{t-1} + \mathbf{C}\mathbf{Z}_t + \mathbf{u}_t^* \tag{10.42}$$

It is possible that $\mathbf{Z}_t$ may contain elements that are lagged exogenous variables, and the explicit account of them may be necessary in some other context. Similarly, this is true for endogenous variables lagged more than one time period.

To observe how various multipliers are formed, we make a few substitutions. We lag (10.42) one time period and substitute the result into the same expression. Thus

$$\begin{aligned} \mathbf{Y}_t &= \mathbf{A}(\mathbf{A}\mathbf{Y}_{t-2} + \mathbf{C}\mathbf{Z}_{t-1} + \mathbf{u}_{t-1}^*) + \mathbf{C}\mathbf{Z}_t + \mathbf{u}_t^* \\ &= \mathbf{A}^2\mathbf{Y}_{t-2} + \mathbf{C}\mathbf{Z}_t + \mathbf{A}\mathbf{C}\mathbf{Z}_{t-1} + \mathbf{u}_t^* + \mathbf{A}\mathbf{u}_{t-1}^* \end{aligned} \tag{10.42a}$$

By applying the similar procedure for (10.42) lagged two periods, we get

$$\begin{aligned} \mathbf{Y}_t &= \mathbf{A}^2(\mathbf{A}\mathbf{Y}_{t-3} + \mathbf{C}\mathbf{Z}_{t-2} + \mathbf{u}_{t-2}^*) + \mathbf{C}\mathbf{Z}_t + \mathbf{A}\mathbf{C}\mathbf{Z}_{t-1} + \mathbf{u}_t^* + \mathbf{A}\mathbf{u}_{t-1}^* \\ &= \mathbf{A}^3\mathbf{Y}_{t-3} + \mathbf{C}\mathbf{Z}_t + \mathbf{A}\mathbf{C}\mathbf{Z}_{t-1} + \mathbf{A}^2\mathbf{C}\mathbf{Z}_{t-2} + \mathbf{u}_t^* + \mathbf{A}\mathbf{u}_{t-1}^* + \mathbf{A}^2\mathbf{u}_{t-2} \end{aligned} \tag{10.42b}$$

It is becoming clear that as further substitution is carried out, say, k times, we obtain a general expression

$$\mathbf{Y}_t = \mathbf{A}^{k+1}\mathbf{Y}_{t-k-1} + \sum_{\alpha=0}^{k}\mathbf{M}_\alpha\mathbf{Z}_{t-\alpha} + \sum_{\alpha=0}^{k}\mathbf{A}^\alpha\mathbf{u}_{t-\alpha}^* \tag{10.43}$$

where $\mathbf{M}_\alpha = \mathbf{A}^\alpha\mathbf{C}$. If we let k go to infinity, we shall get the so-called final form of the model

$$\mathbf{Y}_t = \sum_{\alpha=0}^{\infty}\mathbf{M}_\alpha\mathbf{Z}_{t-\alpha} + \sum_{\alpha=0}^{\infty}\mathbf{A}^\alpha\mathbf{u}_{t-\alpha}^* \tag{10.44}$$

if it can be assumed that $\lim_{\alpha\to\infty} \mathbf{A}^\alpha = 0$. If this last assumption, indeed, holds true, we say that the system is stable. The final form of the model (10.44) allows us to distinguish three types of multipliers.

IMPACT MULTIPLIERS. Letting $\alpha = 0$, and assuming that the expectation of $\mathbf{u}_t^*$ is $\mathbf{0}$ for all t, we can write (10.44) as

$$\mathbf{Y}_t = \mathbf{M}\mathbf{Z}_t = \mathbf{A}^0\mathbf{C}\mathbf{Z}_t = \mathbf{C}\mathbf{Z}_t \tag{10.44a}$$

Since the elements of $\mathbf{C}$ are elements of the reduced form coefficient matrix, say in (10.36b), the partial differentiation of an endogenous variable, say, y_i, with respect to an exogenous variable, say, z_i, gives

$$\frac{\partial y_i}{\partial z_i} = c_{ij} \tag{10.44b}$$

which means that the expected change in y_i as the result of a unit increase in z_i is c_{ij}, all in the current period. We call c_{ij} an impact multiplier. It embodies the contemporaneous joint effect of a one-shot change in z_j on y_i *after* such a change has its force felt throughout the system, which naturally includes all other endogenous variables and exogenous variables. From (10.40c) we notice that the one percentage point increase in i_{-1}^L results in an increased supply of mortgage credit (S) by 989.1 million dollars while causing the mortgage yield (i) to increase by 0.1689 percentage points.

INTERIM OR DELAY MULTIPLIERS. Sometimes it is of interest to know what the one-shot effect of a unit increase in an exogenous variable on an endogenous variable is after a certain number of time periods, say, for $\alpha = n$. Then the elements of the $\mathbf{M}_\alpha$ matrix for $\alpha = n$ become the so-called interim or delay multipliers. That is,

$$\mathbf{M}_n = \mathbf{A}^n\mathbf{C} \tag{10.44c}$$

so that if the typical element of $\mathbf{M}_n$ is $m_{ij.n}$, then the latter gives the effect of increasing the exogenous variable z_j by one unit, holding all other exogenous variables constant, on the variable y_i in n periods hence.

CUMULATED MULTIPLIERS. Thus far we have been concerned with the one-shot unit increase in exogenous variables and with the effect of such an increase on endogenous variables. Now one might also ask what the effect is of a unit increase of an exogenous variable maintained at the new level for, say, l periods. This effect can be measured by the cumulated multiplier matrix

$$\begin{aligned}\mathbf{G}_l &= \sum_{\alpha=0}^{l}\mathbf{M}_\alpha = \sum_{\alpha=0}^{l}\mathbf{A}^\alpha\mathbf{C}\\ &= (\mathbf{I} + \mathbf{A} + \mathbf{A}^2 + \cdots + \mathbf{A}^l)\mathbf{C}\end{aligned} \tag{10.44d}$$

If $\lim_{\alpha \to \infty} \mathbf{A}^\alpha = 0$, then,

$$\begin{aligned} \mathbf{G} &= \sum_{\alpha=0}^{\infty} \mathbf{A}^\alpha \mathbf{C} \\ &= (\mathbf{I} + \mathbf{A} + \mathbf{A}^2 + \cdots)\mathbf{C} \\ &= (\mathbf{I} - \mathbf{A})^{-1}\mathbf{C} \end{aligned}$$

is called the equilibrium multiplier matrix.

Inherent dynamic properties. The matrix $\mathbf{A}$ in (10.42) is of basic significance in analysis of the stability of the system associated with it. Stability of the system requires that $\lim_{\alpha \to \infty} \mathbf{A}^\alpha = \mathbf{0}$ as noted earlier. Indeed, some properties of the matrix $\mathbf{A}$ can tell us about how the time paths of the endogenous variables after a displacement from equilibrium will develop.

To investigate these properties, let $\bar{y}$ be the values of $\mathbf{Y}$ at an initial equilibrium and the displacement made to these values by $\mathbf{y}_0^*$. Similarly, the values of $\mathbf{Y}$ after t periods in the form of deviations from the vector $\bar{y}$ may be denoted by $\mathbf{y}_t^*$. Since we are considering the displacement of endogenous values from an initial equilibrium position, we can assume that there are no changes injected to the exogenous variables or to the disturbances. Hence, if $\mathbf{y}_t^*$ tends to zero monotonically with t, we can tell that the system is stable and if $\mathbf{y}_t^*$ oscillates with increase in t, we may not have a stable system. To see this we can write the relation between the initial displacement ($\mathbf{y}_0^*$) and the displacement after t periods ($\mathbf{y}_t^*$) as follows

$$\mathbf{y}_t^* = \mathbf{A}^t \mathbf{y}_0^* \tag{10.45}$$

where $\mathbf{A}$ is the matrix of coefficients of lagged endogenous variables as shown in (10.42). For the derivation of (10.45), see Goldberger (1964, p. 376). Since $\mathbf{A}$ is a square matrix, it can be written as

$$\mathbf{A} = \mathbf{P}\mathbf{\Lambda}\mathbf{Q} \tag{10.45a}$$

where $\mathbf{\Lambda}$ is a diagonal matrix containing the characteristic roots of $\mathbf{A}$ and $\mathbf{Q} = \mathbf{P}^{-1}$. It is the properties of those matrices that for

$$\mathbf{\Lambda} = \begin{pmatrix} \lambda_1 & 0 & \cdots & 0 \\ 0 & \lambda_2 & \cdots & 0 \\ \cdot & \cdot & \cdot & \cdot \\ \cdot & \cdot & \cdot & \cdot \\ \cdot & \cdot & \cdot & \cdot \\ 0 & 0 & \cdots & \lambda_M \end{pmatrix} \tag{10.45b}$$

it can be shown that

$$\mathbf{A}^t = \mathbf{P}\mathbf{\Lambda}^t\mathbf{Q} \tag{10.45c}$$

$$\mathbf{\Lambda}^t = \begin{pmatrix} \lambda_1{}^t & 0 & \cdots & 0 \\ 0 & \lambda_2{}^t & \cdots & 0 \\ \cdot & \cdot & \cdot & \cdot \\ \cdot & \cdot & \cdot & \cdot \\ \cdot & \cdot & \cdot & \cdot \\ 0 & 0 & \cdots & \lambda_M{}^t \end{pmatrix} \tag{10.45d}$$

Now $\mathbf{A}^t$ may be written as

$$\mathbf{A}^t = \sum_{i=1}^{M} \lambda_i{}^t \mathbf{R}_i \tag{10.45e}$$

where $\mathbf{R}_i = \mathbf{P}_i\mathbf{Q}_i'$ and is the product of the ith column of $\mathbf{P}$ and the ith row of $\mathbf{Q}$†. By combining (10.45) and (10.45e), we obtain

$$\mathbf{y}_t{}^* = \sum_{i=1}^{M} \lambda_i{}^t \mathbf{R}_i \mathbf{y}_0{}^* \tag{10.46}$$

and we observe that the inherent dynamic properties of the system depends entirely on the characteristic roots of $\mathbf{A}$. It is clear then for the stability of the system, where it is required that $\lim_{t\to\infty} \mathbf{A}^t = \lim_{t\to\infty} \sum_{i=1}^{M} \lambda_i{}^t \mathbf{R}_i = 0$, it is necessary that each of the λ_i's be less than one (1) in absolute value. Now, if all λ_i's are positive (and less than one), the time path of $\mathbf{y}_t{}^*$ is monotonically decreasing to zero and, if some λ_i's are negative, we may have an oscillating $\mathbf{y}_t{}^*$ but eventually approaching zero.

† For details about the nature of these matrices, see Goldberger's work just cited and Hadley (1961, pp. 249–251).

APPENDIX A

STATISTICAL TABLES

I am grateful for the permissions granted by the following scholars, journals, and organizations for the tables reproduced in this appendix.

Table A-1. Permission by Professors E. S. Pearson and H. O. Hartley and the *Biometrika* Trustees for reproduction from the *Biometrika Tables for Statisticians*, Vol. I (3rd Ed.), 1966, except for columns under $2Q = 0.2$ and 0.02, the permission for the reproduction of which is given by the publishers and authors of Fisher and Yates: *Statistical Tables for Biological, Agricultural, and Medical Research* (6th Ed.), 1963, published by Oliver and Boyd Ltd., Edinburgh.

Tables A-2, A-3, A-4. Permission by Professors E. S. Pearson and H. O. Hartley and the *Biometrika* Trustees for reproduction from the *Biometrika Tables for Statisticians*, Vol. I (3rd Ed.), 1966.

Table A-5. Permission by the Editor of the *Annals of Mathematical Statistics* and by the Treasurer of the Institute of Mathematical Statistics for publishing calculated values based on Table III, "Tabulation of the Probabilities for the Ratio of the Mean Square Successive Difference to the Variance," B. I. Hart, *Annals of Mathematical Statistics*, Vol. 13, pp. 207–214.

Tables A-6, A-7. Permission by Professors J. Durbin and G. S. Watson and *Biometrika* to reproduce from "Testing for Serial Correlation in Least Squares Regression, II," *Biometrika*, Vol. 38, pp. 173–175.

Table A-8. Permission by Professors H. Theil and A. L. Nagar and *Journal of the American Statistical Association* to reproduce from "Testing the Independence of Regression Disturbances," *Journal of the American Statistical Association*, Vol. 56, pp. 795–806.

Table A-1 Percentage Points of the *t* Distribution[a]

Degrees of Freedom \ Pb	$Q = 0.25$ $2Q = 0.5$	0.1 0.2	0.05 0.1	0.025 0.05	0.01 0.02	0.005 0.01
1	1.000	3.078	6.314	12.706	31.821	63.657
2	0.816	1.886	2.920	4.303	6.965	9.925
3	0.765	1.638	2.353	3.182	4.541	5.841
4	0.741	1.533	2.132	2.776	3.747	4.604
5	0.727	1.476	2.015	2.571	3.365	4.032
6	0.718	1.440	1.943	2.447	3.143	3.707
7	0.711	1.415	1.895	2.365	2.998	3.499
8	0.706	1.397	1.860	2.306	2.896	3.355
9	0.703	1.383	1.833	2.262	2.821	3.250
10	0.700	1.372	1.812	2.228	2.764	3.169
11	0.697	1.363	1.796	2.201	2.718	3.106
12	0.695	1.356	1.782	2.179	2.681	3.055
13	0.694	1.350	1.771	2.160	2.650	3.012
14	0.692	1.345	1.761	2.145	2.624	2.977
15	0.691	1.341	1.753	2.131	2.602	2.947
16	0.690	1.337	1.746	2.120	2.583	2.921
17	0.689	1.333	1.740	2.110	2.567	2.898
18	0.688	1.330	1.734	2.101	2.552	2.878
19	0.688	1.328	1.729	2.093	2.539	2.861
20	0.687	1.325	1.725	2.086	2.528	2.845
21	0.686	1.323	1.721	2.080	2.518	2.831
22	0.686	1.321	1.717	2.074	2.508	2.819
23	0.685	1.319	1.714	2.069	2.500	2.807
24	0.685	1.318	1.711	2.064	2.492	2.797
25	0.684	1.316	1.708	2.060	2.485	2.787
26	0.684	1.315	1.706	2.056	2.479	2.779
27	0.684	1.314	1.703	2.052	2.473	2.771
28	0.683	1.313	1.701	2.048	2.467	2.763
29	0.683	1.311	1.699	2.045	2.462	2.756
30	0.683	1.310	1.697	2.042	2.457	2.750
40	0.681	1.303	1.684	2.021	2.423	2.704
60	0.697	1.296	1.671	2.000	2.390	2.660
120	0.677	1.289	1.658	1.980	2.358	2.617
∞	0.674	1.282	1.645	1.960	2.326	2.576

[a] The column headings, except for the degrees of freedom, are the areas in the tails of the distribution. The smaller probability designated by Q is the area of one tail and the larger probability ($2Q$) is the area of the two tails. In the one-tail test with the level of significance α, set $Q = \alpha$; in the two-tail test let $2Q = \alpha$ and utilize the fact that the t distribution is symmetrical. For example, the critical values of t in a two-tail test with $\alpha = 0.05$ and degrees of freedom $= 11$ are $t = +2.201$. The critical value of t for a one-tail test with $\alpha = 0.05$ and degrees of freedom $= 11$ is either $+1.796$ or -1.796. Notice that the d distribution is identical with the normal distribution when the degrees of freedom approach infinity.

Table A-2 Percentage Points of the F Distribution Upper 5 Percent Points

v_2 \ v_1	1	2	3	4	5	6	7	8	9	10	12	15	20	24	30	40	60	120	∞
1	161.4	199.5	215.7	224.6	230.2	234.0	236.8	238.9	240.5	241.9	243.9	245.9	248.0	249.1	250.1	251.1	252.2	253.3	254.3
2	18.51	19.00	19.16	19.25	19.30	19.33	19.35	19.37	19.38	19.40	19.41	19.43	19.45	19.45	19.46	19.47	19.48	19.49	19.50
3	10.13	9.55	9.28	9.12	9.01	8.94	8.89	8.85	8.81	8.79	8.74	8.70	8.66	8.64	8.62	8.59	8.57	8.55	8.53
4	7.71	6.94	6.59	6.39	6.26	6.16	6.09	6.04	6.00	5.96	5.91	5.86	5.80	5.77	5.75	5.72	5.69	5.66	5.63
5	6.61	5.79	5.41	5.19	5.05	4.95	4.88	4.82	4.77	4.74	4.68	4.62	4.56	4.53	4.50	4.46	4.43	4.40	4.36
6	5.99	5.14	4.76	4.53	4.39	4.28	4.21	4.15	4.10	4.06	4.00	3.94	3.87	3.84	3.81	3.77	3.74	3.70	3.67
7	5.59	4.74	4.35	4.12	3.97	3.87	3.79	3.73	3.68	3.64	3.57	3.51	3.44	3.41	3.38	3.34	3.30	3.27	3.23
8	5.32	4.46	4.07	3.84	3.69	3.58	3.50	3.44	3.39	3.35	3.28	3.22	3.15	3.12	3.08	3.04	3.01	2.97	2.93
9	5.12	4.26	3.86	3.63	3.48	3.37	3.29	3.23	3.18	3.14	3.07	3.01	2.94	2.90	2.86	2.83	2.79	2.75	2.71
10	4.96	4.10	3.71	3.48	3.33	3.22	3.14	3.07	3.02	2.98	2.91	2.85	2.77	2.74	2.70	2.66	2.62	2.58	2.54
11	4.84	3.98	3.59	3.36	3.20	3.09	3.01	2.95	2.90	2.85	2.79	2.72	2.65	2.61	2.57	2.53	2.49	2.45	2.40
12	4.75	3.89	3.49	3.26	3.11	3.00	2.91	2.85	2.80	2.75	2.69	2.62	2.54	2.51	2.47	2.43	2.38	2.34	2.30
13	4.67	3.81	3.41	3.18	3.03	2.92	2.83	2.77	2.71	2.67	2.60	2.53	2.46	2.42	2.38	2.34	2.30	2.25	2.21
14	4.60	3.74	3.34	3.11	2.96	2.85	2.76	2.70	2.65	2.60	2.53	2.46	2.39	2.35	2.31	2.27	2.22	2.18	2.13
15	4.54	3.68	3.29	3.06	2.90	2.79	2.71	2.64	2.59	2.54	2.48	2.40	2.33	2.29	2.25	2.20	2.16	2.11	2.07
16	4.49	3.63	3.24	3.01	2.85	2.74	2.66	2.59	2.54	2.49	2.42	2.35	2.28	2.24	2.19	2.15	2.11	2.06	2.01
17	4.45	3.59	3.20	2.96	2.81	2.70	2.61	2.55	2.49	2.45	2.38	2.31	2.23	2.19	2.15	2.10	2.06	2.01	1.96
18	4.41	3.55	3.16	2.93	2.77	2.66	2.58	2.51	2.46	2.41	2.34	2.27	2.19	2.15	2.11	2.06	2.02	1.97	1.92
19	4.38	3.52	3.13	2.90	2.74	2.63	2.54	2.48	2.42	2.38	2.31	2.23	2.16	2.11	2.07	2.03	1.98	1.93	1.88
20	4.35	3.49	3.10	2.87	2.71	2.60	2.51	2.45	2.39	2.35	2.28	2.20	2.12	2.08	2.04	1.99	1.95	1.90	1.84
21	4.32	3.47	3.07	2.84	2.68	2.57	2.49	2.42	2.37	2.32	2.25	2.18	2.10	2.05	2.01	1.96	1.92	1.87	1.81
22	4.30	3.44	3.05	2.82	2.66	2.55	2.46	2.40	2.34	2.30	2.23	2.15	2.07	2.03	1.98	1.94	1.89	1.84	1.78
23	4.28	3.42	3.03	2.80	2.64	2.53	2.44	2.37	2.32	2.27	2.20	2.13	2.05	2.01	1.96	1.91	1.86	1.81	1.76
24	4.26	3.40	3.01	2.78	2.62	2.51	2.42	2.36	2.30	2.25	2.18	2.11	2.03	1.98	1.94	1.89	1.84	1.79	1.73
25	4.24	3.39	2.99	2.76	2.60	2.49	2.40	2.34	2.28	2.24	2.16	2.09	2.01	1.96	1.92	1.87	1.82	1.77	1.71
26	4.23	3.37	2.98	2.74	2.59	2.47	2.39	2.32	2.27	2.22	2.15	2.07	1.99	1.95	1.90	1.85	1.80	1.75	1.69
27	4.21	3.35	2.96	2.73	2.57	2.46	2.37	2.31	2.25	2.20	2.13	2.06	1.97	1.93	1.88	1.84	1.79	1.73	1.67
28	4.20	3.34	2.95	2.71	2.56	2.45	2.36	2.29	2.24	2.19	2.12	2.04	1.96	1.91	1.87	1.82	1.77	1.71	1.65
29	4.18	3.33	2.93	2.70	2.55	2.43	2.35	2.28	2.22	2.18	2.10	2.03	1.94	1.90	1.85	1.81	1.75	1.70	1.64
30	4.17	3.32	2.92	2.69	2.53	2.42	2.33	2.27	2.21	2.16	2.09	2.01	1.93	1.89	1.84	1.79	1.74	1.68	1.62
40	4.08	3.23	2.84	2.61	2.45	2.34	2.25	2.18	2.12	2.08	2.00	1.92	1.84	1.79	1.74	1.69	1.64	1.58	1.51
60	4.00	3.15	2.76	2.53	2.25	2.25	2.17	2.10	2.04	1.99	1.92	1.84	1.75	1.70	1.65	1.59	1.53	1.47	1.39
120	3.92	3.07	2.68	2.45	2.29	2.17	2.09	2.02	1.96	1.91	1.83	1.75	1.66	1.61	1.55	1.50	1.43	1.35	1.25
∞	3.84	3.00	2.60	2.37	2.21	2.10	2.01	1.94	1.88	1.83	1.75	1.67	1.57	1.52	1.46	1.39	1.32	1.22	1.00

[a] v_1 is the numerator degrees of freedom; v_2 is the denominator degrees of freedom. For F values at different percentage points, other than 1 percent and 2.5 percent, see Biometrika Tables for Statisticians, Vol. I [3rd Ed.], edited by E. S. Pearson and H. O. Hartley, Cambridge, The University Press, 1966.

Table A-3 Percentage Points of the F Distribution Upper 2.5 Percent Points

ν_2 \ ν_1	1	2	3	4	5	6	7	8	9	10	12	15	20	24	30	40	60	120	∞
1	647.8	799.5	864.2	899.6	921.8	937.1	948.2	956.7	963.3	968.6	976.7	984.9	993.1	997.2	1001	1006	1010	1010	1018
2	38.51	39.00	39.17	39.25	39.30	39.33	39.36	39.37	39.39	39.40	39.41	39.43	39.45	39.46	39.46	39.47	39.48	39.49	39.50
3	17.44	16.04	15.44	15.10	14.88	14.73	14.62	14.54	14.47	14.42	14.34	14.25	14.17	14.12	14.08	14.04	13.99	13.95	13.90
4	12.22	10.65	9.98	9.60	9.36	9.20	9.07	8.98	8.90	8.84	8.75	8.66	8.56	8.51	8.46	8.41	8.36	8.31	8.26
5	10.01	8.43	7.76	7.39	7.15	6.98	6.85	6.76	6.68	6.62	6.52	6.43	6.33	6.28	6.23	6.18	6.12	6.07	6.02
6	8.81	7.26	6.60	6.23	5.99	5.82	5.70	5.60	5.52	5.46	5.37	5.27	5.17	5.12	5.07	5.01	4.96	4.90	4.85
7	8.07	6.54	5.89	5.52	5.29	5.12	4.99	4.90	4.82	4.76	4.76	4.57	4.47	4.42	4.36	4.31	4.25	4.20	4.14
8	7.57	6.06	5.42	5.05	4.82	4.65	4.53	4.43	4.36	4.30	4.20	4.10	4.00	3.95	3.89	3.84	3.78	3.73	3.67
9	7.21	5.71	5.08	4.72	4.48	4.32	4.20	4.10	4.03	3.96	3.87	3.77	3.67	3.61	3.56	3.51	3.45	3.39	3.33
10	6.94	5.46	4.83	4.47	4.24	4.07	3.95	3.85	3.78	3.72	3.62	3.52	3.42	3.37	3.31	3.26	3.20	3.14	3.08
11	6.72	5.26	4.63	4.28	4.04	3.88	3.76	3.66	3.59	3.53	3.43	3.33	3.23	3.17	3.12	3.06	3.00	2.94	2.88
12	6.55	5.10	4.47	4.12	3.89	3.73	3.61	3.51	3.44	3.37	3.28	3.18	3.07	3.02	2.96	2.91	2.85	2.79	2.72
13	6.41	4.97	4.35	4.00	3.77	3.60	3.48	3.39	3.31	3.25	3.15	3.05	2.95	2.89	2.84	2.78	2.72	2.66	2.60
14	6.30	4.86	4.24	3.89	3.66	3.50	3.38	3.29	3.21	3.15	3.05	2.95	2.84	2.79	2.73	2.67	2.61	2.55	2.49
15	6.20	4.77	4.15	3.80	3.58	3.41	3.29	3.20	3.12	3.06	2.96	2.86	2.76	2.70	2.64	2.59	2.52	2.46	2.40
16	6.12	4.69	4.08	3.73	3.50	3.34	3.22	3.12	3.05	2.99	2.89	2.79	2.68	2.63	2.57	2.51	2.45	2.38	2.32
17	6.04	4.62	4.01	3.66	3.44	3.28	3.16	3.06	2.98	2.92	2.82	2.72	2.62	2.56	2.50	2.44	2.38	2.32	2.25
18	5.98	4.56	3.95	3.61	3.38	3.22	3.10	3.01	2.93	2.87	2.77	2.67	2.56	2.50	2.44	2.38	2.32	2.26	2.19
19	5.92	4.51	3.90	3.56	3.33	3.17	3.05	2.96	2.88	2.82	2.72	2.62	2.51	2.45	2.39	2.33	2.27	2.20	2.13
20	5.87	4.46	3.86	3.51	3.29	3.13	3.01	2.91	2.84	2.77	2.68	2.57	2.46	2.41	2.35	2.29	2.22	2.16	2.09
21	5.83	4.42	3.82	3.48	3.25	3.09	2.97	2.87	2.80	2.73	2.64	2.53	2.42	2.37	2.31	2.25	2.18	2.11	2.04
22	5.79	4.38	3.78	3.44	3.22	3.05	2.93	2.84	2.76	2.70	2.60	2.50	2.39	2.33	2.27	2.21	2.14	2.08	2.00
23	5.75	4.35	3.75	3.41	3.18	3.02	2.90	2.81	2.73	2.67	2.57	2.47	2.36	2.30	2.24	2.18	2.11	2.04	1.97
24	5.72	4.32	3.72	3.38	3.15	2.99	2.87	2.78	2.70	2.64	2.54	2.44	2.33	2.27	2.21	2.15	2.08	2.01	1.94
25	5.69	4.29	3.69	3.35	3.13	2.97	2.85	2.75	2.68	2.61	2.51	2.41	2.30	2.24	2.18	2.12	2.05	1.98	1.91
26	5.66	4.27	3.67	3.33	3.10	2.94	2.82	2.73	2.65	2.59	2.49	2.39	2.28	2.22	2.16	2.09	2.03	1.95	1.88
27	5.63	4.24	3.65	3.31	3.08	2.92	2.80	2.71	2.63	2.57	2.47	2.36	2.25	2.19	2.13	2.07	2.00	1.93	1.85
28	5.61	4.22	3.63	3.29	3.06	2.90	2.78	2.69	2.61	2.55	2.45	2.34	2.23	2.17	2.11	2.05	1.98	1.91	1.83
29	5.59	4.20	3.61	3.27	3.04	2.88	2.76	2.67	2.59	2.53	2.43	2.32	2.21	2.15	2.09	2.03	1.96	1.89	1.81
30	5.57	4.18	3.59	3.25	3.03	2.87	2.75	2.65	2.57	2.51	2.41	2.31	2.20	2.14	2.07	2.01	1.94	1.87	1.79
40	5.42	4.05	3.46	3.13	2.90	2.74	2.62	2.53	2.45	2.39	2.29	2.18	2.07	2.01	1.94	1.88	1.80	1.72	1.64
60	5.29	3.93	3.34	3.01	2.79	2.63	2.51	2.41	2.33	2.27	2.17	2.06	1.94	1.88	1.82	1.74	1.67	1.58	1.48
120	5.15	3.80	3.23	2.89	2.67	2.52	2.39	2.30	2.22	2.16	2.05	1.94	1.82	1.76	1.69	1.61	1.53	1.43	1.31
∞	5.02	3.69	3.12	2.79	2.57	2.41	2.29	2.19	2.11	2.05	1.94	1.83	1.71	1.64	1.57	1.48	1.39	1.27	1.00

[a] ν_1 is the numerator degrees of freedom; ν_2 is the denominator degrees of freedom. For F values at different percentage points, see the source cited in

Table A-4 Percentage Points of the *F* Distribution Upper 1 Percent Points

v_2 \ v_1	1	2	3	4	5	6	7	8	9	10	12	15	20	24	30	40	60	120	∞
1	4052	4999.5	5403	5625	5764	5859	5928	5981	6022	6056	6106	6157	6209	6235	6261	6287	6313	6339	6366
2	98.50	99.00	99.17	99.25	99.30	99.33	99.36	99.37	99.39	99.40	99.42	99.43	99.45	99.46	99.47	99.47	99.48	99.49	99.50
3	34.12	30.82	29.46	28.71	28.24	27.91	27.67	27.49	27.35	27.23	27.05	26.87	26.69	26.60	26.50	26.41	26.32	26.22	26.13
4	21.20	18.00	16.69	15.98	15.52	15.21	14.98	14.80	14.66	14.55	14.37	14.20	14.02	13.93	13.84	13.75	13.65	13.56	13.46
5	16.26	13.27	12.06	11.39	10.97	10.67	10.46	10.29	10.16	10.05	9.89	9.72	9.55	9.47	9.38	9.29	9.20	9.11	9.02
6	13.75	10.92	9.78	9.15	8.75	8.47	8.26	8.10	7.98	7.87	7.72	7.56	7.40	7.31	7.23	7.14	7.06	6.97	6.88
7	12.25	9.55	8.45	7.85	7.46	7.19	6.99	6.84	6.72	6.62	6.47	6.31	6.16	6.07	5.99	5.91	5.82	5.74	5.65
8	11.26	8.65	7.59	7.01	6.63	6.37	6.18	6.03	5.91	5.81	5.67	5.52	5.36	5.28	5.20	5.12	5.03	4.95	4.86
9	10.56	8.02	6.99	6.42	6.06	5.80	5.61	5.47	5.35	5.26	5.11	4.96	4.81	4.73	4.65	4.57	4.48	4.40	4.31
10	10.04	7.56	6.55	5.99	5.64	5.39	5.20	5.06	4.94	4.85	4.71	4.56	4.41	4.33	4.25	4.17	4.08	4.00	3.91
11	9.65	7.21	6.22	5.67	5.32	5.07	4.89	4.74	4.63	4.54	4.40	4.25	4 10.	4.02	3.94	3.86	3.78	3.69	3.60
12	9.33	6.93	5.95	5.41	5.06	4.82	4.64	4.50	4.39	4.30	4.16	4.01	3.86	3.78	3.70	3.62	3.54	3.45	2.36
13	9.07	6.70	5.74	5.21	4.86	4.62	4.44	4.30	4.19	4.10	3.96	3.82	3.66	3.59	3.51	3.43	3.34	3.25	3.17
14	8.86	6.51	5.56	5.04	4.69	4.46	4.28	4.14	4.03	3.94	3.80	3.66	3.51	3.43	3.35	3.27	3.18	3.09	3.00
15	8.68	6.36	5.42	4.89	4.56	4.32	4.14	4.00	3.89	3.80	3.67	3.52	3.37	3.29	3.21	3.13	3.05	2.96	2.87
16	8.53	6.23	5.29	4.77	4.44	4.20	4.03	3.89	3.78	3.69	3.45	3.41	3.26	3.18	3.10	3.02	2.93	2.84	2.75
17	8.40	6.11	5.18	4.67	4.34	4.10	3.93	3.79	3.68	3.59	3.46	3.31	3.16	3.08	3.00	2.92	2.83	2.75	2.65
18	8.29	6.01	5.09	4.58	4.25	4.01	3.84	3.71	3.60	3.51	3.37	3.23	3.08	3.00	2.92	2.84	2.75	2.66	2.54
19	8.18	5.93	5.01	4.50	4.17	3.94	3.77	3.63	3.52	3.43	3.30	3.15	3.00	2.92	2.84	2.76	2.67	2.58	2.49
20	8.10	5.85	4.94	4.43	4.10	3.87	3.70	3.56	3.46	3.37	3.23	3.09	2.94	2.86	2.78	2.69	2.61	2.52	2.42
21	8.02	5.78	4.87	4.37	4.04	3.81	3.64	3.51	3.40	3.31	3.17	3.03	2.88	2.80	2.72	2.64	2.55	2.46	2.36
22	7.95	5.72	4.82	4.31	3.99	3.76	3.59	3.45	3.35	3.26	3.12	2.98	2.83	2.75	2.67	2.58	2.50	2.40	2.31
23	7.88	5.66	4.76	4.26	3.94	3.71	3.54	3.41	3.30	3.21	3.07	2.93	2.78	2.70	2.62	2.54	2.45	2.35	2.26
24	7.82	5.61	4.72	4.22	3.90	3.67	3.50	3.36	3.26	3.17	3.03	2.89	2.74	2.66	2.58	2.49	2.40	2.31	2.21
25	7.77	5.57	4.68	4.18	3.85	3.63	3.46	3.32	3.22	3.13	2.99	2.85	2.70	2.62	2.54	2.45	2.36	2.27	2.17
26	7.72	5.53	4.64	4.14	3.82	3.59	3.42	3.29	3.18	3.09	2.96	2.81	2.66	2.58	2.50	2.42	2.33	2.23	2.13
27	7.68	5.49	4.60	4.11	3.78	3.56	3.39	3.26	3.15	3.06	2.93	2.78	2.63	2.55	2.47	2.38	2.29	2.20	2.10
28	7.64	5.45	4.57	4.07	3.75	3.53	3.36	3.23	3.12	3.03	2.90	2.75	2.60	2.52	2.44	2.35	2.26	2.17	2.06
29	7.60	5.42	4.54	4.04	3.73	3.50	3.33	3.20	3.09	3.00	2.87	2.73	2.57	2.49	2.41	2.33	2.23	2.14	2.03
30	7.56	5.39	4.51	4.02	3.70	3.47	3.30	3.17	3.07	2.98	2.84	2.70	2.55	2.47	2.39	2.30	2.21	2.11	2.01
40	7.31	5.18	4.31	3.83	3.51	3.29	3.12	2.99	2.89	2.80	2.66	2.52	2.37	2.29	2.20	2.11	2.02	1.92	1.80
60	7.08	4.98	4.13	3.65	3.34	3.12	2.95	2.82	2.72	2.63	2.50	2.35	2.20	2.12	2.03	1.94	1.84	1.73	1.60
120	6.85	4.79	3.95	3.48	3.17	2.96	2.79	2.66	2.56	2.47	2.34	2.19	2.03	1.95	1.86	1.76	1.66	1.53	1.38
∞	6.63	4.61	3.78	3.32	3.02	2.80	2.64	2.51	2.41	2.32	2.18	2.04	1.88	1.79	1.70	1.59	1.47	1.32	1.00

[a] v_1 is the numerator degrees of freedom; v_2 is the denominator degrees of freedom. For *F* values at different percentage points, see the source cited in the footnote of Table A-2.

Table A-5 Values of $D = \delta^2/s^2$ for Different Levels of Significance

	Probability		
n	0.01	0.025	0.05
5	0.6702	0.8300	1.0250
6	0.6730	0.8650	1.0675
7	0.7155	0.9085	1.0915
8	0.7570	0.9420	1.1220
9	0.7970	0.9775	1.1520
10	0.8340	1.0110	1.1795
11	0.8680	1.0430	1.2055
12	0.9030	1.0710	1.2295
15	0.9870	1.1460	1.2910
20	1.0945	1.2385	1.3670
25	1.1735	1.3055	1.4225
30	1.2345	1.3575	1.4660
40	1.3240	1.4385	1.5290
50	1.3890	1.4870	1.5730
60	1.4360	1.4765	1.6070

[a] Interpolated linearly from Table III, "Tabulation of the Probabilities for the Ratio of the Mean Square Successive Difference to the Variance," B. I. Hart, *Annals of Mathematical Statistics*, Vol. 13, pp. 207–214.

Table A-6 The Distribution of Durbin-Watson d
5 Percent Significance Points of d_L and d_U

	$k = 1$		$k = 2$		$k = 3$		$k = 4$		$k = 5$	
n	d_L	d_U	d_L	d_U	d_L	d_U	d_L	d_U	d_L	d_U
15	1.08	1.36	0.95	1.54	0.82	1.75	0.69	1.97	0.56	2.21
16	1.10	1.37	0.98	1.54	0.86	1.73	0.74	1.93	0.62	2.15
17	1.13	1.38	1.02	1.54	0.90	1.71	0.78	1.90	0.67	2.10
18	1.16	1.39	1.05	1.53	0.93	1.69	0.82	1.87	0.71	2.06
19	1.18	1.40	1.08	1.53	0.97	1.68	0.86	1.85	0.75	2.02
20	1.20	1.41	1.10	1.54	1.00	1.68	0.90	1.83	0.79	1.99
21	1.22	1.42	1.13	1.54	1.03	1.67	0.93	1.81	0.83	1.96
22	1.24	1.43	1.15	1.54	1.05	1.66	0.96	1.80	0.86	1.94
23	1.26	1.44	1.17	1.54	1.08	1.66	0.99	1.79	0.90	1.92
24	1.27	1.45	1.19	1.55	1.10	1.66	1.01	1.78	0.93	1.90
25	1.29	1.45	1.21	1.55	1.12	1.66	1.04	1.77	0.95	1.89
26	1.30	1.46	1.22	1.55	1.14	1.65	1.06	1.76	0.98	1.88
27	1.32	1.47	1.24	1.56	1.16	1.65	1.08	1.76	1.01	1.86
28	1.33	1.48	1.26	1.56	1.18	1.65	1.10	1.75	1.03	1.85
29	1.34	1.48	1.27	1.56	1.20	1.65	1.12	1.74	1.05	1.84
30	1.35	1.49	1.28	1.57	1.21	1.65	1.14	1.74	1.07	1.83
31	1.36	1.50	1.30	1.57	1.23	1.65	1.16	1.74	1.09	1.83
32	1.37	1.50	1.31	1.57	1.24	1.65	1.18	1.73	1.11	1.82
33	1.38	1.51	1.32	1.58	1.26	1.65	1.19	1.73	1.13	1.81
34	1.39	1.51	1.33	1.58	1.27	1.65	1.21	1.73	1.15	1.81
35	1.40	1.52	1.34	1.58	1.28	1.65	1.22	1.73	1.16	1.80
36	1.41	1.52	1.35	1.59	1.29	1.65	1.24	1.73	1.18	1.80
37	1.42	1.53	1.36	1.59	1.31	1.66	1.25	1.72	1.19	1.80
38	1.43	1.54	1.37	1.59	1.32	1.66	1.26	1.72	1.21	1.79
39	1.43	1.54	1.38	1.60	1.33	1.66	1.27	1.72	1.22	1.79
40	1.44	1.54	1.39	1.60	1.34	1.66	1.29	1.72	1.23	1.79
45	1.48	1.57	1.43	1.62	1.38	1.67	1.34	1.72	1.29	1.78
50	1.50	1.59	1.46	1.63	1.42	1.67	1.38	1.72	1.34	1.77
55	1.53	1.60	1.49	1.64	1.45	1.68	1.41	1.72	1.38	1.77
60	1.55	1.62	1.51	1.65	1.48	1.69	1.44	1.73	1.41	1.77
65	1.57	1.63	1.54	1.66	1.50	1.70	1.47	1.73	1.44	1.77
70	1.58	1.64	1.55	1.67	1.52	1.70	1.49	1.74	1.46	1.77
75	1.60	1.65	1.57	1.68	1.54	1.71	1.51	1.74	1.49	1.77
80	1.61	1.66	1.59	1.69	1.56	1.72	1.53	1.74	1.51	1.77
85	1.62	1.67	1.60	1.70	1.57	1.72	1.55	1.75	1.52	1.77
90	1.63	1.68	1.61	1.70	1.59	1.73	1.57	1.75	1.54	1.78
95	1.64	1.69	1.62	1.71	1.60	1.73	1.58	1.75	1.56	1.78
100	1.65	1.69	1.63	1.72	1.61	1.74	1.59	1.76	1.57	1.78

Table A-7 The Distribution of Durbin-Watson d
1 Percent Significance Points of d_L and d_U

	$k = 1$		$k = 2$		$k = 3$		$k = 4$		$k = 5$	
n	d_L	d_U	d_L	d_U	d_L	d_U	d_L	d_U	d_L	d_U
15	0.81	1.07	0.70	1.25	0.59	1.46	0.49	1.70	0.39	1.96
16	0.84	1.09	0.74	1.25	0.63	1.44	0.53	1.66	0.44	1.90
17	0.87	1.10	0.77	1.25	0.67	1.43	0.57	1.63	0.48	1.85
18	0.90	1.12	0.80	1.26	0.71	1.42	0.61	1.60	0.52	1.80
19	0.93	1.13	0.83	1.26	0.74	1.41	0.65	1.58	0.56	1.77
20	0.95	1.15	0.86	1.27	0.77	1.41	0.68	1.57	0.60	1.74
21	0.97	1.16	0.89	1.27	0.80	1.41	0.72	1.55	0.63	1.71
22	1.00	1.17	0.91	1.28	0.83	1.40	0.75	1.54	0.66	1.69
23	1.02	1.19	0.94	1.29	0.86	1.40	0.77	1.53	0.70	1.67
24	1.04	1.20	0.96	1.30	0.88	1.41	0.80	1.53	0.72	1.66
25	1.05	1.21	0.98	1.30	0.90	1.41	0.83	1.52	0.75	1.65
26	1.07	1.22	1.00	1.31	0.93	1.41	0.85	1.52	0.78	1.64
27	1.09	1.23	1.02	1.32	0.95	1.41	0.88	1.51	0.81	1.63
28	1.10	1.24	1.04	1.32	0.97	1.41	0.90	1.51	0.83	1.62
29	1.12	1.25	1.05	1.33	0.99	1.42	0.92	1.51	0.85	1.61
30	1.13	1.26	1.07	1.34	1.01	1.42	0.94	1.51	0.88	1.61
31	1.15	1.27	1.08	1.34	1.02	1.42	0.96	1.51	0.90	1.60
32	1.16	1.28	1.10	1.35	1.04	1.43	0.98	1.51	0.92	1.60
33	1.17	1.29	1.11	1.36	1.05	1.43	1.00	1.51	0.94	1.59
34	1.18	1.30	1.13	1.36	1.07	1.43	1.01	1.51	0.95	1.59
35	1.19	1.31	1.14	1.37	1.08	1.44	1.03	1.51	0.97	1.59
36	1.21	1.32	1.15	1.38	1.10	1.44	1.04	1.51	0.99	1.59
37	1.22	1.32	1.16	1.38	1.11	1.45	1.06	1.51	1.00	1.59
38	1.23	1.33	1.18	1.39	1.12	1.45	1.07	1.52	1.02	1.58
39	1.24	1.34	1.19	1.39	1.14	1.45	1.09	1.52	1.03	1.58
40	1.25	1.34	1.20	1.40	1.15	1.46	1.10	1.52	1.05	1.58
45	1.29	1.38	1.24	1.42	1.20	1.48	1.16	1.53	1.11	1.58
50	1.32	1.40	1.28	1.45	1.24	1.49	1.20	1.54	1.16	1.59
55	1.36	1.43	1.32	1.47	1.28	1.51	1.25	1.55	1.21	1.59
60	1.38	1.45	1.35	1.48	1.32	1.52	1.28	1.56	1.25	1.60
65	1.41	1.47	1.38	1.50	1.35	1.53	1.31	1.57	1.28	1.61
70	1.43	1.49	1.40	1.52	1.37	1.55	1.34	1.58	1.31	1.61
75	1.45	1.50	1.42	1.53	1.39	1.56	1.37	1.59	1.34	1.62
80	1.47	1.52	1.44	1.54	1.42	1.57	1.39	1.60	1.36	1.62
85	1.48	1.53	1.46	1.55	1.43	1.58	1.41	1.60	1.39	1.63
90	1.50	1.54	1.47	1.56	1.45	1.59	1.43	1.61	1.41	1.64
95	1.51	1.55	1.49	1.57	1.47	1.60	1.45	1.62	1.42	1.64
100	1.52	1.56	1.50	1.58	1.48	1.60	1.46	1.63	1.44	1.65

Table A-8 Significance Points of the von Neumann Ratio of Least-Squares Estimated Regression Disturbances

n (Number of observations)	k′ (Number of coefficients adjusted): 2		3		4		5		6	
	1%	5%	1%	5%	1%	5%	1%	5%	1%	5%
15	1.07	1.36	1.24	1.53	1.43	1.73	1.65	1.94	1.88	2.16
16	1.08	1.37	1.24	1.53	1.42	1.71	1.62	1.90	1.83	2.11
17	1.10	1.38	1.25	1.53	1.41	1.69	1.59	1.87	1.79	2.06
18	1.12	1.39	1.25	1.53	1.40	1.68	1.57	1.85	1.75	2.02
19	1.13	1.40	1.26	1.53	1.40	1.67	1.56	1.83	1.72	1.99
20	1.15	1.41	1.26	1.53	1.40	1.67	1.54	1.81	1.70	1.96
21	1.16	1.42	1.27	1.53	1.40	1.66	1.53	1.80	1.68	1.94
22	1.17	1.43	1.28	1.54	1.40	1.66	1.53	1.78	1.66	1.92
23	1.19	1.44	1.29	1.54	1.40	1.65	1.52	1.77	1.65	1.90
24	1.20	1.45	1.29	1.54	1.40	1.65	1.51	1.77	1.64	1.89
25	1.21	1.45	1.30	1.55	1.40	1.65	1.51	1.76	1.63	1.87
26	1.22	1.46	1.31	1.55	1.40	1.65	1.51	1.75	1.62	1.86
27	1.23	1.47	1.32	1.55	1.41	1.65	1.51	1.75	1.61	1.85
28	1.24	1.48	1.32	1.56	1.41	1.65	1.51	1.74	1.60	1.84
29	1.25	1.48	1.33	1.56	1.41	1.65	1.50	1.74	1.60	1.83
30	1.26	1.49	1.34	1.57	1.42	1.65	1.50	1.73	1.60	1.82
31	1.27	1.50	1.34	1.57	1.42	1.65	1.50	1.73	1.59	1.82
32	1.28	1.50	1.35	1.57	1.43	1.65	1.50	1.73	1.59	1.81
33	1.29	1.51	1.36	1.58	1.43	1.65	1.51	1.73	1.59	1.81
34	1.30	1.51	1.36	1.58	1.43	1.65	1.51	1.72	1.58	1.80
35	1.31	1.52	1.37	1.58	1.44	1.65	1.51	1.72	1.58	1.80
36	1.31	1.52	1.38	1.59	1.44	1.65	1.51	1.72	1.58	1.79
37	1.32	1.53	1.38	1.59	1.44	1.65	1.51	1.72	1.58	1.79
38	1.33	1.53	1.39	1.59	1.45	1.65	1.51	1.72	1.58	1.79
39	1.34	1.54	1.39	1.60	1.45	1.66	1.51	1.72	1.58	1.79
40	1.34	1.54	1.41	1.61	1.46	1.66	1.52	1.72	1.58	1.78
45	1.37	1.56	1.42	1.61	1.47	1.67	1.53	1.72	1.58	1.78
50	1.40	1.58	1.44	1.63	1.49	1.67	1.54	1.72	1.59	1.77
55	1.43	1.60	1.47	1.64	1.51	1.68	1.55	1.72	1.59	1.77
60	1.45	1.62	1.48	1.65	1.52	1.69	1.56	1.73	1.60	1.77
65	1.47	1.63	1.50	1.66	1.53	1.70	1.57	1.73	1.60	1.77
70	1.49	1.64	1.51	1.67	1.55	1.70	1.58	1.73	1.61	1.77
75	1.50	1.65	1.53	1.68	1.56	1.71	1.59	1.74	1.62	1.77
80	1.52	1.66	1.54	1.69	1.57	1.72	1.60	1.74	1.62	1.77
85	1.53	1.67	1.55	1.70	1.58	1.72	1.60	1.75	1.63	1.77
90	1.54	1.68	1.56	1.70	1.59	1.73	1.61	1.75	1.64	1.78
95	1.55	1.69	1.57	1.71	1.60	1.73	1.62	1.75	1.64	1.78
100	1.56	1.69	1.58	1.72	1.60	1.74	1.63	1.76	1.65	1.78

APPENDIX B

SOME CONCEPTS AND RESULTS OF MATRICES

B.1. Matrix Notations

B.1.1. A matrix, for our purposes, is a rectangular array of numbers which may each be represented by two subscripts, say, a_{ij}, where i indicates the number of the row and j the number of the column. A matrix **A** with the array consisting of m rows and n columns

$$\mathbf{A} = \begin{pmatrix} a_{11} & a_{12} & \cdots & a_{1n} \\ a_{21} & a_{22} & \cdots & a_{2n} \\ \cdot & \cdot & \cdot & \cdot \\ \cdot & \cdot & \cdot & \cdot \\ \cdot & \cdot & \cdot & \cdot \\ a_{m1} & a_{m2} & \cdots & a_{mn} \end{pmatrix}$$

is said to be of the dimensionality $m \times n$ or is referred to as an $m \times n$ matrix. This array also can be written in the typical-element form

$$\mathbf{A} = \{a_{ij}\} \qquad \begin{matrix} i = 1, 2, \ldots, m \\ j = 1, 2, \ldots, n \end{matrix}$$

B.1.2. A special case of a matrix is a vector that is in the form of a row or a column of a matrix. We shall adhere to the notation that a bold-faced lower-case letter represents a column vector. For instance,

$$\mathbf{a} = \begin{pmatrix} a_1 \\ a_2 \\ \cdot \\ \cdot \\ \cdot \\ a_m \end{pmatrix}$$

is an m-element column vector.

B.1.3. The transpose of an $m \times n$ matrix $\mathbf{A}$ is the matrix resulting from interchanging the rows and columns of $\mathbf{A}$. Such an operation is denoted by

$$\mathbf{A}' = \begin{pmatrix} a_{11} & a_{12} & \cdots & a_{1n} \\ a_{21} & a_{22} & \cdots & a_{2n} \\ \cdot & \cdot & \cdot & \cdot \\ \cdot & \cdot & \cdot & \cdot \\ \cdot & \cdot & \cdot & \cdot \\ a_{m1} & a_{m2} & \cdots & a_{mn} \end{pmatrix}'$$

$$= \begin{pmatrix} a_{11} & a_{21} & \cdots & a_{m1} \\ a_{12} & a_{22} & \cdots & a_{m2} \\ \cdot & \cdot & \cdot & \cdot \\ \cdot & \cdot & \cdot & \cdot \\ \cdot & \cdot & \cdot & \cdot \\ a_{1n} & a_{2n} & \cdots & a_{mn} \end{pmatrix}$$

and shows that the result is an $m \times n$ matrix. In the special case of the m-element column vector above

$$\mathbf{a}' = (a_1 \quad a_2 \quad \cdots \quad a_m)$$

Notationally, another form of transpose operation is

$$\mathbf{A}' = \{a_{ij}\}' = \{a_{ji}\} \qquad \begin{matrix} i = 1, 2, \ldots, m \\ j = 1, 2, \ldots, n \end{matrix}$$

B.1.4. If $\mathbf{A}' = \mathbf{A}$, the matrix $\mathbf{A}$ is said to be symmetric.

B.2. Matrix Operations

B.2.1. Two matrices $\mathbf{A}$ and $\mathbf{B}$ are equal if, and only if, the corresponding elements of the two matrices are equal. That is,

$$\mathbf{A} = \mathbf{B}$$

if and, only if, $a_{ij} = b_{ij}$ for arbitrary i and j.

B.2.2. The addition of two matrices $\mathbf{A}$ and $\mathbf{B}$ is defined only if $\mathbf{A}$ and $\mathbf{B}$ have the same dimensionality. The resulting matrix $\mathbf{C}$ is

$$\mathbf{C} = \mathbf{A} \pm \mathbf{B}$$

or

$$\{c_{ij}\} = \{a_{ij}\} \pm \{b_{ij}\} = \{a_{ij} \pm b_{ij}\} \qquad \begin{matrix} i = 1, 2, \ldots, m \\ j = 1, 2, \ldots, n \end{matrix}$$

and has the same dimensionality as $\mathbf{A}$ or $\mathbf{B}$.

B.2.3. Matrices **A** and **B**, in that order, are said to be conformable if the number of columns of **A** is the same as the number of rows of **B**. If **A** and **B** are conformable, the product **AB** is defined as

$$\mathbf{AB} = \mathbf{C} = \{\sum_j a_{ij} b_{jk}\}$$

or the typical element of the product matrix c_{ij} is obtained by multiplying the ith row of **A** "into" the jth column of **B**.

For the m-element column vector $\mathbf{x} = \begin{pmatrix} x_1 \\ x_2 \\ \cdot \\ \cdot \\ \cdot \\ x_m \end{pmatrix}$

we have

$$\mathbf{x}'\mathbf{x} = (x_1 x_2 \cdots x_m) \begin{pmatrix} x_1 \\ x_2 \\ \cdot \\ \cdot \\ \cdot \\ x_m \end{pmatrix}$$

$$= \sum_{i=1}^{m} x_i^{\,2}$$

and

$$\mathbf{x}\mathbf{x}' = \begin{pmatrix} x_1 \\ x_2 \\ \cdot \\ \cdot \\ \cdot \\ x_m \end{pmatrix} (x_1 \quad x_2 \quad \cdots \quad x_m)$$

$$= \begin{pmatrix} x_1x_1 & x_1x_2 & \cdots & x_1x_m \\ x_2x_1 & x_2x_2 & \cdots & x_2x_m \\ \cdot & \cdot & \cdot & \cdot \\ \cdot & \cdot & \cdot & \cdot \\ \cdot & \cdot & \cdot & \cdot \\ x_mx_1 & x_mx_2 & \cdots & x_mx_m \end{pmatrix}$$

The transpose operation of a product of **A** and **B** is

$$(\mathbf{A} \quad \mathbf{B})' = (\mathbf{B}' \quad \mathbf{A}')$$

B.2.4. A scalar product of a matrix **A** may be written $\lambda\mathbf{A}$, where λ is an arbitrary real number, or

$$\lambda\mathbf{A} = \{\lambda a_{ij}\}$$

B.2.5. A linear combination of m vectors $\mathbf{v}_i$, $i = 1, 2, \ldots, m$, may be represented by

$$\boldsymbol{\beta} = k_1\mathbf{v}_1 + k_2\mathbf{v}_2 + \cdots + k_m\mathbf{v}_m$$

where k_i are scalars.

The vectors $\mathbf{v}_i$ are said to be linearly dependent if there exist some k_i's that are not all zero so that $\boldsymbol{\beta} = \mathbf{0}$. If there exists no such k_i's so that $\boldsymbol{\beta} = \mathbf{0}$, then the vectors are said to be linearly independent. One of the implications of linear independence is that, if the $\mathbf{v}_i$'s are the columns of a matrix, no column of the matrix can be written as a constant multiple of another.

B.2.6. The inverse of a square matrix $\mathbf{A}$ is denoted by $\mathbf{A}^{-1}$ and satisfies the property

$$\mathbf{A}^{-1}\mathbf{A} = \mathbf{A}\mathbf{A}^{-1} = \mathbf{I}$$

where $\mathbf{I}$ is an identity matrix. An identity matrix of order k is usually denoted $\mathbf{I}_k$ and represents a $k \times k$ matrix whose diagonal elements are all 1's and the other elements are zero.

The identity matrix has the property that

$$\mathbf{A}\mathbf{I} = \mathbf{I}\mathbf{A} = \mathbf{A}$$

namely, its premultiplication and postmultiplication to any matrix leaves the matrix unaltered.

We also have the results:

$$(\mathbf{A}^{-1})^{-1} = \mathbf{A}$$
$$(\mathbf{A}^{-1})' = (\mathbf{A}')^{-1}$$
$$(\mathbf{A}\mathbf{B})^{-1} = \mathbf{B}^{-1}\mathbf{A}^{-1}$$

If

$$\mathbf{D} = \begin{pmatrix} d_1 & 0 & \cdots & 0 \\ 0 & d_2 & \cdots & 0 \\ \cdot & & \cdot & \cdot \\ \cdot & & \cdot & \cdot \\ \cdot & & \cdot & \cdot \\ 0 & 0 & \cdots & d_m \end{pmatrix}$$

where d_i's are nonzero, then

$$\mathbf{D}^{-1} = \begin{pmatrix} \frac{1}{d_1} & 0 & \cdots & 0 \\ 0 & \frac{1}{d_2} & \cdots & 0 \\ \cdot & \cdot & \cdot & \cdot \\ \cdot & \cdot & \cdot & \cdot \\ \cdot & \cdot & \cdot & \cdot \\ 0 & 0 & \cdots & \frac{1}{d_m} \end{pmatrix}$$

B.2.7. The determinant of an arbitrary square matrix **A** is a scalar and is denoted $|\mathbf{A}|$. If $|\mathbf{A}| = 0$, **A** is said to be singular.

B.3. Related Matrix Concepts

B.3.1. For an arbitrary matrix **A** the rank of **A**, denoted by $\rho(\mathbf{A})$, is the maximum number of linearly independent columns (rows) of **A**. If **A** is $m \times n$, the rank of **A** is less than or equal to the smaller of the dimensionality of **A**, that is,

$$\rho(\mathbf{A}) \leq \min \{m, n\}$$

If the rank of **A** is less than n, and $n < m$, then there are less than n linearly independent columns in **A**. One of the implications here is that, if the columns represent observations on the variables (whose number equals the number of columns), one of the columns may be written as a constant multiple of another. Further, generally speaking, $\rho(\mathbf{A}) < n$ means that one of the columns can be written as a nontrivial linear combination of some other columns, implying redundancy of information in the observations for the variables.

The rank of a product of two matrices **A** and **B** is less than or equal to the rank of **A** and of **B**, namely,

$$\rho(\mathbf{AB}) \leq \min \{\rho(\mathbf{A}), \rho(\mathbf{B})\}$$

Furthermore,

$$\rho(\mathbf{A} + \mathbf{B}) \leq \rho(\mathbf{A}) + \rho(\mathbf{B})$$

$$\rho(\mathbf{A}) = \rho(\mathbf{A}')$$

B.3.2. The trace of a square matrix $\mathbf{A} = \{a_{ij}\}$ is the sum of its diagonal elements, that is,

$$\text{tr}(\mathbf{A}) = \sum_i a_{ii}$$

If $\mathbf{A} + \mathbf{B}$ and **AB** are defined, then

$$\text{tr}(\mathbf{A} + \mathbf{B}) = \text{tr}(\mathbf{A}) + \text{tr}(\mathbf{B})$$

$$\text{tr}(\mathbf{AB}) = \text{tr}(\mathbf{BA})$$

This result can be generalized to a product of more than two matrices. For example

$$\text{tr}(\mathbf{ABC}) = \text{tr}(\mathbf{BCA})$$

B.3.3. A square matrix $\mathbf{A}$ is orthogonal if, and only if, $\mathbf{A}'\mathbf{A} = \mathbf{I}$ or $\mathbf{A}' = \mathbf{A}^{-1}$. That is, if $\mathbf{a}_i$ and $\mathbf{a}_j$ are the ith and jth columns of $\mathbf{A}$,

$$\mathbf{a}_i'\mathbf{a}_j = \begin{cases} 1 & \text{if} \quad i = j \\ 0 & \text{if} \quad i = j \end{cases}$$

If $\mathbf{A}$ is orthogonal, $\mathbf{A}^{-1}$ is also orthogonal.

Column vectors $\mathbf{x}$ and $\mathbf{y}$ are said to be orthogonal if $\mathbf{x}'\mathbf{y} = 0$.

B.3.4. A square matrix $\mathbf{A}$ is idempotent if, and only if, $\mathbf{A}' = \mathbf{A}$ and $\mathbf{A}^2 = \mathbf{AA} = \mathbf{A}$.

B.3.5. If $\mathbf{x}$ is an n-element vector and $\mathbf{A}$ is an $n \times n$ symmetrix matrix, then the scalar $\mathbf{x}'\mathbf{Ax}$ is called a quadratic form in the elements of $\mathbf{x}$. The choice of the elements of $\mathbf{A}$ is crucial in determining the nature of the quadratic form. For example, if $\mathbf{x} = \begin{pmatrix} x_1 \\ x_2 \end{pmatrix}$ and $\mathbf{A} = \begin{pmatrix} a & b/2 \\ b/2 & c \end{pmatrix}$ then $\mathbf{x}'\mathbf{Ax} = ax_1^2 + bx_1x_2 + cx_2^2$. However, if $\mathbf{A} = \begin{pmatrix} a & 0 \\ 0 & c \end{pmatrix}$, then

$$\mathbf{x}'\mathbf{Ax} = ax_1^2 + cx_2^2.$$

The matrix $\mathbf{A}$ is called the matrix of the quadratic form, and the rank of the quadratic form is the same as the rank of $\mathbf{A}$, or $\rho(\mathbf{x}'\mathbf{Ax}) = \rho(\mathbf{A})$.

B.3.6. A quadratic form $\mathbf{x}'\mathbf{Ax}$ is positive definite if $\mathbf{x}'\mathbf{Ax} > 0$, and positive semidefinite if $\mathbf{x}'\mathbf{Ax} \geq 0$. The matrix $\mathbf{A}$ is said to be positive definite (p.d.) or positive semidefinite (p.s.d.) accordingly as the associated quadratic form is p.d. or p.s.d.

If a symmetric matrix $\mathbf{A}$ is p.d., its inverse $\mathbf{A}^{-1}$ is also p.d.

B.3.7. A characteristic root of an $n \times n$ matrix $\mathbf{A}$ is a scalar λ so that $\mathbf{Ax} = \lambda\mathbf{x}$ for some $\mathbf{x} = 0$. The vector $\mathbf{x}$ is called the characteristic vector of the matrix $\mathbf{A}$. This definition implies that λ is a characteristic root of $\mathbf{A}$ if, and only if,

$$(\mathbf{A} - \lambda\mathbf{I}) = \mathbf{0}$$

This is a system of n homogeneous equations in n unknowns, and it has a nontrivial solution if, and only if, $\rho(\mathbf{A} - \lambda\mathbf{I}) < n$ or if, and only if, $|\mathbf{A} - \lambda\mathbf{I}| = 0$. The latter is a polynomial of degree n in $\mathbf{x}$; it follows that $\mathbf{A}$ has n characteristic roots which may or may not be unique.

B.3.8. Diagonalization theorem. If $\mathbf{A}$ is $n \times n$ and $\mathbf{A}' = \mathbf{A}$, there exists an $n \times n$ orthogonal matrix $\mathbf{C}$ so that $\mathbf{C}'\mathbf{AC}$ is a diagonal matrix.

Let the diagonal matrix $\mathbf{C}'\mathbf{AC} = \Lambda$. Then the characteristic roots of $\mathbf{A}$ are the diagonal elements of Λ, and the rank of $\mathbf{A}$ is the number of the nonzero diagonal elements of Λ.

If the matrix $\mathbf{A}$ immediately above is idempotent, then $\rho(\mathbf{A}) = \text{tr}\ (\mathbf{A})$.

Consider a trivial but instructive illustration of these results for matrices $\mathbf{C} = \mathbf{I}$ and $\mathbf{A} = \mathbf{I}$.

B.4. Partitioned Matrices

B.4.1. Let an $m \times p$ matrix $\mathbf{A}$ and a $p \times q$ matrix $\mathbf{B}$ be partitioned as follows

$$\mathbf{A} = \begin{pmatrix} \mathbf{A}_{11} & \mathbf{A}_{12} \\ \mathbf{A}_{21} & \mathbf{A}_{22} \end{pmatrix}$$

$$\mathbf{B} = \begin{pmatrix} \mathbf{B}_{11} & \mathbf{B}_{12} \\ \mathbf{B}_{21} & \mathbf{B}_{22} \end{pmatrix}$$

where $\mathbf{A}_{11}$ is $m_1 \times p_1$, $\mathbf{A}_{12}$ is $m_1 \times p_2$, $\mathbf{A}_{21}$ is $m_2 \times p_1$, $\mathbf{A}_{22}$ is $m_2 \times p_2$, with $m_1 + m_2 = m$ and $p_1 + p_2 = p$, $\mathbf{B}_{11}$ is $p_1 \times q_1$, $\mathbf{B}_{12}$ is $p_1 \times q_2$, $q_1 + q_2 = q$, and so forth. Then

$$\mathbf{AB} = \begin{pmatrix} \mathbf{A}_{11}\mathbf{B}_{11} + \mathbf{A}_{11}\mathbf{B}_{21} & \mathbf{A}_{11}\mathbf{B}_{12} + \mathbf{A}_{12}\mathbf{B}_{22} \\ \mathbf{A}_{21}\mathbf{B}_{11} + \mathbf{A}_{22}\mathbf{B}_{21} & \mathbf{A}_{21}\mathbf{B}_{12} + \mathbf{A}_{22}\mathbf{B}_{22} \end{pmatrix}$$

Notice the row-into-column correspondence to the matrix multiplication defined in paragraph B.2.3.

B.4.2. For a matrix $\mathbf{A}$, as partitioned in the preceding paragraph,

$$\mathbf{A}' = \begin{pmatrix} \mathbf{A}'_{11} & \mathbf{A}'_{21} \\ \mathbf{A}'_{12} & \mathbf{A}'_{22} \end{pmatrix}$$

B.4.3. If an $n \times n$ matrix $\mathbf{A}$ has an inverse and is partitioned as

$$\mathbf{A} = \begin{pmatrix} \mathbf{B} & \mathbf{C} \\ \mathbf{D} & \mathbf{E} \end{pmatrix}$$

where $\mathbf{B}$ is $m \times m$, $m < n$, then the inverse of $\mathbf{A}$ can be calculated by

$$\mathbf{A}^{-1} = \begin{pmatrix} \mathbf{B}^{-1}(\mathbf{I} + \mathbf{C}\mathbf{J}^{-1}\mathbf{D}\mathbf{B}^{-1}) & -\mathbf{B}^{-1}\mathbf{C}\mathbf{J}^{-1} \\ -\mathbf{J}^{-1}\mathbf{D}\mathbf{B}^{-1} & \mathbf{J}^{-1} \end{pmatrix}$$

where

$$\mathbf{J}^{-1} = \mathbf{E} - \mathbf{D}\mathbf{B}^{-1}\mathbf{C}$$

B.5. Matrix Differentiation

B.5.1. Given a product of two vectors **c** and **x**

$$\mathbf{c}'\mathbf{x} = (c_1 \quad c_2 \quad \cdots \quad c_n)\begin{pmatrix} x_1 \\ x_2 \\ \cdot \\ \cdot \\ \cdot \\ x_n \end{pmatrix}$$

The partial derivative of **c′x** with respect to **x** is

$$\frac{\partial(\mathbf{c}'\mathbf{x})}{\partial \mathbf{x}} = \mathbf{c}$$

This means that the derivative is taken with respect to each of the elements of **x**, forming a column

$$\frac{\partial(\mathbf{c}'\mathbf{x})}{\partial x_1} = c_1$$

$$\frac{\partial(\mathbf{c}'\mathbf{x})}{\partial x_2} = c_2$$

$$\cdot$$
$$\cdot$$
$$\cdot$$

$$\frac{\partial(\mathbf{c}'\mathbf{x})}{\partial x_n} = c_n$$

B.5.2. Given a quadratic form in **x**

$$\mathbf{x}'\mathbf{C}\mathbf{x} = (x_1 \quad x_2 \quad \cdots \quad x_n)\begin{pmatrix} c_{11} & c_{12} & \cdots & c_{1n} \\ c_{21} & c_{22} & \cdots & c_{2n} \\ \cdot & \cdot & \cdot & \cdot \\ \cdot & \cdot & \cdot & \cdot \\ \cdot & \cdot & \cdot & \cdot \\ c_{n1} & c_{n2} & \cdots & c_{nn} \end{pmatrix}\begin{pmatrix} x_1 \\ x_2 \\ \cdot \\ \cdot \\ \cdot \\ x_n \end{pmatrix}$$

its partial derivative with respect to **x** is

$$\frac{\partial(\mathbf{x}'\mathbf{C}\mathbf{x})}{\partial \mathbf{x}} = 2\mathbf{C}\mathbf{x}$$

or

$$\frac{\partial(\mathbf{x}'\mathbf{C}\mathbf{x})}{\partial \mathbf{x}} = 2\mathbf{x}'\mathbf{C}$$

To verify this, multiply out **x′Cx** and take the partials of the expanded results with respect to x_i, $i = 1, 2, \ldots, n$.

BIBLIOGRAPHY

Acton, F. S. (1959), *Analysis of Straight Line Data.* New York: Wiley & Sons.

Almon, S. (1965), "The Distributed Lag between Capital Appropriations and Expenditures," *Econometrica*, **33,** 178–196.

Aitken, A. C. (1935), "On Least Squares and Linear Combination of Observations." *Proceedings of the Royal Society of Edinburgh*, **55,** 42–48, (1934–1935).

Anscombe, F. J. (1961), "Examination of Residuals," *Proceedings of the Fourth Berkeley Symposium on Mathematical Statistics and Probability*, Vol. I, edited by J. Neyman. Berkeley: University of California Press.

Barten, A. P. (1962), "Note on Unbiased Estimation of the Squared Multiple Correlation Coefficient," *Statistics Neerlandia*, **16,** 151–163, (1962).

Basmann, R. L. (1957), "A Generalized Classical Method of Linear Estimation of Coefficients in a Structural Equation," *Econometrica*, **25,** 77–83, (January 1957).

Baumol, W. (1959), *Economic Dynamics*, 2nd ed. New York: Macmillan.

Bellman, Richard (1960), *Introduction to Matrix Analysis.* New York: John Wiley and Sons.

Birnbaum, Z. W. (1962), *Introduction to Probability and Mathematical Statistics.* New York: Harper & Brothers.

Booth, G. W., and T. I. Peterson (1960), "Nonlinear Estimation," Mathematics and Applications Department, International Business Machines Corporation, New York.

Box, B. E. P., and G. S. Watson (1962), "Robustness to Nonnormality of Regression Tests," *Biometrika*, **49,** 93–106, (1962).

Chipman, J. S., and M. M. Rao (1964), "The Treatment of Linear Restrictions in Regression Analysis," *Econometrica*, **32,** No. 1–2, 198–207, (January–April, 1964).

Chow, G. (1960), "Tests for Equality Between Sets of Coefficients in Two Linear Regressions," *Econometrica*, **28,** 591–605, (July 1960).

Chow, G. (1964), "A Comparison of Alternative Estimators for Simultaneous Equations," *Econometrica*, **32,** 532–553, (October 1964).

Christ, C. (1966), *Econometric Models and Methods.* New York: Wiley & Sons.

Cramer, H. (1946), *Mathematical Methods of Statistics.* Princeton: Princeton University Press.

David, F. N. (1954), *Tables of the Correlation Coefficient.* Cambridge: University Press.

Draper, N. R., and H. Smith (1966), *Applied Regression Analysis*, New York: Wiley & Sons.

Durbin, J., and G. S. Watson (1950), "Testing for Serial Correlation in Least Squares Regression, I," *Biometrika*, **37**, 409–428, (1950).

Durbin, J., and G. S. Watson (1951), "Testing for Serial Correlation in Least Squares Regression, II," *Biometrika*, **38**, 159–178, (1951).

Ezekiel, M., and K. Fox (1959), *Methods of Correlation and Regression Analysis*, 3rd ed. New York: Wiley & Sons.

Farrar, D. E., and R. R. Glauber (1967), "Multicollinearity in Regression Analysis: The Problem Revisited," *Review of Economics and Statistics*, **XLIX**, No. 1 92–107.

Finney, D. J. (1952), *Probit Analysis*, 2nd ed. Cambridge: University Press.

Freund, R. J. (1963), "A Warning of Roundoff Errors in Regression," *The American Statistician*, 13–15, (December 1963).

Friedman, J., and R. T. Foote (1957), *Computational Methods of Handling Systems of Simultaneous Equations*, Agricultural Handbook No. 94. Washington, D.C.: United States Dept. of Agriculture.

Friedman, Milton (1957), *A Theory of The Consumption Function*, New Jersey: Princeton Univ. Press.

Goldberger, A. S. (1964), *Econometric Theory*. New York: Wiley & Sons.

Goldberger, A. S. (1959), *Impact Multipliers and Dynamic Properties of the Klein-Goldberger Model.* Amsterdam: North-Holland Publishing Co.

Goldberger, A. S. (1968), *Topics in Regression Analysis.* New York: Macmillan Co.

Goldberger, A. S., A. L. Nagar, and H. S. Odeh (1961), "The Covariance Matrices of Reduced Form Coefficients and of Forecasts for a Structural Econometric Model," *Econometrica*, **29,** 556–573, (October 1961).

Graybill, F. A. (1961), *An Introduction to Linear Statistical Models*, Vol. I. New York: McGraw-Hill Book Co.

Graybill, F. A. (1969), *Introduction to Matrices with Applications in Statistics.* Belmont, California: Wadsworth Publishing Co.

Griliches, Z. (1957), "Specification Bias in Estimates of Production Functions," *Journal of Farm Economics*, **39,** 8–20, (February 1957).

Griliches, Z. (1967), "Distributed Lags: A Survey," *Econometrica*, **35,** No. 1, 16–49, (January 1967).

Hadley, G. (1961), *Linear Algebra*, Reading, Massachusetts: Addison-Wesley.

Hart, B. I. (1942), "Tabulation of the Probabilities for the Ratio of the Mean Successive Difference to the Variance," *Annals of Mathematical Statistics*, **13,** 207–214, (1942).

Hartley, H. O., and A. Booker (1965), "Nonlinear Least Squares Estimation," *Annals of Mathematical Statistics*, **36,** 638–650, (April 1965).

Hildreth, C., and J. Y. Lu (1960), *Demand Relations with Autocorrelated Disturbances*, Agricultural Experiment Station Technical Bulletin 276. East Lansing, Michigan: Department of Agricultural Economics.

Hooper, J., and A. Zellner (1961), "The Error of Forecast for Multi-variate Regression Models," *Econometrica*, **29,** 544–555, (1961).

Huang, D. S. (1963), "Initial Stock and Consumer Investment in Automobiles," *Journal of the American Statistical Association*, **58,** 789–798, (1963).

Huang, D. S. (1966), "The Short-Run Flows of Nonfarm Residential Mortgage Credit," *Econometrica*, **34,** 433–459, (April 1966).

Huang, D. S. (1967), "Forecasting Ability of Alternative Least-Squares Estimators of Simultaneous Equations" (mimeographed), Department of Economics, Southern Methodist University, 1967.

Johnston, J. (1963), *Econometric Methods*. New York: McGraw-Hill Book Co.

Kempthorne, O. (1952), *Design and Analysis of Experiments*. New York: Wiley & Sons.

Klein, L. R. (1955), "On the Interpretation of Theil's Method of Estimation of Economic Relations," *Metroeconomica*, **7,** 147–153, (December 1955).

Klein, L. R. (1956), *A Textbook of Econometrics*. New York: Row Peterson & Co.

Klein, L. R. (1962), *An Introduction to Econometrics*, Englewood Cliffs: Prentice-Hall.

Kuh, E., and J. R. Meyer (1955), "Correlation and Regression Estimates When the Data Are Ratios," *Econometrica*, 400–416, (October 1955).

Madansky, A. (1959), "The Fitting of Straight Lines When Both Variables Are Subject to Error," *Journal of the American Statistical Association*, **54,** 173–205, (1959).

Malinvand, E. (1968), *Statistical Methods of Econometrics*. Chicago: Rand McNally.

Mann, H. B. (1949), *Analysis and Design of Experiments*. New York: Dover Publications.

Marquardt, D. W. (1963), "An Algorithm for Least-Squares Estimation of Non-linear Parameters," *Journal of the Society for Industrial and Applied Mathematics*, 431–441, (June 1963).

Meeter, D. (1964), "Problems in the Analysis of Nonlinear Models by Least Squares," unpublished Ph.D. dissertation, Department of Statistics, the University of Wisconsin.

Meyer, J. R., and E. Kuh (1957), "How Extraneous Are Extraneous Estimates?" *Review of Economics and Statistics*, **39,** 380–393, (1957).

Mood, A. M., and F. A. Graybill (1963), *Introduction to the Theory of Statistics*, 2nd ed. New York: McGraw-Hill Book Co.

Morrison, D. F. (1967), *Multivariate Statistical Methods*. New York: McGraw-Hill Book Co.

Orcutt, Guy H. (1960), "Simulation of Economic Systems," *American Economic Review*, **1,** 893–907, (December 1960).

Plackett, R. L. (1960), *Principles of Regression Analysis*. London: Oxford University Press.

Rao, C. R. (1952), *Advanced Statistical Methods in Biometric Research*. New York: Wiley & Sons.

Rosett, R. (1959), "A Statistical Model of Friction in Economics," *Econometrica*, **27,** 263–267, (1959).

Straud, A., A. Zellner, and L. C. Chau (1963), "Program for Computing Two- and Three-Stage Least Squares Estimates and Associated Statistics," Systems Formulation and Methodology Workshop Paper 6308, Social Systems Research Institute, University of Wisconsin, December 1963. (Revised by H. Thornber and A. Zellner, July 1965.)

Theil, H. (1957), "Specification Errors and the Estimation of Economic Relationships," *Review of the International Institute of Statistics*, **25,** 41–51, (1957).

Theil, H. (1958), *Economic Forecasts and Policy*. Amsterdam: North-Holland Publishing Co.

Theil, H. (1965), "The Analysis of Disturbances in Regression Analysis," *Journal of American Statistical Association*, **60,** 1067–1079, (December 1965).

Theil, H. (1966), *Applied Economic Forecasting*. Chicago: Rand McNally.

Theil, H. (1968), "A Simplification of the BLUS Procedure for Analyzing Regression Disturbances," *Journal of the American Statistical Association*, **63,** 242–251, (March 1968).

Theil, H. (1970), *Principles of Econometrics*. New York: Wiley & Sons.

Theil, H., and J. C. G. Bost (1962), "The Final Form of Econometric Equation Systems," *Review of the International Statistical Institute*, **30,** 136–152, (1962).

Theil, H., and A. S. Goldberger (1961), "On Pure and Mixed Statistical Estimation in Economics," *International Economic Review*, **2,** 65–78, (January 1961).

Theil, H., and H. L. Nagar (1961), "Testing the Independence of Regression Disturbances," *Journal of American Statistical Association*, **56,** 793–806, (December 1961).

Tobin, J. (1955), "The Application of Multivariate Profit Analysis to Economic Survey Data," Cowles Foundation Discussion Paper No. 1, 1955.

Toro-Vizcarrondo, C., and T. D. Wallace (1968), "A Test of the Mean Square Error Criterion for Restrictions in Linear Regression," *Journal of the American Statistical Association*, **63,** 558–572.

von Neumann, J. (1941), "Distribution of the Ratio of the Mean Square Successive Difference to the Variance," *Annals of Mathematical Statistics*, **12,** 367–395, (1941).

von Neumann, J. (1942), "A Further Remark Concerning the Ratio of the Mean Square Successive Difference to the Variance," *Annals of Mathematical Statistics*, **13,** 86–88.

Walker, H. (1940), "Degrees of Freedom," *Journal of Educational Psychology*, **31,** 253–269, (1940).

Watts, H. (1960), "An Objective Permanent Income Concept for the Household," Cowles Foundation Discussion Paper No. 99, November 1960.

Wilks, S. S. (1962), *Mathematical Statistics*. New York: Wiley & Sons.

Williams, E. J. (1959), *Regression Analysis*. New York: Wiley & Sons, Inc.

Yule, G. U., and M. G. Kendall (1950), *An Introduction to the Theory of Statistics*, 14th ed. New York: Hafner Publishing Co.

Zellner, A. (1961), "Econometric Estimation with Temporally Dependent Disturbance Terms," *International Economic Review*, **2,** 164–178, (May 1961).

Zellner, A. (1962), "An Efficient Method of Estimating Seemingly Unrelated

Regressions and Tests for Aggregation Bias," *Journal of the American Statistical Association*, **57**, 348–368, (June 1962).

Zellner, A. (1970), *Introduction to Bayesian Inference in Econometrics.* New York: Wiley & Sons.

Zellner, A., and D. S. Huang (1962), "Further Properties of Efficient Estimators for Seemingly Unrelated Regression Equations," *International Economic Review*, **3,** 300–313, (September 1962).

Zellner, A., and T. H. Lee (1965), "Joint Estimation of Relationships Involving Discrete Random Variables," *Econometrica*, **33,** 382–394, (April 1965).

Zellner, A., and H. Theil (1962), "Three-Stage Least Squares: Simultaneous Estimation of Simultaneous Equations," *Econometrica*, **30,** 54–78, (January 1962).

Index